CEREMONIAL LITHICS IN CLASSIC MESOAMERICA

CEREMONIAL LITHICS IN CLASSIC MESOAMERICA

EDITED BY

Zachary Hruby

UNIVERSITY PRESS OF COLORADO
Denver

Published by University Press of Colorado
1580 North Logan Street, Suite 660
PMB 39883
Denver, Colorado 80203-1942

Printed in the United States of America

The University Press of Colorado is a proud member of
Association of University Presses.

The University Press of Colorado is a cooperative publishing enterprise supported, in part, by Adams State University, Colorado School of Mines, Colorado School of Mines, Colorado State University, Fort Lewis College, Metropolitan State University of Denver, University of Alaska Fairbanks, University of Colorado, University of Denver, University of Northern Colorado, University of Wyoming, Utah State University, and Western Colorado University.

∞ This paper meets the requirements of the ANSI/NISO Z39.48-1992 (Permanence of Paper).

ISBN: 978-1-64642-780-2 (hardcover)
ISBN: 978-1-64642-781-9 (ebook)
https://doi.org/10.5876/9781646427819

Cataloging-in-Publication data for this title is available online at the Library of Congress

Cover photograph (*front right*) courtesy of The Cleveland Art Museum. The obsidians (*front left, back*) were replicated with direct percussion and finished with a copper flaker by José Alfonso Amador Domínguez, artisan from the town of Nopalillo in the Sierra de Las Navajas, Hidalgo.

To the memory of Dr. Hattula Moholy-Nagy, a dear friend and colleague responsible for modernizing the study of Tikal lithic artifacts. I also dedicate this volume to all of my advisors and professors in Mesoamerican studies over the years, in chronological order, including Karl Taube, Phil Wilke, Scott Fedick, Stephen Houston, Takeshi Inomata, Wendy Ashmore, Eduardo Douglas, Michael Kearny, John Clark, John Robertson, John Hawkins, Ray Matheny, Donald Forsythe, Rosemary Joyce, Victor Golla, Silvia Bucher, and Tom Jones.

There seems to have been a competition to see who could make the most elaborate design in flint.

J. E. S. Thompson, Maya archaeologist

Contents

List of Figures | xi
List of Tables | xvi
Preface | xvii

Part One: Technological, Typological, and Regional Considerations

1. Introduction: Techno-Morphological Variation in Ceremonial Lithic Artifacts from the Maya Area and Central Mexico
Zachary Hruby | 3

2. Eccentric or Silhouette: A Preliminary Typology of Ceremonial Chipped-Stones from Sierra de las Navajas and Teotihuacan
Alejandro Pastrana and Zachary Hruby | 51

3. An Apologia for the Punch: Ethnographic, Ethnohistoric, and Archaeological Evidence of Indirect Percussion Techniques in North and Central America
Benjamin Eble and Zachary Hruby | 76

Part Two: Studies of Chert and Flint Artifacts

4. Exploring Maya Flint Mastery: An Autopsy of a Ceremonial Scepter

John E. Clark | 131

5. Replicative Experimentation on the Effigy Flints of Copan, Honduras

James C. Woods | 155

6. Was the Elaborate Chert Eccentric from San Andres, El Salvador, Made by the Rosalila, Copan, *El Maestro*?

Payson Sheets | 174

7. Eccentric Production Techniques and Caching Practices from Structure A9 at Xunantunich, Belize

Kelsey J. Sullivan and Jaime J. Awe | 185

8. Comparison of Chert Silhouettes from the Late and Terminal Classic Eastern Maya Lowlands

Zachary Hruby, Christina Halperin, Catharina Santasilia, and Jaime J. Awe | 209

9. Of Water Lilies and Lightning: Technology and Symbolism of Late Classic Lithics Caches in Northwestern Belize

Joshua J. Kwoka | 240

Part Three: Studies of Obsidian Artifacts

10. An Interpretative Framework and Description of Ritualized Obsidian from Caracol, Belize

Lucas R. M. Johnson, Arlen F. Chase, and Diane Z. Chase | 269

11. Obsidian Pressure-Blade Caches from the 8N-11 Structure 69C at Las Sepulturas, Honduras

Jorge Ramos, Zachary Hruby, and Xinwei Li | 295

12. Obsidian for the Dead: The Technology and Patterning of Classic Maya Blade and Debitage in Tomb Contexts at El Peru-Waka'

David Rafael McCormick Alcorta, Zachary Hruby, Damien Marken, Olivia Navarro-Farr, Michelle Rich, and Keith Eppich | 319

13. Volume Summary: Keeping Stone Alive, Active, and Attractive

Lucas R. M. Johnson | 352

Index | 363

About the Authors | 373

Figures

0.1. Map of Mesoamerica and the sites discussed in the volume | xix
1.1. A comparison of naturalistic and stylistic silhouette forms | 6
1.2. Self-referential skeuomorphs in obsidian and chert | 17
1.3. Obsidian silhouettes from Cacaxtla displaying Teotihuacanoid forms and transverse-parallel pressure flaking | 19
1.4. Stylized Maya forms probably borrowed from, or influenced by, Teotihuacan forms | 22
1.5. The five basic technological types of Maya obsidian silhouettes | 23
1.6. The four basic technological types of Maya chert silhouettes | 25
1.7. Map of proposed silhouette traditions in the Maya area | 31
1.8. Comparison of different silhouette forms made from imported "fine brown" bifaces | 38
2.1. Examples of some of large, fine silhouette bifaces from the Pyramid of the Moon excavations | 52

2.2. Example of a Teotihuacan serpentiform zoomorph | 62
2.3. Example of a curved bifacial knife from the Pyramid of the Moon | 62
2.4. Perforated-disc silhouette from excavations at Sierra de las Navajas | 63
2.5. Teotihuacan anthropomorphic silhouettes | 65
2.6. Teotihuacan serpentiform zoomorphic-silhouettes | 66
2.7. Teotihuacan quadrupedal zoomorphic-silhouettes | 67
2.8. Teotihuacan conflated silhouettes | 68
2.9. Teotihuacan geometric silhouettes | 69
3.1. Examples of cut antler and deer ulna punches and pressure flakers used in indirect percussion flaking | 80
3.2. Examples of antler punches and cut tine flakers from Colha | 82
3.3. Early experimenters | 86
3.4. Painting of a flintknapping scene by George Catlin | 93
3.5. Edwin Deming's 1932 illustration of a flintknapping scene | 97
3.6. Thomas Beckwith's 1911 photograph of indirect percussion using the knees and legs | 99
3.7. Hupa or Wintu punch, part of the Ray Collection | 101
3.8. Photographs of Theodore Orcutt's knapping toolkit including an array of antler punches and flakers | 107
3.9. An illustration of Mark Harrington's (1941) knapping process with punch and anvil | 108
3.10. Lacandon punch technique | 110
3.11. Display at the Illinois State Museum showing three phases of knapping | 115
3.12. Sheep horn punches described by Phil Geib (2004) | 116
4.1. Face A of BAB showing light on each edge and a pencil rubbing of the face | 134
4.2. Face B of BAB showing light on each edge and a pencil rubbing of the face | 135
4.3. Close-up photographs of BAB showing the different kinds of flake scars and attributes | 136
4.4. Obsidian biface made by indirect percussion showing Type 1 and Type 2 flake scars | 137
4.5. Drawings of BAB and Copan Biface 90-1 showing the types of flake scars and the transverse cross-section of BAB at 5 cm intervals | 138
4.6. Reference system of the flake scars on BAB and Copan 90-1 | 139

4.7. Flaking sequence of an experimental obsidian biface showing the alternation of platforms on one edge to opposite faces | 140
4.8. Platform placements on the two edges of BAB, showing the alternation of flake removals from Face A to Face B | 141
4.9. Copan 90-1 showing flake types and flaking sequence | 144
4.10. Reconstruction of BAB before the final indirect and pressure retouch | 147
4.11. Modern obsidian bifaces made by direct percussion and indirect percussion showing differences in flake scars | 148
5.1. Examples of oblique or clapboard style notches on replicas | 158
5.2. Examples of direct regular notches on replicas | 159
5.3. Examples of direct notching errors | 160
5.4. Forming facial features on replica showing curved outline of facial features | 161
5.5. Replica of face with smoking celt | 162
5.6. Example of deep expanding notches | 163
5.7. Example of headdress on a replica | 164
5.8. Examples of headdresses on replicas formed by deep element isolation notches | 165
5.9. Element isolations notches by different techniques | 166
5.10. Examples of perverse fractures during forming of deep element isolation notches | 167
5.11. Example of a replica formed primarily by deep element isolation notches | 168
5.12. Replica of Dumbarton Oaks specimen showing example of typical manufacturing error while attempting a deep element isolation notch | 169
5.13. Detail of Copan specimen showing flake scars of varying length | 170
5.14. Holding techniques required for working around delicate elements | 170
5.15. Some tools used in this study | 171
6.1. Map showing locations of San Andres and Copan | 175
6.2. Eccentric chert from San Andres | 176
6.3. Eccentric chert from the Rosalila cache, Copan | 179
6.4. Eccentric chert from the dedication of the Hieroglyphic Stairway, Copan | 182
7.1. Map of ancient Maya sites in the Upper Belize River Valley | 187

7.2. Map of Xunantunich with Structure A9 highlighted | 188
7.3. Cache 1 (Xun-16-A9-CA1) from Structure A9 | 194
7.4. Cache 2 (Xun-16-A9-CA2) from Structure A9 | 196
7.5. Denticulate chert flake eccentrics from Cache 1 (Xun-16-A9-CA1) from Str. A9 | 197
7.6. Bifacially chipped chert eccentrics from Cache 1 (Xun-16-A9-CA1) from Str. A9 | 199
7.7. Obsidian eccentrics from Cache 2 (Xun-16-A9-CA2) | 201
7.8. Close-up of the marginal refit of two obsidian macroblade eccentrics from Cache 2 | 202
7.9. Reconstruction of obsidian blade core from which the obsidian macroblade eccentrics from Cache 2 were removed | 203
8.1. Map of Eastern Petén and Central Belize | 210
8.2. The third-person pronoun *U* glyph in Maya writing | 215
8.3. The basic technological types of chert silhouettes in the Maya | 219
8.4. Archaeological context of Ucanal Cache #1-1 and the remains of Stela 19 in Group J | 220
8.5. Position of the chert silhouettes of Ucanal Cache #1-1 in relation to Stela 19 | 221
8.6. Archaeological context of Xunantunich Cache A3-1 | 222
8.7. Position of the silhouettes in Xunantunich Cache A3-1 | 223
8.8. An example of superimposition from a chert silhouette from British Honduras | 224
8.9. A cross-site comparison of cached chert silhouettes | 226
8.10. A comparison of technological types | 227
9.1. Map of the Three Rivers Region showing location of case study sites | 243
9.2. Location of Cache 37 within K'in T'an *plazuela* group, Blue Creek | 244
9.3. Blue Creek Cache 37 in situ | 245
9.4. Contents of Blue Creek Cache 37 | 246
9.5. Location of Tz'unun Structure 21, Cache TZ18-01, and Cache TZ19-02 | 247
9.6. Large laurel-leaf biface, Tz'unun Cache TZ19-02 | 248
9.7. Tz'unun Cache TZ18-01 in situ | 249
9.8. Contents of Tz'unun Cache TZ18-01 | 250
9.9. Lintel 2 of Tikal Temple 3 | 257

9.10. Portrayals of lightning | 258
9.11. Portrayals of water lilies, with arrows indicating calyx | 259
10.1. Overview of Caracol with causeways and terracing extent | 273
10.2. Obsidian artifacts from SDC117B-6 and SDC177D-9 | 278
10.3. Objects from SDC189B and 1SDC189B-6 | 279
10.4. Objects from SDC204B-1, SDC1K-11, and SDC184B-4 | 281
10.5. Objects from SDC1K-6, SDC210B-10, and SDC4E-9 | 283
10.6. Objects from SDC184B-5, SDC205B-5, and SDC205B-9 | 285
11.1. Plan Map of Group 8N-11 | 296
11.2. Copan Valley with location of Group 8N-11 on the eastern side of urban core | 297
11.3. Location of blade caches in front of Structure 69C | 301
11.4. RC-2-105 | 302
11.5. Location of blade caches, offering RC-2-101 and RC-2-117 in relation to Str. 69C | 303
11.6. Blade offering RC-2-101 in situ | 304
11.7. Refitting process and example of blade pieces from two cores | 305
11.8. Offering RC-2-102: blades and cache vessel in situ | 306
11.9. Blade deposit from offering RC-2-111 in situ | 307
11.10. Blades from offering RC-2-101 | 311
12.1. Map of the Maya area with detail of Western Central Petén indicating the location of El Peru-Waka' in relation to other Maya sites | 320
12.2. Relief map of El Peru-Waka' urban center | 321
12.3. Burial 37 plan view | 330
12.4. M13-12 excavation profile view | 331
12.5. Planview of Burial 39 | 333
12.6. Plan view above burial chamber of Burial 61 | 334
12.7. Pie chart of obsidians recovered at burials discussed in this study | 336
12.8. Bar graph of prismatic cores and core fragments as a percentage of each context | 336
12.9. Bar graph of flakes directly related to prismatic core maintenance | 337
12.10. Bar graph of each burial's blade assemblage in relation to stage of reduction process | 337
12.11. Bar graph of core maintenance strategy applied in obsidian burial assemblages at Perú-Waka' compared with that from Ojo de Agua | 338
12.12. Purposefully broken core from Burial 38 | 339

Tables

4.1. Comparison of the Buenos Aires Biface with Copan biface 90-1 | 143

7.1. Context and contents of eccentric caches from Xunantunich | 190

9.1. Size characteristics of oversized laurel-leaf bifaces | 255

10.1. Summary of cached obsidian by temporal period and technological type as of 2018 | 275

10.2. List of "E," "S," and unifacial styles/forms by period | 286

11.1. Comparison of complete blades from 8N-11 caches that contained obsidian | 310

11.2. Comparison of blade metrics from offering RC-2-101 in color groupings, and individually in Set 2 | 315

12.1. Comparison of the burials discussed in the text | 329

Preface

This book represents the culmination of thirty years of study and interest in Mesoamerican rocks, especially the pieces of art into which they were crafted. In some ways lithic analysis always plays second (or third or fourth) fiddle to the more photogenic fields of epigraphy and iconography in Mesoamerica. Flakes of obsidian or chert don't usually make the headlines, and even a workshop, which is a major discovery for any lithic specialist, isn't a real conversation starter around the water cooler. Eccentric flints, however? They *almost* make the grade, especially if they are of the megalithic type at Cahal Pech, Belize, or the god effigy type from Copan, Honduras. Alas, most eccentrics are much smaller, seem repetitive, and in some cases look a little crude to the untrained eye. But they were often deposited in large groups, in strange patterns, and in very powerful places, which demands our attention as Mesoamericanists. They are almost too strange for modern people to grasp. Complex, arcane forms at times seem to resemble shapes we recognize—a crescent, perhaps for the moon, a disc appearing as the sun, translucent in the light. Others appear, well, eccentric. They are a symbol system we haven't entirely cracked.

https://doi.org/10.5876/9781646427819.c000

Although we haven't "deciphered" the meaning of each symbol chipped into obsidian, chert, and flint, we do know that most Mesoamerican cultures associated these raw materials with fire, or the remnants of it. In most cases chipped stone was and still is believed to be the embodiment, or the material correlate, of celestial fire: lighting, meteors, the sun, and other planetary bodies. Many of the eccentrics, or, as we propose here, *silhouettes*, represent these elements, as well as their godly embodiments in the form of stone effigies. This book decidedly focuses on how these effigies in chipped stone were made; the ancient crafters were playing with fire, probably in what was considered a dangerous and onerous task that required some ritualization unto itself. Many of the objects were likely an essential part of elite divination kits, used for a period of time, and then ritually buried during monument dedications and burials.

An important step in understanding them as a class of objects and how they might have been used in the past is to understand how they were made. All of the contributors to this volume examine some important facet of the *chaînes opératoires*, the different stages of resource acquisition and manufacture, that brought these ritualized chipped-stones into existence. Examining production techniques from local and international levels allows us to understand how different communities of practice interacted with one another, or not, through time and over space. As it turns out, the majority of silhouettes are not technically that hard to make. Getting the size of the notches and proportions correct, not to mention performing any ritual proscription, however, was anything but easy. Even the simplest eccentrics required some esoteric, and probably guarded, knowledge to manufacture.

What many nonspecialists may not be aware of is that these objects were usually buried below some of the most impressive pieces of sculpture, within some of America's most famous temples, and above (and within) many royal tombs. Much like ceramics, they bridge the ordinary and extraordinary worlds, because the ancients couldn't have made their grand monuments without them. The difference of course is that chipped stones were considered deadly, being the most typical implement of human sacrifice and warfare. In a practical sense, sculpture and architecture were but specializations of lithic technology in Mesoamerica. In a deeply religious sense, the construction of their temples wasn't finished without offering the same materials. Even today many Mesoamerican peoples include ritual in the production of architecture. It can be as important to the project as the actual construction process. Interestingly, it seems to have been true for the smaller things as well. At least some chipped-stone production was ritualized. So, ritualized chipped-stones reflect Mesoamerican ontologies, which, for us modern Capitalists, may be difficult to grasp. We go to work, make things, get a paycheck, go home, and move on. That is probably *not* how eccentrics and

FIGURE 0.1. *Map of Mesoamerica and the sites discussed in the volume (map by Hruby).*

tomb obsidians were made. Thus, the social context is important to understand the production process and the *chaîne opératoire*.

While religion is addressed here as an important factor in making ritual instruments, we don't delve too deeply into symbolic interpretation (but see chapters 8, 9, and 12). These topics, including detailed iconographic and epigraphic analyses, are to be addressed in future publications. The finished objects discussed in this volume represent the nexus between iconography, epigraphy, and lithic studies. Here we primarily focus on the technological and social factors that made them possible.

Central Mexico and the Maya area were places of great innovation in chipped-stone manufacture during the Classic period (figure 0.1). Although the book is lopsided toward the Maya area, we cover Teotihuacan-related artifacts directly in two chapters, and address its significance tangentially in four others. As the reader will be made aware, the linkages between the two culture areas extended to the use of chipped stone in monument dedications. Understanding one cannot be done without the other. I would like to thank my mother for supporting my research, my wife Lindsay Powell for her field and lab work, as well as all the contributors to this volume, especially for participating in the original symposium—the very long symposium.

PART ONE

Technological, Typological, and Regional Considerations

1

Introduction

Techno-Morphological Variation in Ceremonial Lithic Artifacts from the Maya Area and Central Mexico

ZACHARY HRUBY

Of the many durable resources in ancient Mesoamerica, obsidian and flint were considered the deadliest. They were also often the most transformative. Without metal, these stones were used to cut and carve the natural world into a human one and, in some cases, into a supernatural one. Civilization building in Mesoamerica involved a reflexive relationship between people and the gods, between chaos and order, and between the quartered realm of the maize field and the wild untamed forest (Taube 2003). Just as the gods created the universe, humans were responsible for maintaining cosmic order by reenacting mythological events, repeatedly over time. It was done directly by transforming mundane spaces into sacred landscapes, by reproducing mythological or distant locales locally, and also through blood sacrifice, which was perhaps the most important part of the production process. Sacrifice was reciprocal. Through great effort, creator deities crafted people, and humans in turn crafted the gods (e.g., Carrasco 2013; Hruby 2007). The caches archaeologists have found by the hundreds over the decades are the material remains of those creative efforts, of which blood was an indelible part. This book is about how obsidian

https://doi.org/10.5876/9781646427819.c001

and flint were transformed into bloodletters and god effigies, which themselves undoubtedly shared in the sanguine offerings that maintained balance in the Mesoamerican universe.

Chipped stone played a central role in both the material and symbolic aspects of creation: carving stones and bodies alike. It was indispensable to human survival and making the civilized world. Like the Mesoamerican storm gods who wielded stones as lightning weapons, however, they were considered dangerous and not entirely predictable. As such, they became metaphor for transformative power and the ability to control it—the very symbol of rulership. Throughout ancient Mesoamerica, rulers were associated with celestial fire, impersonating storm and solar gods who could control lighting, the burning sun, and shooting stars in the form of finely crafted flint, obsidian, jade, and turquoise objects. Given the significance of these items, they were usually created by master craftspeople who could make something fit for the heavens. Thus, the process of production may have also been a religious experience, which heightened the status of the item being produced, and perhaps the producer (Hruby 2007).

Ritually depositing chipped stones is a New World tradition that goes back to Paleoindian times, but in Mesoamerica it gained clear religious and political significance at least by the late Early to Middle Formative periods (1100–900 BCE) (cf. Hirth and Cyphers 2020). The Zapotec and Preclassic Maya, among many other Formative peoples, would ritually deposit finely crafted pressure blades, as well as the cores from which they originated (e.g., Marcus and Flannery 1996). The Paleoindian and Archaic caches undoubtedly were more focused on the more utilitarian notion of *caching*, which is to hide away something for later use. For better or worse, *cache* was adopted by Mesoamerican archaeologists to describe ritual deposits that may not have been meant to be revisited. We use it in this book to refer to purposeful, systematic, and stylized deposition of items largely for ritual purposes, similar to the term ofrenda in Mexican archaeology.

Although undoubtedly a display of conspicuous consumption, the practice of offering chipped stone was also likely tied to blood sacrifice. As chapters 10, 11, and 12 show, the caching of obsidian and chert blades, as well as their inclusion in burial contexts, continued up to the Classic period and beyond. Caches containing chipped stones became a central focus of architectural and monument offerings at the onset of the Classic, and only intensified at most sites into the Late Classic. The Maya and Teotihuacano knappers would take the production of ceremonial chipped stones to heights never again achieved in the New World, and perhaps the Old. This book and its contributors focus on the Classic florescence of lithic technology in Mesoamerica, and the two areas in which it played such a central role in statecraft.

THE SYMBOLIC SIDE OF CHERT, OBSIDIAN, AND CHIPPED STONE AND HOW THEY CAME TO BE

While bedrocks, such as basalt and limestone, had their own unique uses and values, chert, obsidian, and jade in some cases often symbolized potent celestial fires or their aftermath: shooting stars, planets, the sun and moon, and especially lightning before and after it struck (Agurcia Fasquelle et al. 2016; Hruby 2007; Iannone 1993; Taube 1991, 2000). Somewhat paradoxically, they also seem to have been associated with ice and water in some parts of Mesoamerica (Pastrana and Athie 2014) and were subsequently connected in more abstract ways to rain and agricultural fertility, rulership, jaguar claws, castigation, witchcraft and divination, and success in hunting and warfare. In the highlands, obsidian may have been understood as an icy, frozen form of lightning wielded by the Central Mexican storm god, Tlaloc, from his cold mountain abode where obsidian tends to form naturally as a volcanic rock (Pastrana and Athie 2014). As such, these rocks carried a heavy symbolic load that was used to full effect by Mesoamericans in some of their most extreme, elaborate, and likely bloody ceremonies. Perhaps most significant, they were the bringers and even consumers of blood, an essential aspect of the sacrificial covenant between gods and humans. In the ever-escalating strife and warfare of Classic Central Mexico and Yucatan, chipped stone became the symbol of military might, which was ultimately the source of sacrificial victims needed fulfill the obligation (e.g., Carrasco 2013; Taube 2000).

The largest and most finely crafted of these objects required an incredible degree of skill and craft specialization to produce, not to mention large quantities of raw material. Beyond incredible knapping talent and skill, the seemingly arcane forms of some pieces, and perhaps even their measurements (Prado 1999), also required detailed esoteric knowledge to bring off. In a similar way to how royal scribes, painters, and carvers controlled esoteric production knowledge in ancient Mesoamerica (Inomata 2001), the best knappers may have done the same (Hruby 2007; Pastrana and Athie 2014). As Alejandro Pastrana and Silvia Domínguez (2018) have shown for the obsidian workshops at La Sierra de las Navajas, production was ritualized—a practice also shared by the Classic Maya (Iannone 1993; Hruby 2007). Just as the proper construction of architecture required ritual sacrifice and caching, too, the production of smaller ideologically charged items required ritualization. In other words, the product was valuable as a prestige good, but the very *process* of production and the social relationships involved in that process were an important part of social cohesion, competition between artisans, and the material expression of the state ideology (Hruby 2006).

Many Classic lithic offerings do not resemble utilitarian tools but were instead crafted into the shapes of gods and celestial bodies, as well as natural and supernatural forces. The final products could be more stylized or more naturalistic in

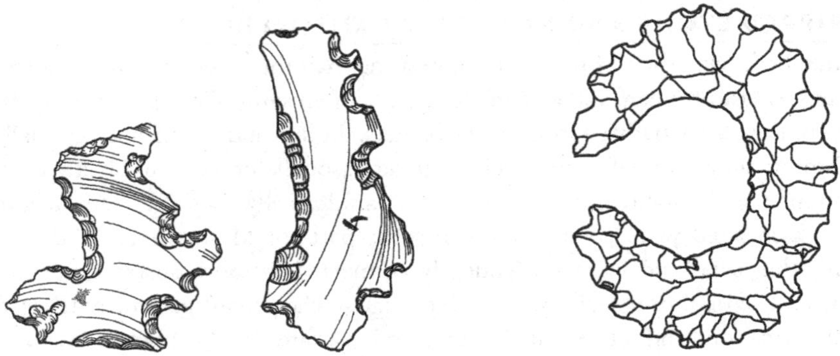

FIGURE 1.1. *A comparison of naturalistic and stylistic silhouette forms (drawings by Hruby).*

their outline, apparently evolving in parallel traditions (figure 1.1). Historically these objects were mislabeled by archaeologists as *eccentrics*, since their meaning and functions were unknown, many were asymmetrical (i.e., eccentric by geometric definition; see chapter 2) and, as the epigraph of the volume illustrates, were thought to have been made by caprice or the wild obsession of their producers. We now know they were made according to a codified symbol system with specific associated meanings.

The symbolisms of eccentrics or silhouettes are treated in depth elsewhere (Hruby in prep.; Hruby et al. in prep.), but it bears noting here that they cannot be understood without the context of their deposition and probable use. They were multivalent objects that drew meaning from their raw material and format of that raw material, technique of production, outline, use before deposition, provenience of deposition, and association with other objects in the final deposit. The Maya seem to have used different colors of chert and technological types of obsidian flakes and cores to evoke subtle contrasts, which included directionality (Hruby 2007). What is more obvious, however, is the stark contrast between obsidian and chert in cache contexts. These materials usually co-occur in caches, apparently to evoke dichotomies between night and day, heavens and underworld, bright celestial fire and burned-out or finished fire, east and west, and likely together as "everything" or completeness, similar to the phrase "the alpha and the omega" (i.e., 24/7, totality, or perhaps eternity). The Maya hieroglyph for *tz'ak*, which means "whole or complete," evokes a similar meaning visually by pairing opposite images, including light and dark, red and green, and sun and moon, among others (Stuart 2003).

As we discuss in chapter 8, and I have argued previously (Hruby 2006, 2007; see also Schele and Miller 1986; Agurcia Fasquelle et al. 2016), many silhouettes represent celestial bodies and forces: sun, moon, constellations, lighting, and others. These concepts were expressed stylistically as simplified icons, such as a crescent representing the moon, or more naturalistically as the anthropomorphic deities that represented these celestial entities (see figure 1.1). Given that some have light use-wear but were simultaneously deposited frequently in cache contexts indicates that they had a use-life beyond the cache deposit. However, that use-life was probably tied to the completion of important periods of time from 1-, 5-, 10-, and 20-year intervals, or reoccurrences of important Calendar Rounds from mythological time, suggesting they were a type of short-term inalienable item (see Kovacevich 2011 for a discussion of inalienability in the archaeological record). The closest approximation to these small portable god-effigies from the ethnographic and ethnohistoric records would be (1) Maya *sastuns*, which are semitranslucent seeing stones used for divination; (2) anthropomorphic or zoomorphic bloodletters; (3) apotheosized game pieces, such as in *patolli*, which moved through calendrical time and the heavens; (4) and elaborate markers for counting days along with the codices, again for divination. Put another way, they seem to be ritual tools used by religious specialists with an array of possible functions. It should be noted that Kwoka (chapter 9) takes a slightly different tack and connects the silhouettes in his sample to a different set of meanings. In his study he emphasizes handheld eccentrics depicted in the iconography at Tikal and many other sites, which were connected to the "water lily jaguar," a being associated with nighttime, the watery underworld, water lilies, and human sacrifice. More mundane versions of these things may have been used by commoners, but silhouettes seem to have been used to reinforce or legitimize elite political ideologies tied to calendrics, monument erection, and mythology.

When the cherts and obsidians were finally cached below stelae, in plazas, and in stairways, they took on yet another set of meanings based on their position in the deposit (cf. Joyce 1992). The ancient Maya and many other Mesoamerican cultures used caches to create a tableau or microcosm of the universe itself, whereby the process itself may have approximated the creation event. Key acts of creation and universe creation gleaned from sources such as the Contact period Popol Vuh (Christenson 2007) and the Late Preclassic San Bartolo murals (Taube et al. 2010) include: (1) setting up the four directions; (2) positioning, sometimes hierarchically, different creator deities in space both at night and during the day (i.e., light and dark); (3) establishing the centrality of the Sun and the Maize God, integrating various sky and lightning deities; and (4) the inclusion of a vessel of some kind that acted as a portal for the sun and the World Tree, but also hold rain or sacrificial blood to feed the gods. Thus, used items and found objects can become physical proof of the recreation of the universe, but in different locations depending on where the present ritual is being carried out and the time of year and day it occurs.

Common tool forms and unmodified production debitage were also frequently offered, but they were placed in symbolically loaded contexts or in specific sequences that betray their characterization as mundane (Hruby and Rich 2014). Some archaeologists attempted to describe the deposition of debitage in funerary contexts in utilitarian ways, for example, as a means to discard dangerous production waste (Moholy-Nagy 1997; but see chapter 10). Recent research has shown that there was a deeper symbolic meaning to these tomb deposits (chapter 12; Hruby and Rich 2014; Johnson and Johnson 2020; Trachman 1999, 2002), and once again the focus must shift to the craft specialists who produced them and the individuals who deposited them.

Over the last three decades studies of ceremonial lithic artifacts have opened untapped avenues of inquiry (Hruby 2007; Hruby and Ware 2009; Iannone 1993; Kwoka et al. 2019; Moholy-Nagy 2008). Early interpretations of the most elaborate creations of the Maya and Teotihuacanos were based on iconography or untested imaginings (e.g., Blom and La Farge 1928; Gann 1914, 1918, 1925, 1929; Gann and Gann 1939; Gruning 1930; Joyce 1932; Heye 1925; Joyce et al. 1928; Kidder 1947; Lane Fox 1871; Linné 1934; Maler 1908; Mason 1935; Price 1897–1899; Rice 1909; Ricketson 1929, 1931; Ricketson and Ricketson 1937; Smith 1972; Stephens 1860; Thompson 1931, 1936, 1939). While ritual lithic caches in the Americas go back to Paleoindian times, and jade and chipped-stone caches are common in Formative Mesoamerica, the creation of chipped-stone effigies seems to have begun in earnest by Teotihuacanos and the Maya during the Terminal Formative and Early Classic (later in this chapter and chapter 2; Agurcia Fasquelle et al. 2016; Iannone 1993, 136–154; Meadows 2001; Willey 1972). After the collapse of Teotihuacan, the creation of related forms continued for a time in Central Mexico at places like Cacaxtla but then also faded away. Similarly, the abandonment of the Late and Terminal Classic Maya kingdoms marked the end of large-biface production that would not appear again until the height of the Aztec Empire. With these caveats in mind, this volume focuses on the Classic period fluorescence of these traditions in the two culture areas they were most developed—Central Mexico and the Maya area—but with the understanding that these two traditions and their practitioners likely influenced one another in more or less direct ways.

Many ceremonial chipped-stones were classified as *eccentrics* over the years, and some of the chapters (2 and 10) in this volume address the ineffectiveness of the term. I briefly discuss problems in classification and typology related to its use, with the proviso that the issue is discussed in more detail elsewhere (see also Johnson 2016; Hruby in prep.; Hruby et al. in prep.). Personally, I have used *eccentric* in most of my previous research. For Mesoamerican archaeologists it is an incredibly useful term, since it is short, is infrequently employed in other contexts, and immediately brings to mind chert and obsidian items that were chipped into nonutilitarian forms. More recently I have used *ceremonial chipped-stone* or *ceremonial lithic artifact* to describe them, which is a bit more cumbersome but certainly more accurate. Under this rubric one may include artifacts with a utilitarian form that were clearly produced for, or in, ritual contexts. Of course, the utilitarian-ceremonial dichotomy implied by most functional classes and types is the source of most of the classificatory problems (Hruby 2003). In Mesoamerica, past and present, ritual and religious practice could be indelibly tied not only to the object being used but also the very production of that object (e.g., Hruby 2003, 2007; Hruby and Flad 2007; Hruby and Rich 2014; Johnson 2016; Trachman 1999, 2002). One issue with *ceremonial lithic artifact* is that it implies that the objects were intentionally produced for a ritual or ceremonial event. We now know that some chipped stones were collected from elsewhere for deposition in ritual contexts. This "elsewhere" could be anywhere, including more mundane production areas, dumps, or even found contexts (see Brown 2000 and 2015 for a discussion of found items used in ritual practice).

Lucas Johnson (2016; also see chapter 10) has recently proposed the term *ritualized lithics*, which is useful for discussing a broad range of artifacts and phenomena. It underscores the oft-shifting identities, or more accurately *identification*, of these objects as a social process. Eccentrics, utilitarian forms, and even debitage neatly fit under this rubric depending on the context. However, it employs a recently nominalized version of *lithic* and by itself is not a descriptor for chipped stone specifically, but rather all stone. In short, the nominalized version of *lithic* is quite vague, and the use of chipped or flaked stone may be more descriptive. *Ritualized lithic* is so broad that we are still left wanting for a specific term for the "eccentrics." I have argued elsewhere that these things had many functions and symbolic meanings, including divining or seeing stones, weapons, bloodletters, and god effigies, none of which make a functional classification clean or simple (Hruby 2007). In the present volume, Alejandro Pastrana and I (chapter 2) propose the use of *silhouette*, which puts the focus on morphology instead of function. *Chipped-stone silhouette* emphasizes the outline of the piece, which makes preliminary sorting cleaner and simpler, at least for Central Mexico. In practice William Coe (1959) focused primarily on outline as well, but the variability in some forms and overall complexity of the Maya system caused him some vexation. Thickness and production technique were important, but one consistent goal seems to have been to create a clear and easily identifiable outline. In this chapter, and throughout the book, we often use the terms *eccentric* and *silhouette* interchangeably depending on the questions we ask of the archaeological record, and the archaeologist doing the asking.

Some chapters of this book focus entirely on lithic technology and the process of production. The simple way to describe how silhouettes and other ceremonial lithic items were made is through direct percussion, indirect percussion, and pressure. As we lay out in chapter 3, this is the trinity of all lithic technology in the

Americas, and likely for much of the rest of the world. All of these basic technologies produce a cone or conchoidal fracture-pattern (see Crabtree 1972). If you can imagine the strange conical break that happens when a BB hits a pane of glass, then you already understand the basics of snapping a flake off a core going back to the Oldowan tool system of *Homo habilis*. You use a portion of that cone of force (i.e., hertzian cone) to remove a flake from a mass of homogeneous stone. However, within each of these basic technological categories there are myriad techniques that people employed the world over, and based on the stone tools and debitage alone (i.e., the flake scars and patterning), one cannot always tell which technique was used. This can make reconstructing *chaînes opératoires* and the stages of production a confusing and contentious endeavor.

It has been hypothesized, but rarely published, that indirect percussion was the technique responsible for making all of the complex notches and volutes we find on silhouettes. Gene Titmus and Jim Woods (2003) were the first to systematically replicate the technique in the lab and publish the results. The idea is that a punch (i.e., a finger-length segment of antler or hardwood) is placed on a hard-to-reach platform and then struck with a hammer. The result is a beautiful, wide flake-detachment with a feathering termination, While I can confirm that indirect percussion must have been used in some cases, it is also clear that ancient stoneworkers were throwing the whole toolkit at these projects. Jim Woods (chapter 5), for example, illustrates in this volume that pressure flaking, in addition to indirect percussion, was an important part of making the large effigy silhouettes from Copan, Honduras. John Clark (chapter 4) moves the focus away from the silhouettes and aims it at how massive bifaces were manufactured. Larger silhouettes were made from these kinds of bifaces. Setting the stage for all of these discussions, however, is Benjamin Eble and my (chapter 3) argument that indirect percussion played as big of a role in the development of New World lithic technology as direct percussion and pressure. In concert with Harry Shafer's observations on Colha and Texas archaeology, we show that tools of indirect percussion have been identified in the Maya world and likely have their origins with early inhabitants of the continent. Chapter 3 demonstrates that indirect percussion was not simply a one-off technique for making rarities, but rather an essential intermediate stage of biface and blade production throughout much of Native American prehistory. The remainder of this introduction and the volume in general focus largely on how these ceremonial lithics were made, with implications for their distribution and deposition. We examine chert and obsidian effigies as well as ritualized prismatic blades and debitage from tomb and cache contexts. We believe that the value of the book extends beyond specialists in Mesoamerica, since the processes of production and replication experiments discussed within may help archaeologists throughout the Americas understand their own samples, chipped or not.

COMMUNITIES OF PRACTICE OR TRADITIONS?

Most of the objects discussed in this volume are, and were, special. They required great skill to produce and esoteric, probably guarded or proscribed, technological and ritual knowledge to finish in a socially acceptable way. In the past I have used practice theory to describe how ancient Maya knappers may have used their knowledge as a form of *symbolic capital* (sensu Bourdieu 1977) to cement their position in society and gain higher status (Hruby 2006, 2007; Hruby and Ware 2009). The situation becomes more complex if we consider the possibility that some knappers were also royal sculptors and architects. In other words, some individuals may have practiced multiple related skills that were considered to be discrete forms of identification in their own city, town, or village. Thus, in discussing how these ideas and practices were shared across great distances, not to mention unfriendly political boundaries, robust theory is a necessity. Most authors in this book question not only how relative homogeneity can occur across contested regions and zones but also how different kinds of knowledge were operationalized *within* polities.

The concepts of *communities of practice* and *constellations of practice* are useful tools to explore local and regional variability in technology and iconography (see chapter 8), but through the lens of individual practitioners who shared production knowledge and histories through practice and situated learning (i.e., legitimate peripheral participation). Situated learning, originally explored by Jean Lave and Etienne Wenger (1991; see also Wenger 1998), helps archaeologists theorize: (1) different kinds of skills—ranging from simple to complex—as they were shared in more or less restricted ways; (2) the possibility of membership in more than one community of practice; and (3) how technologies and design elements were transmitted between individuals and between communities of practice through brokers and boundary objects (e.g., Roddick 2016; Roddick and Stahl 2016). Archaeological models that focus on different artifact traditions and phases have the effect of removing the artisans, as individual agents, from the processes of change vis-à-vis relative stability, or essentializing all this complexity into a master-apprentice relationship. Originally a bit nebulous, these ideas have been refined and used successfully by archaeologists to postulate how different suites crafting practice change over time (e.g., Mills et al. 2013; Sassaman and Rudolphi 2001), especially in the realm of ceramic crafting (Roddick and Stahl 2016).

For this chapter, in which I make some broad and sweeping generalizations about ceremonial lithic artifacts from throughout Mesoamerica, identifying communities or constellations of practice is a bit out of reach. The reason is that detailed technological and material analyses are rare, and some are not published or well illustrated. Such a study would also be a gargantuan task of the highest order, and would require the involvement of many scholars in a sort of "Pecos

Conference"–style information exchange. Nevertheless, beginning the endeavor from a practice theory perspective directs our future analyses in potentially useful ways. The data required for practice-based studies go beyond simply looking at the final product (i.e., outline and material) and toward detailed technological, morphological, and iconographic analyses. The examination of *chaînes opératoires* requires that we consider the social relations at each stage of production and resource acquisition at each site. Ultimately context, and thus the function and use-life of the objects, must play a role as well. The reality is that some sites have received this kind of attention, especially at Teotihuacan, Tikal, Piedras Negras, Caracol, Copan, and some sites in Belize, but overall, they are quite rare, especially from a site-wide and regional perspective.

Theories of communities and constellations of practice encourage more of a "bottom-up" approach and are especially helpful in the study of cache goods, since so little is known about their purpose and creation. In this volume, chapter 8 examines possible constellations of knapping practice in the Eastern Maya Lowlands, specifically between the sites of Ucanal and Xunantunich. The sites are relatively close to one another, and their caches have been analyzed in great detail, revealing similar but divergent traditions. More conservative communities of practice seem to have existed at nearby Tikal, where some types of silhouettes were produced for hundreds of years, albeit using a variety of techniques. Things get more complicated with elaborate effigy flints from Copan (Agurcia Fasquelle et al. 2016; Clark et al. 2012; Payson Sheets, chapter 6) that were made by a group of master artisans whose pieces are found as far afield as Belize and Campeche. Were these artisans mobile, did they share their knowledge across vast distances, or were the final products simply traded long distances and then copied? The issue of the centralization of communities of practice, and the possibility of maintaining communities of practice over large distances, which is not a problem in today's world, is addressed in chapter 8. We are learning now that people were moving quite frequently between centers as distant as Tikal and Teotihuacan (Stuart 2000, 2024). The question is whether these intermittent but politically charged points of contact and forms of situated learning can be considered a *community of practice* in ancient Mesoamerica.

Given the issues mentioned earlier, the divisions I examine in this chapter are better addressed as archaeological *traditions*. Gordon Willey's (1945) concept of tradition had, and still does, enjoy widespread use in American archaeology. It is useful for conceptualizing massive quantities of data over a long period of time in a particular region. Emil Haury and coauthors (1956, 36) concisely define tradition as "a socially transmitted culture form which persists in time." As noted earlier in this section, it creates an interpretation rather bereft of people and individual agents, and it does not explain *how* practices were transmitted between crafters and communities of practice. On the other hand, it is excellent for looking at regional variations over time and summarizing temporal and spatial changes in artifact types and varieties. It was in this spirit that William Coe made his still-useful eccentric typology, starting at Piedras Negras in the 1950s and then refined with the much larger sample from Tikal (Coe 1959, 1965; Moholy-Nagy and Coe 2008; typology of specific forms). It is complementary to the *horizon* concept, which consists of a culture trait—technological, morphological, symbolic—that appears over a wide area, indicating some kind of cultural continuity or transmission, in a relatively short period of time.

The Early Classic in Mesoamerica seems to have given rise to a widespread horizon in ceremonial lithic artifacts, which may have developed through increased direct interaction between Central Mexico and the Maya area. Each of these culture areas had already been developing their own traditions in the Preclassic (i.e., Formative), with increased specialization and experimentation in the Late Preclassic. After that point, regional traditions seem to emerge and crystalize in different parts of Central Mexico and the Maya area. The bulk of this horizon consists of the adoption of geometric, sinuous, and naturalistic forms produced for ritual caches, burials, and other high-profile religious ceremonies. I use *tradition* rather loosely for the moment to push forward the agenda of understanding large-scale changes in lithic artifact types over the entirety of the Classic period. The initial types presented in this chapter (see also chapters 2, 8, and 9) are based on technological traits, morphological traits, size, and material type, and will undoubtedly be modified over time or replaced by more useful theoretical approaches. These different traditions may or may not reflect communities or constellations of practice.

DISTRIBUTION OF TECHNOLOGICAL TYPES AS REGIONAL TRADITIONS

Aside from the variation in contexts mentioned, there is also a great deal of variety in technological types regionally. Further, different kinds of ceremonial chipped stones performed different social functions and cannot be conceptualized as a single class of artifact. Large debitage deposits tend to be associated with so-called termination deposits and above burial deposits. Big silhouettes (>20 cm) with clearly chipped handles were used as ceremonial weapons, scepters, and regalia elements used in deity impersonation and blood sacrifice. They were commonly wielded by elites, often in scenes of blood sacrifice, and seem to be a distinct emic group of objects from the smaller cache eccentrics. Chipped-handle silhouettes are usually not found in caches, with the most obvious exceptions being the Maize God–K'awiil scepters found at Copan (see chapters 5 and 6 here) and the famous perforated axe head from Altar de Sacrificios (Willey 1972). In the Maya area chipped-handle silhouettes likely developed out of the stemmed macroblade tradition in chert and obsidian. These kinds of handled "daggers" have been found in obsidian at Kaminaljuyu, Guatemala (Hruby 2004), in the Preclassic, as well as at Copan, Honduras, in the Classic period (Hruby and Tokovinine n.d.). It is unclear whether the chert macroblade and obsidian macroblade manufacture informed one another in any way, but it is likely, especially given that practices as complex as monumental carving were shared between the two regions. Smaller bifaces, unifaces, and modified flakes tend to be found in dedicatory caches, and more rarely as burial furniture. They are almost always found in royal and elite contexts, and, as I mentioned earlier, probably served simultaneously as divining tools (Hruby 2007, 2020) and small god effigies (Hruby 2007, in prep.; Iannone 1993). However, these are general trends, and, as is so common in Mesoamerica, there are always exceptions to the rule (see chapter 10; Johnson 2016). In this section I review some general technological patterns found in Mesoamerica, with a special focus on the silhouette forms of Classic period Teotihuacan and Maya areas.

Ritualized lithic deposits were undoubtedly important to the religious life of Classic Oaxacan and Isthmian peoples, but they did not play as large a role as in Central Mexico and the Maya zone. That is not to say that chipped stones were never ritualized by the Zapotec, but rather that many of the forms seem to be imported as finished objects or were largely unmodified before interment. One exception is a Middle Formative blade that appears to have been chipped into a centipede or stingray spine effigy (Marcus and Flannery 1996, 127), but this was made previous to the time period discussed here. Some of the crescent and biface forms found in Oaxaca were imported from Teotihuacan but do not seem to be used in the same ways. Bifaces have been found in temple contexts and have been interpreted as knives used for heart sacrifice (167). Prismatic blades seem to be a common aspect of Classic period tombs at Monte Alban and were likely part of the bloodletting tools of the dead, or perhaps used by those reentering the tombs, long after the initial interment of the deceased. Caches are present at Monte Alban, of course, but often contain ceramic effigy censers and jade (e.g., Marcus and Flannery 1996, 104). In other words, the Classic period Zapotec tradition contrasts with those of Central Mexico and the Maya area, where ritualized lithic artifacts are found by the hundreds or thousands.

The production of chipped-stone silhouettes undoubtedly began in the Preclassic, with a vague correlation between iconography and artifact. In early Maya art, lightning was depicted as sinuous and serpentine, which recalls the Pan-American conceptualization of lighting, rain, and storms in general, as related to the snake (Agurcia Fasquelle et al. 2016; Hruby 2007, 2020; Looper 2003). At the same time iconographic conventions depicting sharp, shiny, or fiery elements began to overlap with lightning forms, particularly for undulation an notching that recalls the serrated or jagged margins of chipped stones. Hence, a few Preclassic ceremonial bifaces and blades feature notches. These notches did not have a utilitarian function but were rather mimicking an iconography that emphasized notions of celestial fire and sharpness. In more precise terms, some Mesoamerican craftspeople created chipped-stone *skeuomorphs* that represented idealized, or stylistic, representations *of* chipped-stone materials. I failed to find an art historical term that adequately describes this tradition.

The term *skeuomorph* usually refers to an object made of one material that is meant to look like an object made of another. A Maya example would be a ceramic vessel that was made to look like a cut gourd or woven basket (Houston 2014). True skeuomorphs also existed in flint and obsidian, as is the case with the famous obsidian axe from Yaxchilan (Kaneko 2003), where a hafted celtiform axe was rendered completely in one piece of obsidian. In the case of the idealized skeuomorphic artifacts, however, it is difficult to find an analogy in Western culture. A poor one might be a Duraflame log: it is made of wood, performs the same function as wood, and is meant to look like an idealized log, but it isn't *necessary* that it look like a log to perform its function. Another analogy might be the McDonalds McRib sandwich. It is made of meat but is transformed, apotheosized if you will, into an idealized, almost supernatural version of meat to increase its social value and iconicity. What I term a *self-referential skeuomorph* tradition was not terribly widespread until the Classic period, and the symbolic load seems to have been carried by the basic utilitarian forms in the Preclassic. The idea of producing an object to fit the iconographic representation of the material, however, had been born.

CENTRAL MEXICO

The tradition of chipped-stone silhouettes seems to have emerged in Central Mexico around the same time it did in the Maya area but with some key differences. Some of the earliest examples of Central Mexican silhouettes, dating to the Terminal Formative (100 BCE–150 CE), were discovered by David Carballo (2014) and his team at La Laguna, Tlaxcala. They boast some forms that were never seen before nor have been seen since. However, the starting point—the preform from which the silhouettes were made—was the laurel-leaf or lenticular biface. Interestingly, the preforms seem to have been fully formed, Teotihuacan-style bifaces, complete with transverse parallel pressure flaking that became a cultural marker for Teotihuacan and later Cacaxtla. One type resembles centipede and stingray forms (Carballo 2014, 207–212), which would not appear in the Maya area, at least in chipped stone, until the Classic period. The other more numerous type in the La Laguna cache is probably a crude or prototypical version of a plumed serpent, which is the most common type at Teotihuacan in the Early Classic. Roughly contemporaneous offerings from the Pyramid of the Sun (0–100 CE), although not terribly complex, contained anthropomorphs and miniature atlatl points (Carballo 2014, 198; Millon and Drewitt 1961, 375–376; Millon et al. 1965).

By the Late and Terminal Formative, Central Mexican knappers began producing chipped obsidians that would influence the rest of Mesoamerica until the Colonial period. Teotihuacano stone workers would later knap blades into serpent forms, recalling the lighting snake traditions in the Maya area, but added other variations on that theme. They also developed a silhouette tradition that arose out of the projectile point industries Teotihuacan is famous for: tanged dart points finished with the hallmark transverse parallel pressure-flaking. Interestingly, the tradition of transverse-parallel pressure flaking seems to have begun around the same time that the first silhouettes were being created for architectural caches (Pastrana and Carballo, personal communication 2020). The Teotihuacan pattern is reverse that of the Archaic tradition in the Great Basin (e.g., Elko points) in which the direction of pressure flaking travels from the nearest margin diagonally and up to the right (see Hirth et al. 2003 for an experimental approach).

There is less variety in symbolic forms at Teotihuacan than in the Maya area, with forms including laurel-leaf forms, serpents, serrated serpents, crescents, discs, trefoils, hooked batons, zoomorphs, anthropomorphs, and conflations with atlatl darts (see Parry and Kabata n.d.; see also chapter 2). While one could argue that there are hundreds of types found in the Maya area, which is partially true depending on how one constructs their typology, the reality is that the majority of smaller Maya eccentrics are based on these prototypical forms. I have argued that Maya eccentrics can best be characterized as small god effigies (also see Iannone 1993), which were probably used as divining stones (Hruby 2007). Although their function is not well understood, Teotihuacan eccentrics do not seem to carry as high a degree of multivalency. The serpents and anthropomorphs, the most common in caches and burials aside from the stemmed atlatl dart, are also technologically and formally far more standardized than their Maya counterparts. Teotihuacan silhouettes were made using an array of techniques, the most basic of which included the notching and pressure-flaking of prismatic blades (i.e., third series, or 3s, blades in the Clark and Bryant [1997] typology). Even some of these small examples are finished with transverse-parallel flaking, and in rare cases fully bifaced. The same forms are present in miniature as they are in large sizes and were often accompanied by unworked blades and double-notched blades. For example, they were found in great numbers in a massive obsidian cache found on the centerline of the Temple of the Feathered Serpent, and a large circular mass of them were found in the Pyramid of the Sun (Sugiyama 2005; Sugiyama and Cabrera Castro 2018; Sugiyama and López Luján 2006).

The most high-profile silhouettes, however, are of a medium and large size, with the most spectacular pieces ranging above 30 cm in length. Some of the largest examples fall into the category mentioned above: chipped-handle silhouette.

FIGURE 1.2. *Self-referential skeuomorphs in obsidian and chert: (a) Teotihuacan example (drawing by Hruby; see obverse in figure 2.3); (b) artistic representation of obsidian blade in Teotihuacan art (drawing by Hruby); (c) Classic Maya depiction of a hafted obsidian blade (drawing by Hruby); (d) chert flake or silhouette with notches to show sharpness (Palenque Tablet of the Slaves, drawing by Hruby); (e) chert silhouette symbolically emphasizing sharpness (Altar de Sacrificios, Petén, drawing by Hruby).*

The most typical form is the fully bifaced curved knife, which had a chipped handle and is among the few obsidian objects, other than atlatl darts, prismatic blades, and blade cores, that are depicted in Teotihuacan iconography. In artistic renderings of sacrificial knives, the curved "daggers" often feature an excised human heart pierced at the end (figure 1.2). I believe this form is an example of the self-referential skeuomorph outlined earlier. Over time, the exaggerated iconographic representation of an obsidian blade, a black pointed curve, was replicated as a bifacial silhouette of that icon. Circumstantial evidence for this idea is that the chipped-handle curved daggers are rare in the archaeological record but ubiquitous in the iconography. I postulate that most of the sacrificial knives in the iconography were probably indexing macroblades, or perhaps even simple laurel-leaf bifaces.

In any case, the knapper's goals do not seem to include extreme thinness, as some examples do from the Maya world but rather a good lenticular cross-section that allowed for transverse-parallel pressure flaking. It is unclear whether there is any great benefit of this kind of flaking pattern, other than aesthetic concerns. It can shimmer in the light and give a biface an almost iridescent shine—an effect that is magnified by the "cat's eye" or chatoyant green obsidian often used to make them. This flaking tradition is a true cultural marker for Central Mexico, however, since it was never adopted by the Maya. It does not seem to have been adopted by the Zapotec either, although there is some evidence of this pattern in the Middle Formative of San José Mogote (Marcus and Flannery 1996, 127). When the Maya attempted to imitate this technique, they only got as far as the rough outline and not the pressure patterning itself. When the Lowland Maya finally began creating fully bifaced projectile points and eccentrics in obsidian, it seems to have lasted from the end of the Late Classic and Terminal Classic periods, with the exception the Southeastern Maya periphery, where it continued into the Postclassic. It should also be noted that Teotihuacan silhouettes were traded, or brought personally, all over Mesoamerica. They can be found in Oaxaca, as well as virtually every corner of the Maya world during the Early Classic period (Kidder 1985). If Teotihuacan was proselytizing, as well as creating economic and political connections, then these objects seem to have been an important part of this missionary work.

There seem to have been two basic tracks knappers took to produce midsized silhouettes: proceeding from a fully or mostly bifaced blank, or beginning from a lightly modified percussion blade (see chapter 2 for a complete discussion). In both cases the blanks were created by modifying or partially bifacing macroblades and other percussion debitage derived from blade-core reduction. Even those examples that retain the original ventral surfaces of the macroblades were finished with transverse parallel pressure flaking. The largest silhouettes represent a much different trajectory and planning process, which has been explored in detail recently by John Clark, Alejandro Pastrana, and James Woods (personal communication) but involves bifacing large natural slabs or clasts (see chapter 2). In the case of big silhouettes (>30 cm), the knappers must have begun with massive clasts or "megaflakes" taken from the largest clasts available at their respective sources. Clark (chapter 4) has shown that wooden tools can be used effectively not only to produce the large preforms but also the very large pressure flakes removed to finish them. While the large serpent forms often retain very little of the original percussion scarring, the much wider anthropomorphs retain a sizable remnant of the original percussion work on their central faces.

Epiclassic Central Mexico

After the collapse of the Teotihuacan state, their lithic traditions lived on. Their religious and military practices continued at sites in Central Mexico—such as Cacaxtla, Xochicalco, and Tula—and all over the Maya world at most major Late and Terminal Classic sites. Atlatl darts and silhouettes continued in production at these centers that in some sense never let the dream of Teotihuacan die, but crafters at each site materialized their vision of Teotihuacan ancestry in slightly different formats and media. For example, the Maya never mastered the transverse-parallel flaking technique but continued to produce Teotihuacan forms, while at Cacaxtla they seem to have continued transverse-parallel flaking and similar outlines, but in different proportions (figure 1.3).

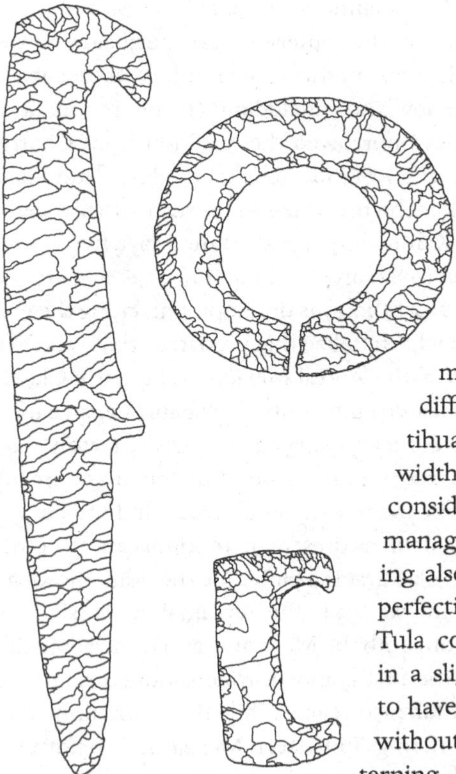

FIGURE 1.3. *Obsidian silhouettes from Cacaxtla displaying Teotihuacanoid forms and transverse-parallel pressure flaking (drawing by Hruby).*

The Cacaxtla examples are truly masterworks, with a higher degree of difficulty than many of the earlier Teotihuacan forms. They seem to have small width-to-thickness ratios while retaining considerable length, which is difficult to manage in volcanic glass. The pressure flaking also seems to be of the same level of perfection produced at Teotihuacan. Later, Tula continued producing dart points but in a slightly different style, which appears to have been shared with Chichen Itza, and without systematic transverse-parallel patterning. As far as the author is aware, neither the Toltecs nor the Chichen Maya continued the large silhouette traditions of their Central Mexican forerunners. In rare cases, when they are discovered, they seem to reflect a different symbol system. Prestige items appear to have gone in a different direction, with a heightened emphasis on turquoise mosaic crafts and dabbling in gold and copper metallurgy.

THE MAYA AREA

In Western Mesoamerica the laurel-leaf, or lenticular, biface seems to have been the basic preform for Late Preclassic silhouettes, while Eric Gibson (1989) and Gyles Iannone (1993) argue Maya silhouette traditions originated with the partially bifaced, stemmed macroblade (Hester 1982; Shafer 1991). It is unclear whence the laurel-leaf form was first created, but its use in rituals, and perhaps even as a utilitarian form, likely began in the west. Many of the earliest silhouettes and atlatl darts in Central Mexico were made from laurel-leaf bifaces. It may be significant that the laurel-leaf biface, as well as its many modified forms, do not become a common cache item in the Maya world until the "Entrada" era of the mid-fourth century (Stuart 2000, 2024).

In an unpublished master's thesis, Gyles Iannone (1993, 136–154) summarized the debate on the origin of eccentrics in Mesoamerica, and specifically in the Maya region. Early archaeologists had formulated a number of hypotheses about the genesis of Maya eccentrics in the lowlands, including: (1) the Teotihuacan hypothesis, which argued that the Maya borrowed the tradition from Central Mexico (Kidder 1985, 108; Willey 1972, 201; 1978, 104–105); (2) the Highland Maya hypothesis (Willey 1972, 207; 1978, 104), which was based on scanty evidence of an early eccentric axe-head found at Kaminaljuyu; and (3) the Maya hypothesis, arguing for local development. Gibson (1989) and Iannone (1993, 136–154) argued that Maya eccentrics represented an autochthonous development, citing the existence of Preclassic eccentrics from Northern Belize that predated what was then understood to have been first contact with Central Mexico in the Early Classic. We now know that contact between the two culture areas probably began earlier, perhaps as much as 200 or 300 years earlier to varying degrees (Clayton 2005; Coe and Houston 2015; Laporte 1989, 2003). Furthermore, the cited zoomorph and anthropomorph from El Pozito, Belize (Hester et al. 1991), and the curved scepter from Colha (Gibson 1989) are all forms present at Teotihuacan. Whether they show up first in Teotihuacan or Belize first is a bit beside the point, however, since it is unknown who was influencing whom in the Terminal Preclassic.

What we do know is that crafters from both the Maya area and Central Mexico began seriously experimenting with the first chipped-stone silhouettes around the start of the Common Era, give or take 150 years. While they were only intermittently used as part of dedicatory offerings in Eastern Mesoamerica, however, by 200–250 CE they had been formalized, replicated, and placed in sacred numbers within the largest temples at Teotihuacan (López Luján and Sugiyama 2018). Burials 2 and 6, for example, are some of the most elaborate and impressive ritual deposits in all of Mesoamerica, including sacrificial victims, birds, jaguars, shells, and of course the obsidian silhouettes (López Luján and Sugiyama 2018). These were undoubtedly marking the pyramids as central places and tying them to local creation mythologies.

The Maya only started producing caches that combined silhouettes with shells, jade, and birds just before the *entrada*, when actual relatives of the king of Teotihuacan took control of the throne at Tikal (Martin and Grube 2008; Stuart 2024). It may be significant that the Mundo Perdido complex at Tikal, which features buildings in a quasi-Teotihuacan architectural style that date to well before the *entrada*, did not produce any caches with silhouettes. Nevertheless, the earliest silhouette caches at Tikal date to around 340 CE (8.15.0.0.0 in the Maya Long Count; Moholy-Nagy and Coe 2008; Moholy-Nagy 2020), about the time when direct political and economic ties between the two sites was firmly established (Stuart 2024; Sugiyama and Sugiyama 2020; Sugiyama et al. 2020). Given these data, it may seem as though I am advocating Tristram Kidder's hypothesis for a

Teotihuacan origin for eccentrics in the Maya Lowlands (Kidder 1985). Instead, I argue that the Early Classic Maya caching tradition, which was essentially a shift to much smaller silhouettes, was the product of cultural syncretism between the two areas. The analogy of Maya iconography and its treatment of foreign symbols seems to bear this out, since the Maya seemed to enjoy integrating Teotihuacan art and ideologies into its system, but while maintaining its own traditions. The Maya likely combined flint and obsidian in its caches on purpose to reflect this binding of east and west with, in the case of Tikal, Tikal in the center.

The formalization of Maya silhouette traditions did not start in earnest until regular direct contact with Teotihuacan was established, and they appear to be most common at sites that feature a sustained relationship with Teotihuacan, or in most cases Tikal. The silhouettes were likely syncretic and largely inspired by Central Mexican cultures. In contrast to those of Teotihuacan, however, the silhouettes of the Maya appear to take on a more complex range of meanings, uses, and morphologies. I have recently argued that the engine behind the increasing complexity of Maya silhouettes resulted from the co-option of the Teotihuacan forms and their integration into elite Maya daykeeping, astronomical, and divinatory practices. One possibility is that the Maya integrated extant Maya silhouette forms (i.e., deity portraits, scorpion, and trident scepter forms) with those of Teotihuacan (i.e., the crescent, disc, ring, trilobes, and laurel-leaf forms) and then proceeded to attach specific Maya deities to each type. The Maya eccentrics were then deposited in numerically significant ways and in different sequences, again reflecting a possible borrowing from Central Mexico. This manner of syncretism occurred in a similar fashion during the Conquest period in the Maya area, where foreign gods and symbols were given characteristics of Maya deities (see Christenson 2016 for a detailed discussion). Understanding the vast complexity of Maya forms is addressed in a separate forum (Hruby in prep), but suffice it to say, this amazing complexity can be reduced to roughly twenty different symbolic forms that were reused over the remainder of the Classic period.

The majority of Maya eccentrics are small to medium size; that is, they were between 5 and 12 cm in the longest dimension. They would have been easily carried in handbags or bundles. In the iconography, most ritualized lithics would fit into a small plate, bowl, or basket and then ostensibly bundled. It is important to remember that even modern Maya bundles may be quite large but contain a very small object that was considered to be a god or god effigy. In other words, size does not always matter vis-à-vis ritual or religious significance. The small Maya silhouettes fall into two basic categories, which I have termed *stylistic* and *naturalistic* forms (figure 1.4).

The stylistic forms largely mimic those from Teotihuacan but also include more "utilitarian" forms, such as tanged spearheads and laurel-leaf knife and point forms, which may have Central Mexican origins. The naturalistic forms

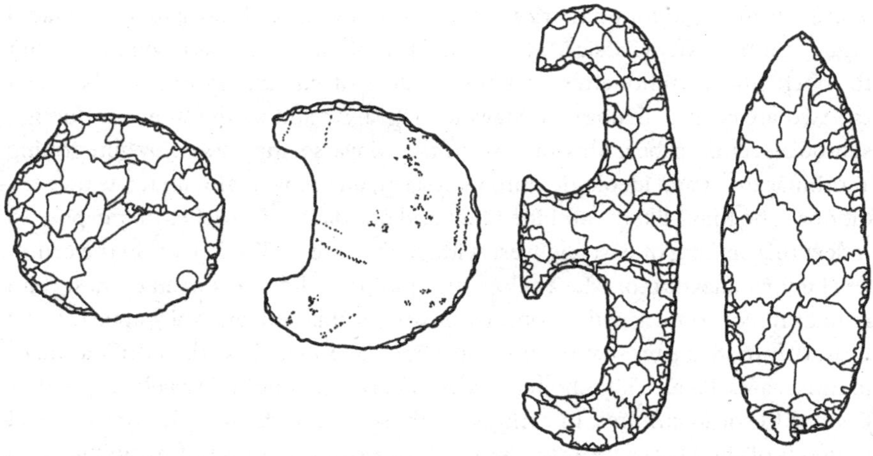

FIGURE 1.4. *Stylized Maya forms probably borrowed from, or influenced by, Teotihuacan forms: Disc, Crescent, Trident, and Laurel-Leaf types (drawings by Hruby after Moholy-Nagy and Coe 2008, Appendix 13). Trident example roughly 8–10 cm long.*

are chipped into beings and objects that appear as those beings and objects in Maya art and writing. A K'awiil lightning deity (God K), for example, looks like a K'awiil, and a Maize God appears as a Maize God silhouette. In fact, most of what were considered "eccentric" forms were misinterpreted naturalistic forms. One of the major stumbling blocks to their interpretation was that the outline of a silhouette was only *part* of the finished product. Often a rough outline of a form was chipped, and then the details of the piece were finished with painting (Hruby and Ware 2009), which unsatisfyingly makes the final symbolic analysis of the piece difficult. With only the rough outline of a particular deity, and the paint long since eroded or washed away, it is virtually impossible to determine the symbolism of all the silhouettes.

The Technology of Maya Obsidians

The technologies of ritualized lithics are partially accessible to archaeologists through experimentation and replication. The exact techniques may never be completely reconstructed, since different methods can produce similar results, but they are nevertheless useful for identifying basic parameters of production. Production included a combination of direct percussion, pressure flaking, and blade making, as well as indirect percussion, which Eble and I (chapter 3) argue is underrecognized in New World archaeology in general. While there is some overlap between the techniques used to reduce obsidian and flint, they were for the most part very different (see chapter 10 for obsidian). These differences result from the apparent distinction the Classic Maya made between blade and biface

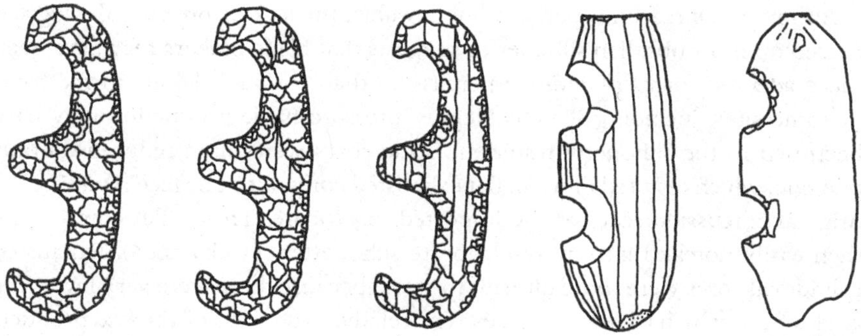

FIGURE 1.5. *The five basic technological types of Maya obsidian silhouettes: (a) fully bifaced blade cores; (b) mostly bifaced blade cores; (c) partially bifaced blade cores; (d) notched blade core; (e) modified flake or blade (drawings by Hruby).*

technologies, not to mention the format of the blank or preforms used to create the silhouettes in each material (Hruby 2006, 2007). I have argued elsewhere that for the majority of the Classic period, the Maya must have had distinct crafting guilds, or *oficios*, which may have been exclusive (see earlier). At some sites, however, near the end of the Late Classic and Terminal Classic, some obsidians were fully bifaced, and finished by pressure similar to chert. I interpreted this pattern at Piedras Negras as a disintegration of boundaries between obsidian- and chert-knapping *oficios* (Hruby 2007). It is unclear whether this pattern exists at other sites. Interestingly, some sites in the Highlands, such as Kaminaljuyu, never seem to embrace biface technologies, and even when commissioned, knappers do not seem to have comprehended, much less practiced, biface technology.

The vast majority of obsidian silhouettes were made from percussion flakes and blades derived from blade-core reduction, as well as exhausted blade cores in various formats. Most of these were not reduced extensively as bifaces but rather partially bifaced, or simply notched via indirect percussion with occasional pressure finishing. There are only a few known cases of obsidian bifaces that were transformed into silhouettes (see Kaneko 2003 for an example from Yaxchilan). The blade-core silhouettes can be separated into four basic types: (1) *fully bifaced blade cores*, which can be identified by miniscule remnant blade-scars, are the least-common type; (2) *mostly bifaced blade cores* are almost as rare as the fully bifaced examples but can be found at virtually all sites that have a robust silhouette tradition; (3) *partially bifaced blade cores* feature at least a portion of their outline that has been reduced bifacially; the most common type; (4) *notched blade-core* silhouettes, which seem to have been notched using indirect percussion of some kind; and finally (5) *modified flakes* and *blades* that were made with rough pressure or indirect percussion (figure 1.5).

Although the reason is perhaps unknowable, the lack of pressure flaking witnessed on most obsidian silhouettes suggests that blade makers may have been more adept at forms of indirect percussion than pressure flaking in the hand. At some sites, including Piedras Negras, pressure blade production may have been tied to the silhouette tradition. Blade cores were often reduced on only one side, which resulted in wide, flat exhausted cores with one face retaining the original percussion scars of the imported macrocore. These "flat cores" were then easily notched into more elaborate silhouettes (Hruby 2006). Exhausted cylindrical cores were more difficult to work, because there were very few acute angles from which to remove flakes. The relative crudeness of the end products bears this out. Even when only subtle outlines could be achieved due to the morphology of the blank, knappers persevered and successfully evoked some basic silhouette forms, which were more easily recognizable in chert or obsidian blades. The average size of obsidian eccentrics can also vary from site, along with pressure blades and exhausted cores, based in part on the availability of imported blade cores (Hruby 2024).

Flakes and blades of obsidian were more easily worked but rarely fully bifaced. The *modified flake* and *blade* traditions also parallel the *incised obsidian* traditions of Tikal and Uaxactun. In fact, these incised obsidians can act as a substitution cypher for interpreting the other more opaque symbolic forms of both the stylistic and naturalistic types (Moholy-Nagy and Coe 2008; Hruby 2007; Iannone 1993). While some of the processes of selection appear to be *ad hoc*, other sites with strong silhouette traditions in obsidian seem to have correlated technological types to specific symbolic forms. I (2007) have recorded some of these patterns at Piedras Negras and Tikal, and argued that this pattern indicates a connection between technology, production, and symbolism. This research shows that flakes and blades were sometimes removed in a sequence from a blade core and then systematically transformed into silhouettes (see also Kelsey Sullivan and Jaime Awe, chapter 7). Thus, the final form of the eccentric is only one facet of the symbolic load they carried. They record the very processes of production and should themselves be understood as processual. Modified flakes and blades were created using a combination of indirect percussion, and rough pressure flaking, but not with the intent of running pressure flakes across the surfaces of the flake (see chapter 10 for examples from Caracol; see also Johnson 2016). Pressure was often used only insomuch as knappers could complete the intended outline. Finally, at some sites, such as Piedras Negras, obsidians were fully bifaced with fine pressure finishing. These rival the eccentric flints in their level of skill, symmetry, and thinness, and from a Western perspective are among the most beautiful obsidian artifacts in the Maya world.

Although beyond the scope of this chapter, one wonders whence the macroflakes, macroblades, and exhausted blade cores used as silhouette blanks

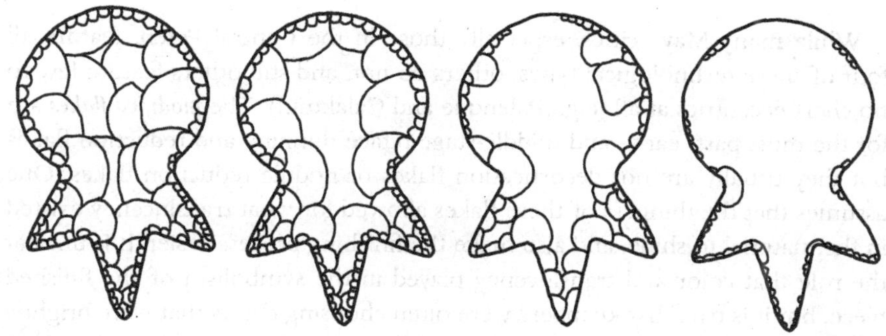

FIGURE 1.6. *The four basic technological types of Maya chert silhouettes: (a) fully bifaced; (b) mostly bifaced; (c) partially bifaced; (d) and modified flake silhouettes (drawings by Hruby).*

originated. The obvious conclusion is that they were produced by the local blade makers at each polity capital (Hruby 2007; Johnson 2016; Moholy-Nagy 2008), and this remains the best hypothesis. The reality, however, is that we don't really know. The problem is that as of this writing, there are only three probable blade workshops so far identified in the entirety of the Maya Lowlands, and one of these is from Copan, which can only vaguely be identified as a Lowland site. These workshops are likely difficult to find precisely because the debitage was considered to be so dangerous and workshops cleaned so fastidiously. Interestingly two of these workshops—one at Busilja, Chiapas (Roche Recinos 2021), and the other on the outskirts of El Pilar, Petén (Ford 2004; Hintzman 2000)—are not located in polity capitals but rather in small subsidiary sites. Complicating matters is that the production in these workshops seems to be of small, partially reduced blade cores. As a result, the macrodebitage used for eccentrics at the major centers must have come from somewhere else. The secondhand cores from these subsidiary centers were too small for making the modified core eccentrics listed earlier. Finally, there is very little evidence of obsidian silhouette production anywhere, for any time period, in the Maya area. Thus, while local traditions at silhouette-producing sites clearly existed, they may also have been importing already-exhausted cores as silhouette blanks. This pattern occasionally occurs with flint silhouettes (see later in chapter). Only detailed future research can resolve this issue entirely.

The Technology of Maya Flints

The production techniques for chert, flint, jasper, and chalcedony silhouettes are wide and varied but can be broken down into a few basic categories: *fully bifaced*, *mostly bifaced*, *partially bifaced*, and *modified flake* silhouettes (figure 1.6; also see chapter 8).

While many Maya cities, especially those in the Central Petén, feature all four of these technological types, others do not, and still others feature few to no chert eccentrics at all (e.g., Palenque and Calakmul). The *modified flakes* are, for the most part, early- and middle-stage biface thinning and reduction flakes, but they usually are not decortication flakes or nodule reduction flakes. One assumes that the thinness of these flakes allowed for what translucency existed in the material to shine, and also make the finishing process easier. It is unclear the role that color and translucency played in the symbolism of the finished piece, but it is true that knappers were often choosing cherts that were brightly colored and partially translucent. The problem with identifying the original colors and translucency of Maya ritualized cherts is that, depending on the soil chemistry, the material becomes easily hydrated and oxidized. A bright white patina can form over centuries, and color changes can happen in a matter of moments. I have personally witnessed a chert transform, right before my eyes, from a red color to a light pink in a matter of minutes. If the symbolic importance of this transformation is disregarded for the moment, the flakes were then modified with a combination of indirect percussion and pressure techniques, and due to their inherent lack of symmetry, they were often used to create more naturalistic silhouette forms.

Partially bifaced examples required more time to produce than the modified flakes and were made from larger nodule reduction flakes. One of the hallmarks of this technological type is that they were not terribly symmetrical along the margin. Similar to the way things were done with modified flakes, the focus appears to have been on the outline of the form from a plan view, with less interest in the symmetry of the horizontal or vertical cross-sections. All of the most elaborate silhouette forms could be made using this technique, including those that feature holes or empty spaces (see chapter 7). In most cases it appears that the Maya capitalized on naturally occurring depressions in the surface of the cortex, vugs, and shell inclusions in the raw material to start this interior reduction process. If this interpretation is correct, then the natural layout of the stone may have played a role in how the final form of the eccentric was determined. As Sullivan and Awe discuss in this volume (chapter 7), the Maya may have drilled some of these holes to help create the empty space needed to start the internal reduction process. Not all of the indirect-percussion notching techniques evidenced on the *full bifaces* were used to produce the *partially bifaced* pieces, since they were usually quite a bit smaller than the other types. Small flake size likely made using large punches problematic, since immobilizing the flake during production would have been difficult. Sometimes these were finished with light pressure flaking, but interestingly they were just as likely to retain the rough perimeter left behind by percussion work. In fact, on even the largest silhouettes, it appears that fewer flake scars was a desired trait of the finished product. The only real differences between the three bifaced types lay in the lack of marginal symmetry and the retention of either the ventral or dorsal flake scars from the original nodule-reduction flake.

Mostly bifaced silhouettes are closer in morphology to the fully bifaced examples. All of the same large-scale, indirect percussion techniques as those found in the full bifaces seem to have been used to make them. The reason for a different technological category is that the blank or preform used to make mostly bifaced silhouettes was derived from either a smaller nodule, or more frequently a macroflake that was too small or thin to completely biface. In some cases, they may reflect a lack of skill, and thus some kind of hierarchical arrangement among the group of knappers producing the pieces. For most groups of cached silhouettes, teams of knappers, instead of a single knapper, are hypothesized. The variation in styles among the pieces and that some production techniques required more than one person to bring off (see chapter 3). Small portions of cortex or large percussion scars from the original blank are retained on one or both faces, but they are usually symmetrical from the point of view of the margin.

From a Western, and possibly emic, perspective, the *fully bifaced* silhouettes are considered the most beautiful, and perhaps awe inspiring, of Mesoamerican ritualized lithics. They required the most time, skill, and high-quality raw material to produce. The knappers were likely considered the masters of their craft, and probably spent much of their lives acquiring the skills to make these items. There are economic parameters to consider as well, since access to high-quality and sizable pieces of raw material was not necessarily a given, except perhaps near the Maya world's largest chert sources in Northern Belize and South Central Campeche. The full biface silhouettes were often larger than the other technological types, and they required a greater number of stages to produce properly. Most significant, a large thin preform must have been produced. In most cases these preforms were either massive celtiform or laurel-leaf bifaces (see chapter 4). These preforms, that is, their basic outline, undoubtedly added yet another level of symbolism to the final form of the eccentric. The largest and most elaborate eccentrics discussed in this volume are the massive effigy flints from Copan and elsewhere in the Southeastern Maya region, but for all of their incredible detail, each one began life as a giant celt. At their core, the effigy flints were axes. The same cannot be said for the smaller effigy flints of Altar de Sacrificios (Willey 1972), which were based on a laurel-leaf form.

Now, while these differences in silhouette techniques may seem insignificant at first glance, the analysis of these objects in their cache contexts, not as isolated objects, certain patterns emerge. For example, in chapters 7 and 8 it is clear that some of the silhouettes were produced at the same time, perhaps not for the cache ritual per se but for use in that particular group of eccentrics. The clearest examples in this volume come from Ucanal (chapter 8), where at least two silhouettes were produced from the same nodule. This pattern is also found at Piedras Negras, where a single nodule was reduced, likely not much bigger than 20 cm in length. A large, fully bifaced silhouette was produced, but the largest of the nodule reduction flakes, and perhaps some of the larger well-formed biface thinning flakes, were reserved for smaller silhouettes (Escobedo and Hruby 2002). These pieces all fit into a larger, preplanned group of eccentrics that usually numbered from 7, 9, 13, 18, or 21 chert eccentrics. The number of nodules used to create the final group of silhouettes probably added an additional symbolic component. It is unknown at this point whether the larger, more elaborate bifaces were considered to be more important than their smaller counterparts.

Regional Variation in the Maya Area

The primary factor in understanding morphological and technological variation throughout the Maya world is resource type and availability. Unlike obsidian, ritualized cherts were usually made out of locally available materials. I have argued elsewhere that the use of local cherts may have been a statement of local identification, a marker of the centrality of the ancient cities and the stones on and with which they were built (see also Iannone 1993). Similar to how the ancient Maya made cache bowls, plates, and cups—almost always locally—they seem to have embraced the local materials for ritual caching. Strangely, Guatemalan obsidians may have been perceived to be "local" at some Lowland centers, since green Mexican obsidians and artifacts are rarely found in Maya lithic caches. There are some outliers at Altun Ha (Pendergast 1979, 1990, 2003) and Tikal (Moholy-Nagy 1999, 2020), but these cannot be characterized as typical. Mexican obsidians in the Maya area seem to have functioned differently than eccentrics for the most part, serving as prestige goods and status markers, as well as special tools and weapons.

The situation becomes more complex when we consider that many of the Maya cities that embraced Central Mexican religion went to great lengths to bring Central Mexico *to* the Maya area. Many temples from some of the largest Maya sites were built, however misguidedly in some cases, to recall and evoke Teotihuacan structures and landscape. I have noted that the great acropolis at Piedras Negras, for example, was decorated with Teotihuacan-style mountains, which, symbolically at least, connected the royal line to that distant place (other examples include Acanceh, Dzibanche, and Tikal). In some cases, then, the production and use of medium- and small-sized eccentrics can be understood as an extension of syncretized Central Mexican religion in the Maya world, which further legitimized these structures as having a western origin or connection (see Escobedo and Hruby 2002 for an example of a Teotihuacan-themed cache at Piedras Negras, Temple O-13).

Raw material availability has a massive impact on what can be produced with subtractive technologies. Simply put, if there are no large, locally available rocks, then you can't make large bifaces. If you don't have hundreds of large nodules at your disposal, then it will be very difficult to impossible to learn how to make giant eccentrics. On average, a place like the Central Petén does not yield very large chert nodules, and consequently the eccentrics found at Tikal, Piedras Negras, and even Altar de Sacrificios are not much bigger than a human hand and usually much smaller (5–15 cm in length). The largest eccentrics, from Lamanai and Northern Belize, can be as big or bigger than a human torso (i.e., nearly a meter in the maximum dimension). To a large extent, this is why the production of silhouettes reflect local traditions: (1) homogeneity in material and eccentric size per cache; (2) similar proportions and production technique; and (3) coherent styles and symbol types continued over time. Even in cases where bifaces are imported for caches, they appear to have been modified by local knappers into local style (Meza Rodríguez 2008).

That being said, some silhouettes were clearly produced for exchange. Exported silhouettes were usually of the chipped-handle type, and often made from the famous *fine brown* flint (see later in chapter). These include the large trefoil *manople*, or knuckleduster, common in the iconography (see Kwoka, chapter 9), laurel-leaf forms (referred to as scepters by Clark, chapter 4), the perforated axe head (Willey 1972), and possibly the elaborate effigy flints from Copan (see chapters 5, 6, and later in chapter). Although the source for this material has not been identified with certainty, I postulate that it was located in Northern Belize based on artifact distributions, raw material characteristics, and a number of road cut surveys in the Chert Bearing Zone (CBZ) carried out by the author, Mike McBride, Ben Eble, and Lindsay Powell. I use the past tense, because its products appear in the archaeological record for a short period of time and thus may have been mined out of existence in ancient times. It was famous throughout the Maya world, and may have even been understood emically as a category of tool stone distinct from that of obsidian and chert. The large laurel-leaf bifaces made from this material seem to be restricted to the Middle and Late Classic periods and can be found as far afield as the Chiapas and Guatemala Highlands (Clark, chapter 4), to Copan and El Salvador in the southeast (chapters 5 and 6), the Merida area in the Northern Yucatan, to El Peru-Waka', Piedras Negras, and El Tigre (Itzamkanac) on the frontier of the Western Lowlands.

This hypothesized source and associated workshop also traded smaller projectile and spear points, which are found at all Maya sites to one degree or another (see Moholy-Nagy 2003 for a discussion of imported weapons of *fine brown*; and Moholy-Nagy and Coe 2008 for a discussion of the use of this material in caches). At a few sites, most clearly at Altar de Sacrificios, at El Tigre, and for a short time at Tikal (Meza Rodríguez 2008; Moholy-Nagy and Coe 2008; Willey 1972), these smaller laurel-leaf bifaces were transformed into the unhafted-type silhouettes most common in cache contexts. Since each of these sites feature different styles in this material, my preliminary assessment is that the bifaces were exchanged long distance and then reduced locally. The implication is that the knappers making the large handheld forms and laurel-leaf bifaces for exchange, even the massive effigy flints, may not have had specific knowledge of how to make the sequences of forms so common to sites like Altar de Sacrificios or Tikal. A question to be addressed with future research is whether the larger effigy eccentrics (e.g., the Rosalila flints) were finished at the workshop, or instead that large bifaces, and perhaps partially reduced tabular-nodules, were traded and the silhouettes finished elsewhere (see chapter 6 for a discussion).

The most widely distributed chert formations in the Maya area occur primarily in the Petén of Guatemala, but they overlap with the surrounding regions as well (i.e., Usumacinta, Pasion-Petexbatun, Campeche–Quintana Roo, and Central-Northern Belize). The nodules are relatively small, from 10 to 20 cm in the longest dimension, and are usually multicolored chert-chalcedony mixes. They differ from Campeche sources, which form as massive bedded cherts that can produce blocks into the hundreds of pounds. Further to the east into Northern Belize, the chalkier substrates produce massive tabular flint nodules that resemble bedded chert, as well as the super *fine brown* material from which the massive effigy flints were likely produced. To the west and south, sites such as Altar de Sacrificios, Piedras Negras, and perhaps even El Tigre feature a wider range of stones, including opaque cherts, jaspers, and other diverse stone, thanks to the alluvial action of the Usumacinta and Chixoy Rivers. El Tigre may represent its own unique geologic zone, perhaps associated with massive alluvial deposits of what appear to be typical Central Petén and Southern Campeche cherts that accumulated further down the San Pedro Mártir River. In the Northern Yucatan, the eastern Puuc region is much richer in chert, but the Maya also likely imported much of their material from South Central Campeche. In contrast, the far north, and especially the east, is virtually bereft of chert. The Northeast Yucatan Maya relied on other stones to get by, especially hard limestones, which may have been incorrectly described as silicified limestone by me (Dahlin et al. 2011) and also discussed by Hearth and Fedick (2011). This dense limestone could be used for making flakes for cutting and crude choppers, but due to their porous nature they often erode and are virtually impossible to identify in the archaeological record.

Interestingly, in the far western lowlands of Palenque, they do not deem to have much chert, and no recorded eccentric caches, but just northeast at the lower San Pedro Mártir, El Tigre (Itzamkanac), eccentric caches are common, and the chert so abundant that they used it as structure fill (Meza Rodríguez 2008). It is unclear why Palenque did not attempt to acquire these chert materials, but suffice it to say Palenque appears to have sunk most of its wherewithal into producing enormous ceramic censers for their caching rituals (Carrasco

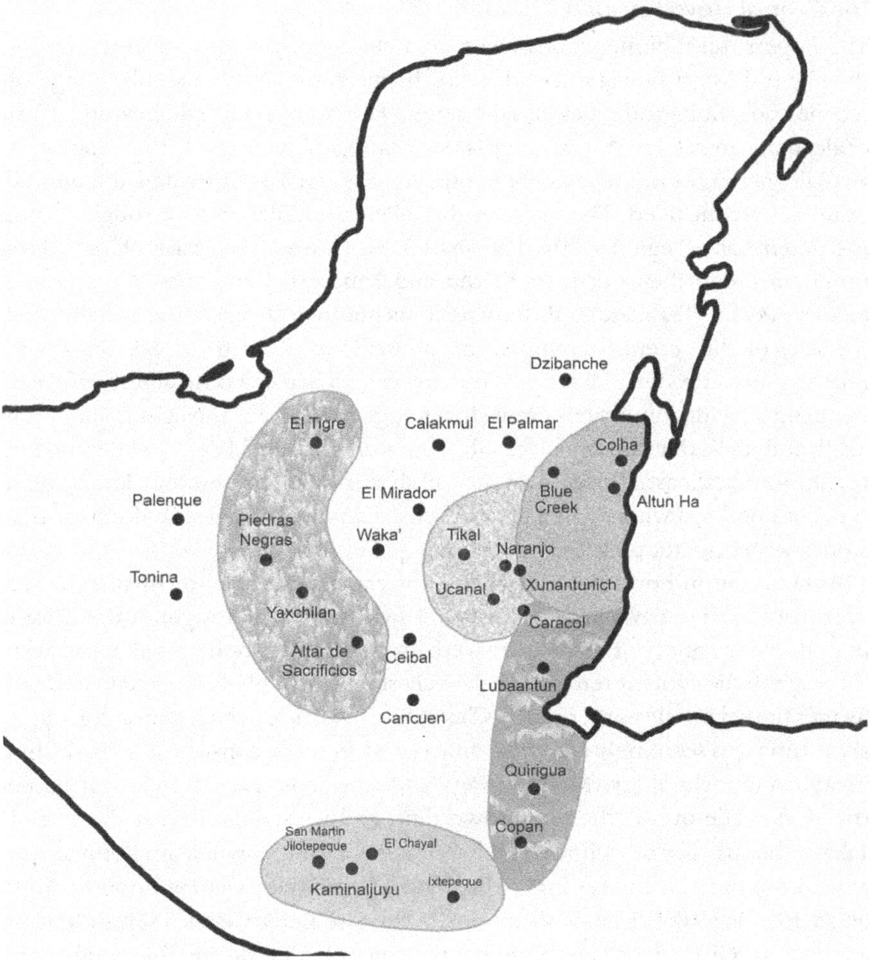

FIGURE 1.7. *Map of proposed silhouette traditions in the Maya area (map by Hruby).*

2005; Cuevas García 2000, 2004, 2007). In other words, many Maya religious specialists were communicating similar messages and doing so with the same materials, but the execution of these symbols and the media they used varied greatly. Other Southern Maya Lowland sites where eccentrics are markedly scarce are El Peru-Waka', Ceibal, Calakmul, La Corona, and Cancuen, even though lithic objects, such as bifaces, flakes, and blades, were used ceremonially. In the following sections I outline different regional silhouette traditions from the Maya area (figure 1.7). These different traditions seem to correlate, at least partially, with different resource zones.

The Central Maya Lowland Tradition

The largest, semi-homogenous chert- and chalcedony-bearing zone is that of the Central Maya Lowlands, and it also happens to feature a similar range of techno-morphological types of silhouettes. The cherts, chalcedonies, and chert-chalcedony mixes are not in themselves homogenous; in fact, they appear in virtually every color of the rainbow except for green if heat-treated and burned examples are included. The range of these Petén nodules extends roughly from the Petén Lakes region to the Petexbatun, eastward to the base of the Maya mountains, and then north to Xultun and Southern Campeche. Central and Eastern Petén sites feature all four basic technological types (see earlier), and the sizes of the eccentrics range from miniscule to about hand size. Sites with related silhouette styles all seem to borrow from Tikal as a central place and can be further divided into subregions, including: (1) El Zotz, Uaxactun, and Tikal itself; and (2) Yaxha, Naranjo, Ucanal, Xunantunich, Cahal Pech, and Baking Pot to the east-southeast. The eastern sites all share similar numerology and symbol types but perhaps with a wider array of chalcedonies or at least chalcedony-rich stones with brighter pink and red colors.

As far as the author is aware, Tikal has the earliest examples of small unhafted eccentrics used in royal architectural and stela caches. Throughout the Classic period a wide variety of techniques were used to make silhouettes at Tikal, even if the symbolic content remained relatively stable (Moholy-Nagy and Coe 2008). As mentioned briefly earlier, Coe's Class EF2 eccentrics,[1] which persist for a very short time and seemingly comprise only two structures, consists of a major shift in style and technology (Moholy-Nagy and Coe 2008, 19). This class of eccentrics was made out of the fine brown flint, and were deposited at the end of Tikal's "hiatus" period before the reign of Jasaw Chan K'awiil from Long Count 9.11.0.0.0–9.12.0.0.0. Interestingly, this kind of tradition was common at Altar de Sacrificios at the time, with origins in the late Early Classic. Simon Martin and Nikolai Grube (2008, 41) note that the ruler of Altar during this period may have been the son of the Tikal Ruler "Animal Skull," which may explain a greater similarity in material culture between the polities. The fine brown eccentrics from both sites seems to have been imported as finished bifaces, and then transformed locally into regionally distinct silhouette forms. The common eccentric types are discussed in detail by Coe, who created a useful typology for them based on Tikal and Piedras Negras examples, and are revisited in chapter 8 of this volume.

1 Although William Coe's typology is described in chapter 8, it should be noted here that he created three basic categories of eccentrics, including eccentric flints (EF), eccentric obsidians (EO), and incised obsidians (IO).

Northern Belize Tradition

This region, often described as the Northern Belize CBZ, features an amazing variety of cherts, flints, and chalcedonies and extends over the modern border and into Quintana Roo before the central plateau rises to the west, and also likely into the northeastern corner of Guatemala. As Joshua Kwoka (2014) has shown, Xultun enjoys one of the largest chert sources in Northeastern Guatemala, but it is unclear the degree to which the Xultun chert formations relate geologically to those of the CBZ. A little farther north at El Palmar, across the border on southern Quintana Roo, a large chert source is being analyzed by Sullivan, but it again it is unclear how geologically similar the source is to the CBZ. In the south, the CBZ gives way to more of a Central Maya Lowlands–type geology and embodied by the Upper Belize River Valley lithic traditions. Silhouettes and ritualized lithics in general tend to be larger in the CBZ thanks to a higher frequency of large chert and flint nodules.

While attempts have been made to replicate the famous effigy eccentrics (chapter 5) and the >40 cm, ultrathin laurel-leaf bifaces (chapter 4) found scattered throughout the Maya world, no one, as far as I know, has attempted to replicate the megalithic eccentrics of Northern Belize (e.g., Pendergast 1981; Stemp et al. 2014). These pieces often come from burial contexts as well as caches, and it can be assumed that they were created using a combination of direct and indirect percussion. Beyond this speculation, however, the techniques and strategies employed in making chipped-stone rings close to a meter in diameter are unknown. One must assume they were knapped in teams. The very largest examples are quite thick, and thinness and a fine pressure-flaked finish were not the intended goal but rather awe-inspiring monumentality and symmetry. The megalithic silhouettes were not finished with pressure, although some of "smaller" 30–40 cm pieces from Lamanai and Altun Ha were to a small degree. According to the material characteristics, as well as the existence of failed production attempts in the debitage dumps of Colha (Hester 1982; Shafer and Hester 1983), it is fairly certain that many of these monsters were made there. However, as the samples from Lamanai, Altun Ha, and Blue Creek demonstrate, the Maya of Northern Belize used a wide array of materials to produce silhouettes, including flint, chert, and chalcedony.

The majority of the silhouettes from Northern Belize appear to be fully bifaced, and as Joshua Kwoka discusses in this volume (chapter 9), they take a wide variety of forms, many of which are shared with those found elsewhere in the Maya area. At sites like El Tigre in Campeche, it is abundantly clear that the laurel-leaf form had incredible symbolic and economic importance for the Maya, since objects with this form are cached in large numbers there. Kwoka (chapter 9) and team (Kwoka et al. 2019) have shown that this is also true for Northern Belize, with the addition of the stemmed spearhead. Further, Kwoka notes that a fine brown biface was cached at Blue Creek, again showing that it was acceptable to include that material in some local caches. There are indicators that silhouettes were used in different ways in Northern Belize. Unlike other areas, eccentrics appear to have been important burial furnishings, and the sheer size of some pieces undoubtedly made it difficult to cache them in the same way as the smaller ones. The use of eccentrics in burials may mark a heightened status for lithic craft production in this region.

As Jason Barrett (2011) has argued, however, the oft-plentiful fine chert at many sites was not an inexhaustible resource. Sites like Blue Creek began to resort to smaller nodules of chalcedony and chalcedony-chert mix, which are so common in the Petén, into the Late and Terminal Classic. There seem to be at least two basic technological types made in Northern Belize: more massive, thicker silhouettes that were not finished with pressure and retained very wide and untouched notching scars, and a thinner type that was finished with pressure and finer indirect percussion work requiring a smaller, semicylindrical punch (see chapter 3; Shafer 2023; Titmus and Woods 2003). In terms of ritualized lithics, broadly construed, the Maya of Northern Belize were one of the few groups to produce chert macroblades for caching rituals. It has not been proved that macroblade production was the genesis of the eccentric flint tradition in the Maya world (Gibson 1989; Iannone 1993). However, it is clear that large ceremonial cherts got their start in this region. After the Preclassic, the large trefoil *manoples*, laurel leafs, and perforated axe heads were in high demand at all of the largest cities in the Maya Lowlands, and it appears that they originated in Belize.

Southeastern Lowland Tradition

These zones are interesting because they feature some rather large alluvial deposits of river-rolled jaspers and cherts, and there is a river in Izabal, Guatemala, named Rio Pedernales by the Spaniards, which is rife with the material. However, they do not seem to have been used extensively at nearby sites, neither for large quantities of systematically chipped, formal tools nor for ritualized lithics. Southern Belize sites, such as Lubaantun and Nimli Punit, seem to have imported preforms or perhaps finished bifaces from Northern Belize for their ceremonial lithic instruments. The sandstones of Lubaantun did not produce copious amounts of high-quality tool stone, and others located more closely to the Maya Mountains found chert difficult to come by (e.g., Caracol; Johnson 2016). The Maya Mountains are largely igneous and metamorphic in nature, and only small pockets of chert can be found in the limestone and sandstone there, which may partially account for the focus on obsidian for the creation of silhouettes at places like Caracol. Sites further south and west in Belize, such as Uxbenka, may have traded with the Petén for their raw materials, and there are copious examples of large, fine brown flints at Lubaantun.

Moving to the Rio Dulce, Motagua, and Copan River drainages, the stones are largely metamorphic and volcanic, and the limestone that does exist does not produce copious chert nodules. Sites such as Quirigua and Copan seemed to have relied on obsidian for the majority of their formal tools but had a vibrant flake core and projectile point industry (Meissner and Marino n.d.). At Copan, even with massive jasper nodules and wonderstone cobbles present in the river, local materials were not often used for caches and burials. However, massive jasper boulders were used for *yunques* (i.e., anvil stones) for working other lithic materials (chapter 11), and smaller cobbles were transformed into multidirectional flake cores. Instead of making biface celts for agriculture and architecture, the Copanecos reduced tough, volcanic river cobbles into rough choppers. A cache of production flakes of these materials was discovered at Las Sepulturas (chapter 11). It may be that the jaspers cherts, and wonderstones were simply too tough to work easily, or that Copanecos were more interested in using obsidian for their finer tools. The result is that local materials at Copan were used for some projectile points and small spear heads, but they were not elaborated into silhouettes. The closest thing at Copan to a chert celt tradition, one of the hallmarks of Maya civilization, are is general outline of the elaborate effigy flints found there.

Compared to the rest of the Maya area, there is an intensive use of elaborate effigy eccentrics, as well as very long, thin laurel-leaf bifaces, which Clark (chapter 4) and Woods (chapter 5) discuss here. These are found at Quirigua and Copan, but also, as Sheets examines (chapter 6), they are also found further south into El Salvador. It is important to remember, however, that one of these effigy eccentrics was found at El Palmar in Southern Campeche (Thompson 1936; Tsukamoto et al. 2021). The large laurel-leaf forms, which were clearly made in the same style and with the same material as the Copan and Quirigua examples, are found as far afield as Solola, Chiapas, and El Peru-Waka', Petén. They were made from very large macroflakes, or more likely massive tabular flints that probably formed in a chalk substrate, indicating they are true flint. The preforms too must have been huge, perhaps measuring over 50 cm in length, and then were reduced in a more refined way with indirect percussion. The final flaking was done with extremely precise indirect-percussion and pressure flaking (chapter 5). Virtually all of the fine brown bifaces retain some cortex at their distal end, perhaps another calling card of the workshop.

The vexing question of where fine brown effigy flints were produced remains. Even though the greatest number of them come from Copan, there is no local geology that produces such massive nodules of high-grade flint in the Copan pocket. Furthermore, the archaeology of Copan does not support the idea that there was a large biface tradition of any kind in the Copan valley. After decades of excavation, no thinning flakes of the so-called fine brown material have been discovered. Three basic scenarios are possible: (1) There is an as-yet-undiscovered source of fine brown flint in the Southeastern Lowlands, and the knappers were local to Copan; (2) the knappers were local, or moved to Copan from elsewhere, and large blanks or preforms of fine brown were imported from my hypothesized source in Northern Belize (or less likely as far afield as Southwestern Campeche near El Tigre or Altar de Sacrificios); or (3) the pieces were imported finished to Copan from its trading allies in the Eastern Lowlands and the knappers lived in Northern Belize. The latter would make the most sense, except for the extreme difficulty of transporting such delicate works of art. One can only assume that knappers who reached that level of skill must have had absolutely unfettered access to large, high-quality material to become that good. To verify any of these hypotheses a proper source must be identified, along with an associated workshop, as well as broken rejects from failed production attempts at making them. As a coda, during a brief analysis of chert artifacts from El Peru-Waka', I identified decortication flake of fine brown material from general excavations. This identification opens up the possibility that there is a fine brown source between Waka' and El Tigre somewhere, or that small tabular nodules of the material were traded partially reduced from a distantly located source.

Central Campeche Resource Zone

This geologic zone cannot be said to feature a distinct silhouette tradition. It overlaps with the Central Lowlands zone but with the addition of some of Mesoamerica's largest-known bedded-cherts. In roadcuts, some of the blocks are roughly the size of a wheelbarrow or a tuk-tuk and weigh over a ton. Some are so large that reducing them may have been impossible using ancient technologies. Furthermore, the quality of the material is often low, or at least contains fractures and inclusions. This material was used to create utilitarian items, some of which are quite large. It is possible that the massive chert picks (50 cm+) found in the quarry areas of El Mirador (Woods and Titmus 1996) were imported from this Central Campeche and Quintana Roo area during the Preclassic.

As far as ritualized lithics are concerned, however, the massive chert sources do not seem to have translated into a silhouette industry. Interestingly, there is no robust or even notable eccentric tradition at Calakmul or Becan, suggesting that it may not have been important in Southern Campeche. El Peru-Waka' similarly did not participate in a silhouette tradition, although some very large fine brown bifaces were discovered by Héctor Escobedo and Juan Carlos Meléndez in 2007. In other words, there is no systematic silhouette tradition north of Waka', perhaps even including many of the Puuc sites. It is important to note, however, that the chipped-handle types are featured in the iconography as far north as Dzibilchaltun (Martin and Grube 2008), and a fine skeuomorphic flint axe is on file in the Peabody as hailing from "Northern Yucatan." On monuments from Calakmul to the Puuc, silhouettes appear to have been hafted as crescent axe heads and sometimes large handheld trefoils, which may have been imported from elsewhere.

To the south of Calakmul, there is a chain of sites on the central axis of the Southern Lowlands, including El Peru-Waka', La Corona, Cancuen, and perhaps Ceibal, where a robust or even extant silhouette tradition cannot be identified. All of these sites had close relationships with Calakmul, including intermarriage between royal families, which from a macroscale seem to reflect the constant push and pull between Tikal and Calakmul to control trade across the Yucatan Peninsula. An interesting hypothesis to test is that Calakmul influenced ritual caching along this corridor in the same way Tikal may have done in the western rivers and Central Petén regions flanking it.

The Western Rivers Tradition

A high degree of variability and conflation of different symbolic forms characterizes this tradition, as well as a wide variety of stone types used in manufacture. I divide this region into three subregions: (1) Middle Usumacinta, including Piedras Negras, Yaxchilan, and their subsidiary sites; (2) Lower Pasion, including Altar de Sacrificios and Dos Pilas; and (3) the Lower San Pedro Mártir, which includes El Tigre-Itzamkanac. In terms of archaeologically recovered silhouettes, after Tikal, Piedras Negras and Altar de Sacrificios have produced the largest samples of these objects, and they feature all four of the basic technological types. Stylistically, however, there are some differences between the objects from Altar de Sacrificios and Piedras Negras. While Piedras Negras seems to include deity portraits made from painted and modified flakes (Hruby 2007; Hruby and Ware 2009), those of Altar de Sacrificios are often more complex, fully bifaced, and derived from a laurel-leaf preform. Thus, the Altar forms, although much smaller, are more akin to the massive effigy flints of the Southeastern Maya Lowlands and Northern Belize. The Piedras Negras–style portraits are more similar to the Tikal tradition in the sense that the head is the focus of the portrait, and not the entire body.

The Middle Usumacinta and Lower Pasion subregions conflate different forms than those commonly found at Tikal, and could be considered a bit more creative. There is some adherence to the Tikal pattern of 9 flints per cache, although 13, 17, and 21 cherts are also common. The concept of conflation is discussed in greater detail with specific examples in chapter 8. In general, Altar de Sacrificios seems to have had access to different kinds of stones than those available at Piedras Negras, even though both samples are geologically heterogenous as a whole. It seems that Altar had access to larger chert nodules and focused to a greater extent on eccentrics that were based on the laurel-leaf format. The eccentrics from Piedras Negras are smaller and tend to focus perhaps more on partial biface and modified flake traditions than Altar. Generally speaking, Piedras Negras can be characterized as being poor in chert resources, or at least large, high-quality nodules. Yaxchilan did not produce many eccentrics as far as the author knows; however, their existence at the site was expected given their frequent use at El Zotz (Hruby 2018), and likely political connections to the Pa' Chan polity (Garrison and Houston 2018).

The tradition of the Lower San Pedro Mártir, typified by El Tigre, seems most similar to that of Altar de Sacrificios, although the sample is rather small. As at Altar, El Tigre boasts a rather extensive use of a fine brown flint in the production of eccentrics. Both of these sites transformed medium and large laurel-leafs into elaborate forms and in a few cases seem to have identical types (figure 1.8; Meza Rodríguez 2008; Willey 1972).

From available information, it appears that there were master knappers at El Tigre who likely produced laurel-leaf bifaces for export. Hence, the most elaborate eccentrics form the site are large and very thin, and based mostly on the laurel-leaf tradition. It is unknown whether there is another fine brown source located in Southwestern Campeche along the San Pedro Mártir, or whether they were importing finished laurel-leaf

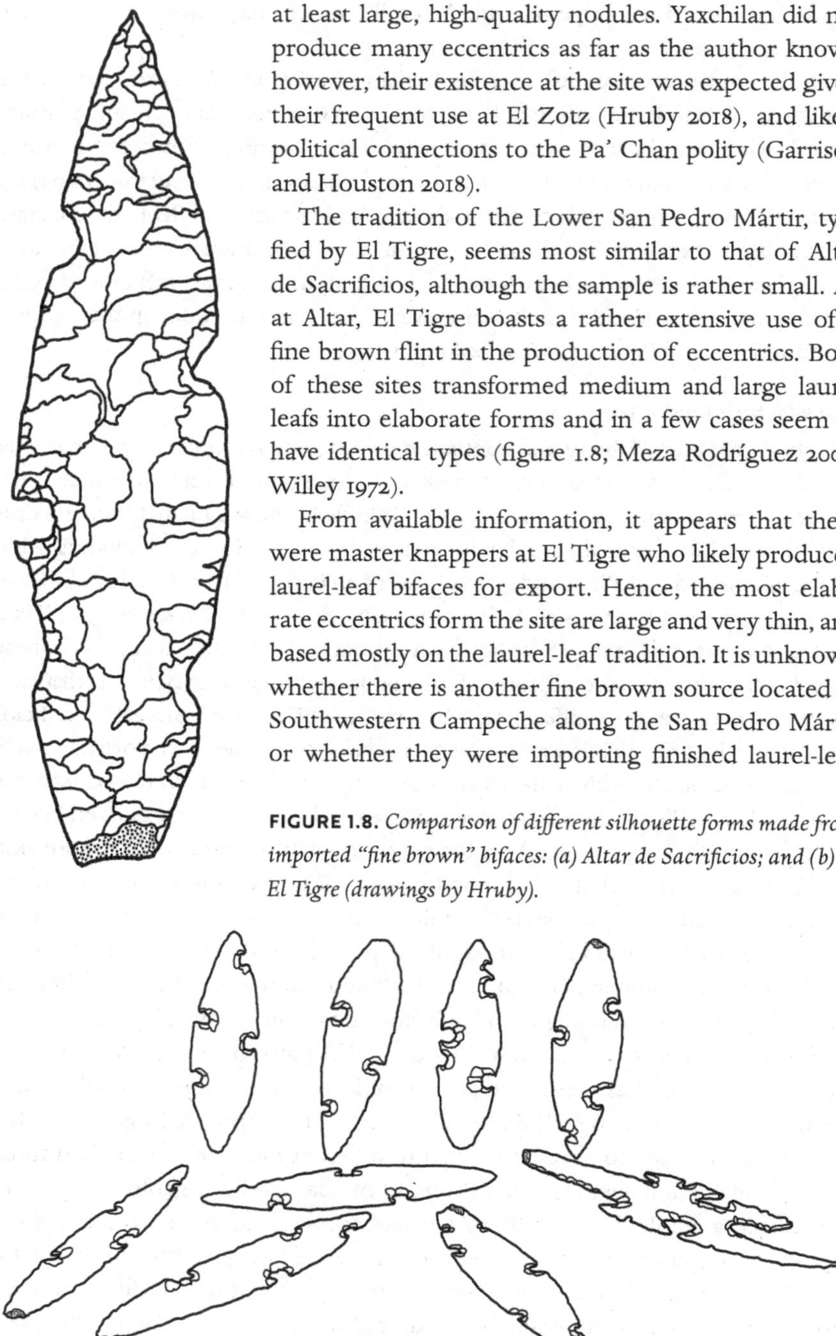

FIGURE 1.8. *Comparison of different silhouette forms made from imported "fine brown" bifaces: (a) Altar de Sacrificios; and (b) El Tigre (drawings by Hruby).*

bifaces from Belize. The most important factor in understanding the fine brown eccentrics from El Tigre and Altar de Sacrificios is that they only minimally transform the surface of the original biface. In other words, they chipped the silhouettes *after* the laurel leaf had been completely finished by indirect percussion and pressure. This fact alone indicates that the production process is of a completely different sort from that of the elaborate effigy flints. The bifaces used to make the effigy flints were probably rougher bifaces or preforms before extensive notching by indirect percussion began. It is important to note that the Piedras Negras Maya never used fine brown flint in their caches, for whatever reason, and focused entirely on locally available stones.

The Highlands

The Highlands boast few eccentrics that are verifiably local. I mention this region because there is a stemmed macroblade tradition, as well as a smattering of locally made crescent pieces of the chipped handle type but no cohesive silhouette tradition (Iannone 1993). In general, the region is similar to Copan, with a doubtful history in eccentric production but a massive blade-centered economy. It is important to note that even at centers where silhouettes were not produced, there nevertheless existed rich religious traditions that *included* ritualized lithics. Similar concepts were expressed via general material and utilitarian technological types. For example, at Copan, where we find some of the most elaborate flint-effigies, they did not transform their obsidian in a similar way. As chapter 11 examines, the Copanecos instead focused on blades and cores for their caches. In many ways Copan and Kaminaljuyu are sister cities with regard to lithic traditions, focusing, at least in elite spheres, on obsidian blade-core reduction as the source of most economic wherewithal and prestige associated with chipped stones. As Iannone has pointed out, there may be some connection between the Preclassic focus on macroblades at both Kaminaljuyu in obsidian and chert at Colha. At places like Copan, pressure blades and macroblades were the most common ceremonial lithic offerings well into the Late Classic period (Hruby and Tokovinine n.d.)

ORGANIZATION OF THE VOLUME

The book is organized roughly by technology, with the first half dedicated primarily to biface technologies, and the latter half to blade technologies, and more generally into chert and obsidian studies. In chapter 2, Pastrana continues the introductory theme with a detailed historical investigation of how some ceremonial lithic artifacts came to be known as *eccentrics* and why this term is inaccurate. Pastrana, Clark, Eble, and Woods (chapters 2, 3, and 4) focus on the possible range of direct, indirect percussion, and pressure techniques that went into producing very large bifaces. The next two chapters, by Woods and Sheets (chapters 5 and 6), narrow the focus to a very specific "knapping school" that was responsible for creating the elaborate Rosalila flints of Copan and how widely these pieces were distributed. Chapters 7, 8, and 9 center on the rich lithic traditions of Belize and surrounding zones in Northeastern Guatemala and Southern Mexico. A comparative approach is taken to examine the technological and morphological variability from caches in the same site, to those in satellite relationships of a region. Sullivan continues the replicative approach to understand the manufacture of much smaller effigy bifaces of both obsidian and chert.

The volume then shifts gears to examine blade-core technologies and the ceremonial lithic artifacts produced from them. Lucas R. M. Johnson et al. (chapter 10) summarize and analyze decades of obsidian collection at Caracol, Belize. They study production techniques and argue that blade-production artifacts were "ritualized" during calendrical and death-related events. Relatively heterarchical distributions at Caracol emphasize those materials essential for daily life and reflect a shared ritual identity. Jorge Ramos and colleagues (chapter 11) continue the examination of ritualized obsidian at the Las Sepulturas neighborhood of Copan, Honduras. Although elaborate effigies were not created in obsidian, the blades were removed in different sequences, as well as being notched and used in ways that resemble distant and abandoned Teotihuacan. Finally, blade-core debitage recovered from four different tombs at Waka' is investigated to look at the similarities and differences in production and deposition between them (chapter 12). The question of the relationship of the technology and symbolism between Maya cache rituals and burials is addressed. A detailed analysis of the symbolism of these objects is to be examined in future publications.

REFERENCES

Agurcia Fasquelle, Ricardo, Payton Sheets, and Karl Andreas Taube. 2016. *Protecting Sacred Space: Rosalila's Eccentric Chert Cache at Copan and Eccentrics Among the Classic Maya*. Monograph 2. San Francisco: Precolumbia Mesoweb Press.

Barrett, Jason W. 2011. "Ancient Maya Exploitation of Non-Renewable Resources in the Eastern Maya Lowlands." In *The Technology of Maya Civilization: Political Economy and Beyond in Lithic Studies*, edited by Zachary X. Hruby, Geoffrey E. Braswell, and Oswaldo Chinchilla Mazariegos, 57–68. Sheffield, UK: Equinox.

Blom, Frans, and Oliver La Farge. 1928. *Tribes and Temples*. New Orleans, LA: Tulane University.

Bourdieu, Pierre 1977. *Outline of a Theory of Practice*. Cambridge: Cambridge University Press.

Brown, Linda A. 2000. "From Discard to Divination: Demarcating the Sacred Through the Collection and Curation of Discarded Objects." *Latin American Antiquity* 11(4):319–333.

Brown, Linda A. 2015. "When Pre-Sunrise Beings Inhabit a Post-Sunrise World: Time, Inanimate Objects, and Contemporary Tz'utujil Maya Ritual Practitioners." In *The Measure and Meaning of Time in the Americas*, edited by Anthony Aveni. Dumbarton Oaks, Washington, DC.

Carballo, David. 2014. "Obsidian Symbolism in a Temple Offering from La Laguna, Tlaxcala." In *Obsidian Reflections: Symbolic Dimensions of Obsidian in Mesoamerica*, edited by Marc Levine and David Carballo, 195–222. Boulder: University Press of Colorado.

Carrasco, David. 2013. *Religions of Mesoamerica*, 2nd ed. Long Grove, IL: Waveland Press.

Carrasco, Michael. 2005. "The Mask Flange Iconographic Complex: The Art, Ritual, and History of a Maya Sacred Image." PhD diss., University of Texas, Austin.

Christenson, Allen J. 2007. *Popol Vuh: The Sacred Book of the Maya*. Norman: University of Oklahoma Press.

Christenson, Allen J. 2016. *The Burdon of the Ancients: Maya Ceremonies of World Renewal from the Pre-Colombian Period to the Present*. Austin: University of Texas Press.

Clark, John E., and Douglas Donne Bryant. 1997. "A Technological Typology of Prismatic Blades and Debitage from Ojo de Agua, Chiapas, Mexico." *Ancient Mesoamerica* 8(1):111–136.

Clark, John E., Fred W. Nelson, and Gene L. Titmus. 2012. "Flint Effigy Eccentrics." In *Ancient Maya Art at Dumbarton Oaks*, edited by Joanne Pillsbury, Mariam Doutriaux, Reiko Ishihara-Brito, and Alexandre Tokovinine, 273–287. Pre-Columbian Art at Dumbarton Oaks, Number 4. Washington, DC: Dumbarton Oaks.

Clayton, Sarah C. 2005. "Interregional Relationships in Mesoamerica: Interpreting Maya Ceramics at Teotihuacan." *Latin American Antiquity* 16(4):427–448.

Crabtree, Don. 1972. "The Cone Fracture Principle and the Manufacture of Lithic Materials." *Tebiwa* 15(2):29–42.

Coe Michael D., and Stephen Houston. 2015. *The Maya*, 9th ed. New York: Thames and Hudson.

Coe, William R. 1959. *Piedras Negras Archaeology: Artifacts, Caches, and Burials*. Museum Monographs. Philadelphia: University Museum, University of Pennsylvania.

Coe, William R. 1965. "Caches and Offertory Practices of the Maya Lowlands." In *Archaeology of Southern Mesoamerica, Part One*, edited by Gordon R. Willey, 462–468. Handbook of Middle American Indians 2. Austin: University of Texas Press.

Cuevas García, Martha. 2000. "Los incensarios del Grupo de las Cruces, Palenque." *Arqueología Mexicana* 45:54–61.

Cuevas García, Martha. 2004. "The Cult of Patron and Ancestor Gods in Censers at Palenque." In *Courtly Art of the Ancient Maya*, edited by Mary Miller and Simon Martin, 250–252. London: Thames and Hudson.

Cuevas García, Martha. 2007. *Los incensarios efigie de Palenque: Deidades y rituales mayas*. Mexico City: Universidad Nacional Autónoma de México, Instituto Nacional de Antropología e Historia.

Dahlin, Bruce, Marjukka Bastamow, Timothy Beach, Zachary Hruby, Scott Hutson, and Daniel Mazeau. 2011. "The Case of Phantom Lithics from Chunchucmil, Yucatán, Mexico." In *The Technology of Maya Civilization: Political Economy and Beyond in Lithic Studies*, edited by Zachary Hruby, Geoffrey Braswell, and Oswaldo Chinchilla Mazariegos, 76–90. London: Equinox Press.

Escobedo, Héctor L., and Zachary X. Hruby. 2002. "El contexto y producción del Escondite O-13-57 de Piedras Negras, Guatemala." In *XV Simposio de Investigaciones Arqueológicas en Guatemala, 2001*, Vol. 2, edited by Juan Pedro Laporte, Héctor Escobedo, and Bárbara Arroyo, 927–935. Guatemala City: Museo Nacional de Arqueología y Etnología.

Escobedo, Héctor L., and Juan Carlos Meléndez. 2007. "Excavaciones en la Estructura M12-32." In *Proyecto Arqueológico El Perú-Waka': Informe no. 4, temporada 2006*, edited by Héctor L. Escobedo and David A. Friedel, 89–107. Guatemala City: Instituto de Antropología e Historia.

Ford, Anabel. 2004. "Maya Subsistence, Settlement Patterns, and the Influence of Obsidian on the Maya Economy Around El Pilar, Belize." In *Archaeological Investigations in the Eastern Maya Lowlands: 2003 Belize Archaeology Symposium*, edited by Jaime Awe, J. Morris, and S. Jones, 61–81. Belize City: Institute of Archaeology.

Gann, Thomas W. F. 1914. "Report on Some Excavations in British Honduras." *Annals of Archaeology and Anthropology* 7:28–42.

Gann, Thomas W. F. 1918. *The Maya Indians of Southern Yucatan and Northern British Honduras*. Bureau of American Ethnology Bulletin 64:655–692. Washington, DC: Smithsonian Institution.

Gann, Thomas W. F. 1925. *Mystery Cities*. London: Camelot Press.

Gann, Thomas W. F. 1929. *Discoveries and Adventures in Central America*. New York: Scribner.

Gann, Thomas W. F., and Mary Gann. 1939. *Archeological Investigations in the Corozal District of British Honduras*. Bureau of American Ethnology, Anthropological Papers 7. Washington, DC: Smithsonian Institution.

Garrison, Thomas, and Stephen Houston. 2018. *An Inconstant Landscape: The Maya Kingdom of El Zotz, Guatemala*. Boulder: University Press of Colorado.

Gibson, Eric C. 1989. "The Organization of Late Preclassic Maya Lithic Economy in the Eastern Lowlands." In *Research in Economic Anthropology, Supplement* 4, edited by Patricia A. McAnany and Barry L. Isaac, 115–138. London: JAI Press.

Gruning, E. L. 1930. "Report on the British Museum Expedition to British Honduras, 1930." *Journal of the Royal Anthropological Institute* 60:477–483.

Haury, Emil W., Robert L. Rands, Albert C. Spaulding, Walter W. Taylor, Raymond H., Thompson, Robert Wauchope, and Marian E. White. 1956. "An Archaeological

Approach to the Study of Cultural Stability." In *Memoirs of the Society for American Archaeology 11, Seminars in Archaeology*, edited by Robert Wauchope, 31–57. Cambridge: Cambridge University Press.

Hearth, Nicholas, and Scott L. Fedick. 2011. "Defining the Chert Paucity Problem in the Northern Maya Lowlands: A First Approximation." In *The Technology of Maya Civilization: Political Economy and Beyond in Lithic Studies*, edited by Zachary Hruby, Geoffrey Braswell, and Oswaldo Chinchilla Mazariegos, 69–75. London: Equinox Press.

Hester, Thomas R. 1982. "The Maya Lithic Sequence in Northern Belize." In *Archaeology at Colha, Belize: The 1981 Interim Report*, edited by Thomas R. Hester, Harry J. Shafer, and Jack Eaton, 39–59. San Antonio: Center for Archaeological Research, University of Texas.

Hester, Thomas R., Harry J. Shafer, and Then Berry. 1991. "Technological and Comparative Analyses of the Chipped Stone Artifacts from El Pozito, Belize." In *Maya Stone Tools: Selected Papers from the Second Maya Lithic Conference*, edited by Thomas R. Hester and Harry Shafer, 67–83. Monographs ln World Archaeology 1. Madison, WI: Prehistory Press.

Heye, George G.1925. "Eccentric Chipped Objects from British Honduras." Museum of the American Indian, Heye Foundation. *Indian Notes* 2:99–102.

Hintzman, Mark W. 2000. "Scarce-Resource Procurement and Use: The Technological Analysis of an Obsidian Blade Workshop in the Lowlands of Belize." Master of Science, University of California, Riverside.

Hirth, Kenneth, and Anne Cyphers. 2020. *Olmec Lithic Economy at San Lorenzo*. Boulder: University Press of Colorado.

Hirth, Kenneth G., Gene L. Titmus, J. Jeffrey Flenniken, and Jaques Tixier. 2003. "Alternative Techniques for Producing Mesoamerican-Style Pressure Flaking Patterns on Obsidian Bifaces." In *Mesoamerican Lithic Technology: Experimentation and Interpretation*, edited by Kenneth G. Hirth, 147–152. Salt Lake City: University of Utah Press.

Houston, Stephen D. 2014. *The Life Within: Classic Maya and the Matter of Permanence*. New Haven, CT: Yale University Press.

Hruby, Zachary. 2003. "Re-evaulación de las categorías utilitarias y ceremoniales de artefactos mayas de piedra tallada." In *XVI Simposio de Investigaciones Arqueológicas en Guatemala*, edited by Juan Pedro Laporte, Héctor L. Escobedo, and Barbara Arroyo, 507–512. Guatemala City: Museo Nacional de Arqueología y Etnología.

Hruby, Zachary. 2004. "Observaciones Preliminares del Análisis de Obsidiana del Proyecto Arqueológico Parque Kaminaljuyú." In *El Primero Informe del Proyecto Arqueológico de Kaminaljuyú*, edited by Matilde Ivic. Technical Report. Instituto de Antropologia e Historia de Guatemala (IDAEH).

Hruby, Zachary X. 2006. "The Organization of Chipped-Stone Economies at Piedras Negras, Guatemala." PhD diss., University of California, Riverside.

Hruby, Zachary X. 2007. "Ritualized Chipped-Stone Production at Piedras Negras, Guatemala." In *Rethinking Craft Specialization in Complex Societies: Archaeological Analysis of the Social Meaning of Production*, edited by Zachary Hruby and Rowan K. Flad, 68–87. Archaeological Papers of the American Anthropological Association 17. Arlington, VA: American Anthropological Association.

Hruby, Zachary. 2018. "Lithic Technologies and Economies at El Zotz." In *An Inconstant Landscape: The Maya Kingdom of El Zotz, Guatemala*, edited by Thomas Garrison and Stephen Houston, 229–254. Boulder: University Press of Colorado.

Hruby, Zachary. 2020. "The Classic Maya Atlatl." *Mexicon* 42(3): 59–66.

Hruby, Zachary. 2024. "Toward an Understanding of Obsidian Values Among the Ancient Maya Through a Comparative Approach." In *Realizing Value in Mesoamerica: The Dynamics of Desire and Demand in Ancient Economies*, edited by Scott Hutson and Charles Golden, pp. 139–162. Palgrave McMillan, Cham. Switzerland.

Hruby, Zachary. In prep. "The Symbolism of Classic Maya Solar Caches and Cache Goods." Manuscript in possession of editor.

Hruby, Zachary, Héctor Escobedo, Francisco Estrada-Belli, Stephen Houston, Juan Carlos Meléndez, and Luis Romero. In Prep. "A Comparative Study of Classic Caches from the Maya Lowlands." Manuscript in possession of editor.

Hruby, Zachary X., and Rowan Flad, eds. 2007. *Rethinking Craft Specialization in Complex Societies: Archaeological Analysis of the Social Meaning of Production*. Archaeological Papers of the American Anthropological Association 17. Arlington, VA: American Anthropological Association.

Hruby, Zachary X., and Michelle Rich. 2014. "Flint for the Dead: Ritual Deposition of Production Debitage from El Perú-Waka', Burial 39." In *Archaeology at El Perú-Waka': Ancient Maya Performances of Ritual, Memory, and Power*, edited by O. C. Navarro-Far and M. Rich, 167–183. Tucson: University of Arizona Press.

Hruby, Zachary, and Alexandre Tokovinine. N.d. "Exploring the Symbolic Aspects of the Stromsvik Macroblade Cache from the Grand Plaza of Copan, Honduras." In *The Rise, Collapse, and Rebirth of a State*, edited by Cameron McNeil and Alexandre Tokovinine. Gainesville: University of Florida Press.

Hruby, Zachary X., and Gene Ware. 2009. "Painted Lithic Artifacts from Piedras Negras, Guatemala." In *Maya Archaeology*, vol. 1, edited by Charles Golden, Stephen Houston, and Joel Skidmore, 76–85. San Francisco: Mesoweb Press.

Iannone, Gyles J. 1993. "Ancient Maya Eccentric Lithics: A Contextual Analysis." MA thesis, Department of Anthropology, Trent University, Peterborough, ON.

Inomata, Takeshi. 2001. "The Power and Ideology of Artistic Creation: Elite Craft Specialists in Classic Maya Society." *Current Anthropology* 42:321–349.

Johnson, Lisa M., and Lucas Martindale Johnson. 2020. "Sealing with Stone: Assessing an Assemblage of Lithic Debitage from a Funerary Context at the Lowland Maya City of Caracol, Belize." *Latin American Antiquity* 32(1):1–18.

Johnson, Lucas Martindale. 2016. "Toward an Itinerary of Stone: An Investigation of Obsidian from Caracol, Belize." PhD diss., University of Florida, Gainesville.

Joyce, Rosemary A. 1992. "Ideology in Action: Classic Maya Ritual Practice. In Ancient Images, Ancient Thought: The Archaeology of Ideology." In *Proceedings of the 23rd Annual Chacmool Conference*, edited by A. Sean Goldsmith, Sandra Garvie, David Selin, and Jeannette Smith, 497–506. Calgary: Archaeological Association, University of Calgary.

Joyce, Thomas A. 1932. "Presidential Address: The 'Eccentric Flints' of Central America." *Journal of the Royal Anthropological Institute of Great Britain and Ireland* 62:xvii–xxvi.

Joyce, Thomas A., Thomas Gann, E. L. Gruning, and R. C. E. Long. 1928. "Report on the British Museum Expedition to British Honduras, 1928." *Journal of the Royal Anthropological Institute of Great Britain and Ireland* 258:323–350.

Kaneko, Akira. 2003. *Artefactos lítico de Yaxchilán*. Serie Arqueología, Colección científica, v. 455. Mexico City: INAH.

Kidder, Alfred V. 1947. *Artifact of Uaxactun, Guatemala*. Carnegie Institution of Washington, Publication 576. Washington, DC.

Kidder, Tristram R. 1985. "Artifacts." In *A Consideration of the Early Classic Period in the Maya Lowlands*, edited by Gordon R. Willey and Peter Mathews, 95–110. Publication lo. Albany: Institute for Mesoamerican Studies, State University of New York.

Kwoka, Joshua J. 2014. "Ideological Presentism and the Study of Ancient Technology: Preclassic Lithic Production at San Bartolo, Guatemala." PhD diss., State University of New York at Buffalo, Buffalo.

Kwoka, Joshua J., H. Colleen Hanratty, and Thomas H. Guderjan. 2019. "From Crafting to Caching: Technological and Iconographic Analyses of Blue Creek Cache 37." *Journal of Field Archaeology* 43(6):352–366.

Kwoka, Joshua J., and M. Steven Shackley. 2019. "Technological Analysis and Source Provenance of Obsidian Artifacts from a Sun Pyramid Substructure Cache, Teotihuacan, Mexico." *Latin American Antiquity* 30:205–210.

Lane Fox, A. H. 1871. "A Flint Implement Brought from Honduras." *Proceedings of the Society of Antiquaries of London* 5:93–95.

Laporte, Juan Pedro. 1989. "Alternativas del clásico temprano en la relación Tikal-Teotihuacan: Grupo 6C-XVI, Tikal, Petén, Guatemala." PhD diss., Universidad Nacional Autónoma de México, Mexico City.

Laporte, Juan Pedro. 2003. "Architectural Aspects of Interaction Between Tikal and Teotihuacan." In *The Maya and Teotihuacan: Reinterpreting Early Classic Interaction*, edited by Geoffrey Braswell, 199–216. Austin: University of Texas Press.

Linné, Sigvald. 1934. *Archaeological Researches at Teotihuacan, Mexico*. Publication, 1. Stockholm: Ethnographical Museum.

Looper, Matthew G. 2003. *Lighting Warrior: Maya Art and Kingship at Quirigua*. Austin: University of Texas Press.

López Luján, Leonardo, and Saburo Sugiyama. 2018. "Ritual Deposits in the Moon Pyramid at Teotihuacan." In *Teotihuacan: City of Water, City of Fire*, edited by Matthew Robb, 82–89. Oakland: Fine Arts Museums of San Francisco, de Young and University of California Press.

Maler, Teobert. 1908. *Explorations in the Department of Petén, Guatemala, and Adjacent Regions*. Memoirs of the Peabody Museum, 4(2). Cambridge: Peabody Museum, Harvard University.

Marcus, Joyce and Kent V. Flannery. 1996. *Zapotec Civilization: How Urban Society Evolved in Mexico's Oaxaca Valley*. London: Thames and Hudson.

Martin, Simon, and Nikolai Grube. 2008. *Chronicle of the Maya Kings and Queens: Deciphering the Dynasties of the Ancient Maya*. New York: Thames and Hudson.

Mason, J. Aldon. 1935. "Preserving America's Finest Sculptures." *National Geographic Magazine* 68(5):537–570.

Meadows, Richard. 2001. "Crafting K'awil: A Comparative Analysis of Maya Symbolic Flaked Stone Assemblages from Three Sites in Northern Belize." PhD diss., University of Texas, Austin.

Meissner, Nathan J., and Marc D. Marino. N.d. "Producers on the Margins: Chert Crafting in the Copán Hinterland of Río Amarillo, Honduras." In *Copán: The Rise, Collapse, and Aftermath of a State*, edited by Cameron L. McNeil and Alexandre Tokovinine. Gainesville: University of Florida Press.

Meza Rodríguez, Carolina. 2008. "Los Instrumentos de Pedernal (tuk 'tuunich) en El Tigre, Campeche: Procesos de explotación y consumo." Tesis Licenciatura, en archivo INAH.

Millon, Rene, and Bruce Drewitt. 1961. "Earlier Structures Within the Pyramid of the Sun at Teotihuacan." *American Antiquity* 26(3):371–380.

Mills, Barbara J., Jeffery J. Clark, Matthew Peeples, W. R. Haas Jr., John M. Roberts Jr., J. Brett Hill, Deborah L. Huntley, Lewis Borck, Ronald L. Breiger, Aaron Clauset, and M. Steven Shackley. 2013. "Transformation of Social Networks in the Late Prehispanic US Southwest." *Proceedings of the National Academy of Sciences* 110(15):5785–5790.

Moholy-Nagy, Hattula. 1997. "Middens, Construction Fill, and Offerings: Evidence for the Organization of Classic Period Craft Production at Tikal, Guatemala." *Journal of Field Archaeology* 24:293–313.

Moholy-Nagy, Hattula. 1999. "Mexican Obsidian at Tikal, Guatemala." *Latin American Antiquity* 10(3):300–313.

Moholy-Nagy, Hattula. 2003. *The Artifacts of Tikal: Utilitarian Artifacts and Unworked Material, Tikal Report 27B*. Philadelphia: University of Pennsylvania Museum of Archaeology and Anthropology.

Moholy-Nagy, Hattula. 2008. *The Artifacts of Tikal: Ornamental and Ceremonial Artifacts and Unworked Material, Tikal Report 27A*. Philadelphia: University of Pennsylvania Museum of Archaeology and Anthropology.

Moholy-Nagy, Hattula. 2020. "Problematical Deposits at Tikal, Guatemala: Content, Context, and Intent." *Ancient Mesoamerica*:1–17.

Moholy-Nagy, Hattula, and William R. Coe. 2008. "Appendix 11: Typology and Description of Flint and Obsidian Eccentrics and Incised Obsidians." In *The Artifacts of Tikal: Ornamental and Ceremonial Artifacts and Unworked Material, Tikal Report 27A*. Philadelphia: University of Pennsylvania Museum of Archaeology and Anthropology. Accessed February 1, 2018. https://doi.org/10.6067/XCV8CJ8D43.

Parry, William J., and Shigeru Kabata. N.d. "Chronology of Obsidian Artifacts from the Moon Pyramid." In *The Moon Pyramid, Teotihuacan*, edited by Saburo Sugiyama and David Carballo. Albuquerque: University of New Mexico Press.

Pastrana, Alejandro, and Ivonne Athie. 2014. "The Symbolism of Obsidian in Postclassic Central Mexico." In *Obsidian Reflections: Symbolic Dimensions of Obsidian in Mesoamerica*, edited by Marc Levine and David Carballo, 75–110. Boulder: University Press of Colorado.

Pastrana, Alejandro, and Silvia Domínguez. 2018. *Proyecto Yacimientos de Obsidiana, Temporada de Campo 2016–2017. Informe técnico al Pastrana Alejandro y Silvia Domínguez 2017, Proyecto Yacimientos de Obsidiana, Temporada de Campo 2016–2017*. Informe técnico al Archivo del Consejo de Arqueología, INAH. Archival manuscript, Consejo de Arqueología, INAH, Mexico City.

Pendergast, David M. 1979. *Excavations at Altun Ha*. Vol. 1. Toronto: Royal Ontario Museum.

Pendergast, David M. 1981. "Lamanai, Belize: Summary of Excavation Results, 1974–1980." *Journal of Field Archaeology* 8:29–53.

Pendergast, David M. 1990. *Excavations at Altun Ha, Belize, 1964–1970*. Vol. 3. Toronto: Royal Ontario Museum.

Pendergast, David M. 2003. "Teotihuacan at Altun Ha: Did It Make a Difference?" In *The Maya and Teotihuacan: Reinterpreting Early Classic Interaction*, edited by Geoffrey E. Braswell, 235–247. Austin: University of Texas Press.

Prado, Antonio. 1999. *El creador maya*. Guatemala City: Editorial Galaría Guatemala.

Price, W. H. 1897–1899. "Excavations on Sittee River, British Honduras." *Proceedings, Society of Antiquarians, London* 17:339–344.

Rice, R. Garraway. 1909. "Notes on a Flint in Human Shape Found in the Thames." *Proceedings of the Society of Antiquaries of London* 22:359–360.

Ricketson, Oliver G. 1929. "Excavations at Baking Pot, British Honduras." In *Contributions to American Archaeology*, 1(1):1–28. Washington, DC: Carnegie Institute of Washington.

Ricketson, Oliver G. 1931. "The Present Status of Archaeological Investigation in Guatemala." *Year Book* 30:112–114. Washington, DC: Carnegie Institution of Washington.

Ricketson, Edith B., and Oliver G. Ricketson. 1937. *Uaxactun, Guatemala: Group E-1926–1931*. Washington, DC: Carnegie Institution of Washington.

Roche Recinos, Alejandra. 2021. "Regional Production and Exchange of Stone Tools in the Maya Polity of Piedras Negras, Guatemala." PhD diss., Brown University.

Roddick, Andrew P. 2016. "Scalar Relations: A Juxtaposition of Craft Learning in the Lake Titicaca Basin." In *Knowledge in Motion: Making Communities and Constellations of Practice Across Time and Place*, edited by Andrew P. Roddick and Ann B. Stahl, 126–154. Tucson: University of Arizona Press.

Roddick, Andrew P., and Ann B. Stahl. 2016. " 'Knowledge in Motion': Making Communities and Constellations of Practice Across Time and Place." In *Knowledge in Motion: Making Communities and Constellations of Practice Across Time and Place*, edited by Andrew P. Roddick and Ann B. Stahl, 3–35. Tucson: University of Arizona Press.

Sassaman, Kenneth E., and Wictoria Rudolphi. 2001. "Communities of Practice in the Early Pottery Traditions of the American Southeast." *Journal of Anthropological Research* 57(4):407–425.

Shafer, Harry J. 1991. "Late Preclassic Formal Tool Production at Colha, Belize." In *Maya Stone Tools: Selected Papers from the Second Maya Lithic Conference*, edited by Thomas R. Hester and Harry Shafer, 11–44. Monographs ln World Archaeology 1. Madison, WI: Prehistory Press.

Shafer, Harry J. 2023. "Legacy of the Ancient Maya Tranchet Flake: Or How Two Texas Archaeologists Ended Up in the Maya World." *Journal of Texas Archaeology and History* 7(1):1–24.

Shafer, Harry J., and Thomas R. Hester. 1983. "Ancient Maya Chert Workshops in Northern Belize, Central America." *American Antiquity* 48:5l9–543.

Smith, A. L. 1972. *Excavations at Altar de Sacrificios: Architecture, Settlement, Burials, and Caches*. Papers of the Peabody Museum of Archaeology and Ethnology 62(2). Cambridge, MA: Harvard University.

Stemp, James W., Jaime T. Awe, and Christophe Helmke. 2014. "The Macrolith of El Chiquero, Belize." *Mexicon* 36(5):154–150.

Stephens, John Lloyd. 1860. *Incidents of Travel in Yucatan*. New York: Harper & Bros.

Stuart, David. 2000. " 'The Arrival of Strangers': Teotihuacan and Tollan in Classic Maya History." In *Mesoamerica's Classic Heritage: From Teotihuacan to the Aztecs*, edited by David Carrasco, Lindsay Jones, and Scott Sessions, 465–513. Boulder: University Press of Colorado.

Stuart, David. 2003. "On the Paired Variants of TZ'AK." Mesoweb. www.mesoweb.com/stuart/notes/tzak.pdf.

Stuart, David. 2024. *Spearthrower Owl: A Teotihuacan Ruler in Maya History*. Studies in Pre-Columbian Art and Archaeology. Washington, DC: Dumbarton Oaks, Trustees for Harvard University.

Sugiyama, Saburo. 2005. *Human Sacrifice, Militarism, and Rulership: Materialization of State Ideology at the Feathered Serpent Pyramid, Teotihuacan*. Cambridge: Cambridge University Press.

Sugiyama, Saburo, and Ruben Cabrera Castro. 2018. "The Moon Pyramid and the Ancient State of Teotihuacan." In *Teotihuacan: City of Water, City of Fire*, edited by Matthew Robb, 74–81. Oakland: Fine Arts Museums of San Francisco, de Young and University of California Press.

Sugiyama, Saburo, and Leonardo López Luján, eds. 2006. *Sacrifices of Consecration in the Pyramid of the Moon* Mexico City: INAH.

Sugiyama, Saburo, and Nawa Sugiyama. 2020. "Interactions Between Ancient Teotihuacan and the Maya World." In *The Maya World*, edited by Scott Hutson and Traci Ardren, 689–711. New York: Routledge.

Sugiyama, Saburo, Nawa Sugiyama, William L. Fash, and Barbara Fash. 2020. "The Maya at Teotihuacan? New Insights into Teotihuacan-Maya Interactions from Plaza of the Columns Complex." In *Teotihuacan: The World Beyond the City*, edited by Kenneth Hirth, Barbara Arroyo, and David Carballo, 139–172. Dumbarton Oaks Pre-Columbian Symposia and Colloquia series. Washington, DC: Dumbarton Oaks Research Library and Collection.

Taube, Karl A. 1991. "Obsidian Polyhedral Cores and Prismatic Blades in the Writing and Art of Ancient Mexico." *Ancient Mesoamerica* 2(1):61–70.

Taube, Karl A. 2000. "The Turquoise Hearth: Fire, Self Sacrifice, and the Central Mexican Cult of War." *In Mesoamerica's Classic Heritage: From Teotihuacan to the Great Aztec Temple*, edited by David Carrasco, Lindsay Jones, and Scott Sessions, 269–340. Niwot, University Press of Colorado.

Taube, Karl A. 2003. "Ancient and Contemporary Maya Conceptions About the Field and Forest." In *The Lowland Maya Area: Three Millennia at the Human-Wildland Interface*, edited by Arturo Gómez-Pompa, Michael F. Allen, Scott Fedick, and J. Jiménez-Osornio, 461–492. New York: Haworth Press.

Taube, Karl A., William Saturno, David Stuart, Heather Hurst, and Joel Skidmore. 2010. "The Murals of San Bartolo, El Petén, Guatemala. Part 2. The West Wall." *Ancient America* 10.

Thompson, J. Eric S. 1931. *Archaeological Investigations in the Southern Cayo District, British Honduras*. Field Museum of Natural History Anthropology Series 17(3). Chicago: Field Museum.

Thompson, J. Eric S. 1936. "An Eccentric Flint from Quintana Roo, Mexico." *Maya Research* 3: 316–318.

Thompson, J. Eric S. 1939. *Excavations at San Jose, British Honduras*. Carnegie Institution of Washington, Publication 506. Washington, DC: Carnegie Institution of Washington.

Thompson, J. Eric S. 1963. *Maya Archaeologist*. Norman: University of Oklahoma Press.

Titmus, Gene L., and James C. Woods. 2003. "The Maya Eccentric: Evidence for the Use of the Indirect Percussion Technique in Mesoamerica from Preliminary Experiments Concerning Their Manufacture." In *Mesoamerican Lithic Technology: Experimentation and Interpretation*, edited by Kenneth G. Hirth, 132–146. Salt Lake City: University of Utah Press.

Trachman, Rissa M. 1999. "An Additional Technological Perspective on Obsidian Polyhedral Core Platform Rejuvenation." *Lithic Technology* 24(2):119–125.

Trachman, Rissa M. 2002. "Early Classic Obsidian Core-Blade Production: An Example from the Site of Dos Hombres, Belize." In *Pathways to Prismatic Blades: A Study in Mesoamerican Obsidian Core-Blade Technology*, Monograph 45, edited by K. Hirth and B. Andrews, 105–120. Los Angeles: Cotsen Institute of Archaeology Press at UCLA.

Tsukamoto, Kenichiro, Octavio Esparza Olguin, Daniel Salazar Lama, and Luz Evelia Campana Valenzuela. 2021. "Upakal K'inich: A Late Classic Period Ruler of El Palmar, Mexico." *The PARI Journal* 22(2):1–20.

Willey, Gordon R. 1945. "Horizon Styles and Pottery Traditions in Peruvian Archaeology." *American Antiquity* 11:49–56.

Willey, R. Gordon. 1972. *The Artifacts of Altar of Sacrifices, Papers of the Peabody Museum* 64(1). Cambridge: Peabody Museum.

Willey, Gordon R. 1978. "Artifacts." In *Excavations at Seibal*. Memoirs of the Peabody Museum of Archaeology and Ethnology, Harvard University, vol. 14(1):1–189.

Woods, James C., and Gene L. Titmus. 1996. "Stone on Stone: Perspectives on Maya Civilization from Lithic Studies." In *Eighth Palenque Round Table, 1993*, edited by Martha J. Macri and Jan McHargue, 479–489. San Francisco: Pre-Columbian Art Research Institute.

2

Eccentric or Silhouette

A Preliminary Typology of Ceremonial Chipped-Stones from Sierra de las Navajas and Teotihuacan

ALEJANDRO PASTRANA AND ZACHARY HRUBY

This chapter is based on a long-term study of Teotihuacan-style artifacts from the site of La Sierra de las Navajas, Hidalgo, also known as the Pachuca obsidian source. All of the artifact types discussed here were not present at Sierra de las Navajas, so we have included a sample from the great center of Teotihuacan as a point of comparison. Pastrana and colleagues have analyzed and experimentally replicated many of the production techniques used at the source, including those used to make tools, weapons, and status objects, as well as the religious items discussed here (Pastrana and Domínguez 2010, 2011, 2012, 2013, 2014, 2016, 2018; Pastrana et al. 2018). These instruments were crafted from the most-coveted obsidian in Mesoamerica, boasting a wide array of colors and textures: beautiful translucent green, gold, chatoyant, and dark green colors in banded and solid configurations. A small portion of the Sierra de las Navajas workshop debitage is clearly related to the production of Teotihuacan eccentrics, likely for transport directly to the elites of that city. Some of these eccentrics are well known at Teotihuacan, while others are not, and we wondered whether they may be eccentrics at all. What differentiates these items from other utilitarian

https://doi.org/10.5876/9781646427819.c002

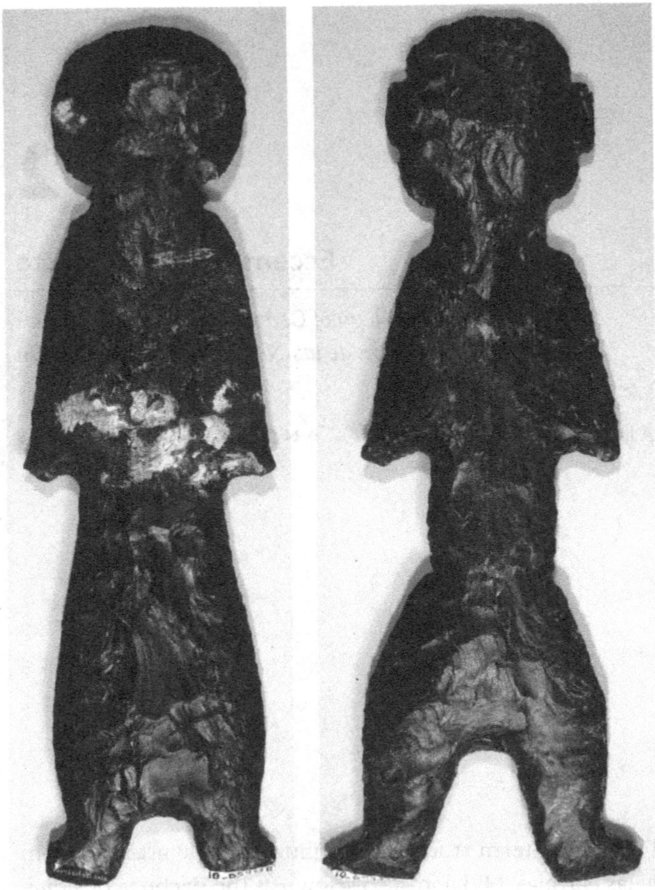

FIGURE 2.1. *Examples of some of large, fine silhouette bifaces from the Pyramid of the Moon excavations. © Proyecto Pirámide de la Luna.*

instruments is that they have strong symbolic aspects beyond those associated with utilitarian forms. We present here a morpho-technological typology to begin the process of understanding the variety of forms from Classic Central Mexico.

In ancient Mesoamerica so-called eccentrics are most commonly associated with the Central Mexican Highland and Maya areas. There is some debate about where these unique objects were first invented, specifically those that were audaciously cached in elite and public architectural settings (chapter 1). The most elaborate of these caches often contained the finest eccentrics (figure 2.1) and were deposited in some of the most important temples in all of Mesoamerica, such as the Pyramid of the Moon at Teotihuacan and Tikal Temple 1.

Interestingly, eccentrics in the Maya area are not common before direct and sustained contact was made with Teotihuacan. Zachary Hruby (chapter 1) argues that some eccentric traditions may have resulted from a reciprocal exchange of ideas between the two cultures, but their deposition, at least as a systematic form of ritual architectural offering, likely began at Teotihuacan. The smaller Maya eccentrics, those most commonly placed in architectural and substela caches, seem to have arisen through a syncretic combination of Teotihuacano and Maya signs. It could be argued that the Maya eccentric tradition began at Early Classic Tikal, the epicenter for contact between the two peoples in the Maya Lowlands (Stuart 2020, 2024). It should be noted that influence went the other direction as well (Taube 1999), if not in terms of the style of eccentrics, then in the inclusion of Guatemalan obsidian in high-profile lithic caches at Teotihuacan. We believe that a systematic study of these objects can lead to a better understanding of how production and ritual evolved over time.

A TERMINOLOGICAL NIGHTMARE: ECCENTRIC

Creating this typology led us to ask, from an archaeological perspective, what exactly is an *eccentric* and why was the term used in the first place? We argue that the use of the term *eccentric* for certain chipped lithic objects is eccentric to the original idea of their form. *Eccentric* was first used in Mesoamerican archaeology by British and American Mayanists who were at a loss to describe certain elaborately chipped artifacts. It was probably first employed in the Southern United States where some flint *silhouettes or figures* were referred to as *totem figures* and also as *amulets* (Hattula Moholy-Nagy 1999; Joyce 1932; Smith 1934). W. H. Holmes (1917, 294) refers to what would now be called a *quadrupedal zoomorphic eccentric* as a "large, thick, animal-shaped figure of flint, probably specialized exclusively with the hammerstone." In the Maya area it may have been first used by Thomas Gann, but generally speaking, by the 1920s and '30s it had already been passed around the early Mayanists working in Southern Mexico, Guatemala, and Belize (for the earliest accounts of Maya flints, see Blom and la Farge 1928; W. F. Gann 1918, 1925, 1929; W. F. Gann and M. Gann 1939; Gruning 1930; Joyce 1932; Joyce et al. 1928; Lane Fox 1871; Linné 1934; Maler 1908; Mason 1935; Price 1897–1899; Rice 1909; O. Ricketson 1929, 1931; O. Ricketson and E. Ricketson 1937; Stephens 1860; Thompson 1939).

The definitions of *eccentric* in English and Spanish are similar to one another:

Eccentric ("eccentric" 2019)

2b: located elsewhere than at the geometric center

also: having the axis or support so located

technical: not following a perfectly circular path

Merriam-Webster Dictionary

Excéntrico, ca ("excéntrico" 2014)

De ex-[1] y céntrico.

2. adj. Geom. Que está fuera del centro, o que tiene un centro diferente.

Diccionario de la Real Academia Española

Thus, it refers to a shape whose axis is outside the center, that is, something asymmetrical. There is no good archaeological definition of the term, however, partially due to the word itself, which is simultaneously too general, ambiguous, and inaccurate in this context. It has no defined limits and therefore does not contribute at all to any morphological, technological, or functional description of the artifacts. By 1959 William Coe had already mentioned the inappropriateness of the term, since several artifacts clearly featured symmetry. The term captures the essence of the ignorance that early researchers could not quite escape. Very few reasonable interpretations were even attempted. J. Eric Thompson posited, perhaps out of frustration, "there seems to have been a competition to see who could make the most elaborate design in flint" (1963, 265). In some ways he was not wrong. There was likely competition between knappers, though this competition likely did not cause the genesis of the eccentric traditions, only their refinement and perhaps elaboration.

Although *eccentric* is a poor descriptor for artifacts, up to this point it has been too ingrained in the literature to be discarded (e.g., Coe 1959; Joyce 1932; Kidder 1947; Willey 1972), and no adequate replacements have been proposed (but see also chapter 10 here). Supplementary qualifiers, like *elaborately chipped*, do not provide consistent guides amid any sorting procedure either. As Coe and Moholy-Nagy (2008, appendix 11, 1) state: "Normally in a cache or burial context, items that would be classified as eccentrics are either representational or symbolic, figural and non-figural. Three major categories are perfunctorily recognized: Eccentric Flints, Eccentric Obsidians, and Incised Obsidians. More often than not, two if not the three were assembled for collective deposition, a fact sustaining their corporateness." Thus, they argue that there is some symbolic equivalence between these general classes, even though they were made in very different ways. In contrast to the other sense of the word, they represent coherent things that were not the product of eccentric caprice.

Virgil Y. Russel (1951–1957, 146), used the term *effigy* to refer to what is now known as an eccentric for an artifact with the silhouette of an eagle. In English *effigy* means an image or representation, especially of a person ("effigy" 2020a). This term as well as *shaped figure*, *silhouette*, *perimeter shape*, and perhaps *contour* are preferable to *eccentric*, simply because they are more accurate and relevant to these objects. Somehow *eccentric* is the term that has survived over time, roughly a hundred years hence. At best, *eccentric* could be viewed as a morphological designation with some connotation as a graphically representational instrument; yet without an explicit definition and without greater precision, it remains of little use. Among archaeologists using the term, there is an implicit agreement that these artifacts were made of flint, chalcedony, and obsidian. They are not working instruments and were not used as functional weapons either, although they may represent weapons, various objects, plants, animals, celestial bodies, landscape reliefs, humans—in general any object, being, idea, or concept that is significant for a particular culture or cultures.

By the particular shape of the contour, or perimeter, of some artifacts, we know they are figurative or symbolic representations. Based on the often monumental contexts of ritual offerings (i.e., caches and burials), in which sets of eccentrics have been found in the Maya area and Central Mexico, most archaeologists have assigned them ceremonial-religious functions. One of, if not the, most important aspects of these artifacts is their shape, or "silhouette," that is, the contour or perimeter of the piece. It can be an explicit graphic representation of symbols or perhaps even elements of pictographic writing, ideographic, symbolic, or phonetic (cf. Taube 1999, 2011). Based on the foregoing information, the following questions arise: (1) Are these objects individual units, or are they parts of composite artifacts; (2) did they represent scenes in the cache context; (3) why were they made in obsidian, chert, flint, or chalcedony; (4) were they archetypal conceptions associated with lightning and war (Eliade 1983), because of the hardness, durability, and meaning of flint and obsidian stones; (5) were they made in other types of perishable materials; (6) were there other representations, such as drawings, paintings, reliefs, or sculptures with contours similar to the shapes of the silhouettes? The answer to all these questions, to some degree, is affirmative.

According to V. Gordon Childe (1958, 36–40), the sets of elements called eccentrics are only *potential* archaeological documents and are difficult to place in an archaeological classification, since they are not defined as types or as products of a common pattern of behavior. From our perspective, the classification of lithic artifacts is related to the type of raw material and can be morphological, technological, and functional. Morphological classification is based on geometric elements and dimension parameters, without necessarily knowing the technique used to produce it or its function. Technological classifications are based on characteristics of the raw material, the techniques and strategies used to chip them, and the instruments needed to do the work.

Functional classifications are one of the main factors in the definition of types, for which one should ideally understand the function, form, production technique, and raw material. For example, in Central Mexico there is the well-known type, "spoon-shaped obsidian unifacial scraper," which can be summarized as "maguey scraper." Another, the "curved, green-obsidian sacrificial knife," is colloquially known as a "Teotihuacan sacrificial knife." The distribution and chronology of the types may vary in regions, but what all of these monikers have is *utility*. The terms and characterizations say something useful about the artifact, no matter what the typology may focus on. The term *eccentric* does not refer to any characteristic or classificatory attribute, which thus does not allow progress toward useful classification based on the definition of morphological, technological, and functional types. It could be that *eccentric* has persisted in use by exclusion; that is, they are a group of artifacts that were different and that had some special but little understood connotation. They were culturally distinct from the types of artifacts commonly known as utilitarian or practical tools.

In the development of science new terms arise, are modified, and then sometimes discarded according to the advancement of knowledge. We, and some others in this volume, argue that *eccentric* has lost its utility and should be discarded. To move toward a useful and orderly classification, we propose the term *silhouette* for artifacts that do not have a clear utilitarian function and are most often found in ritual contexts. It refers specifically to the contour or perimeter as the main attribute of the artifact. Secondary attributes can include the description of the technological and morphological characteristics, which ultimately leads to the identification of the represented figure and its meaning. The term *silhouette* directs us to ask what the silhouette represents and implies the need for refined definitions of their various shapes.

The definition of *silhouette* is ("silhouette" 2020b):

> **1**: a likeness cut from dark material and mounted on a light ground or one sketched in outline and solidly colored in
>
> **2**: the outline of a body viewed as circumscribing a mass
>
> *Merriam-Webster Dictionary*.

We argue that creating morphological and technological types for lithic silhouettes, by culture or by cultural region, either as individual pieces or as sets, could generate a useful basis for archaeologists to address the variety of silhouettes in a more precise and useful way. While this chapter focuses on silhouettes from Central Mexico, it should be understood that this study should overlap in a useful way with a parallel study of Maya silhouettes (chapter 1 and 8 here). To better explain how we use the term *silhouette* for Teotihuacan artifacts, we first provide some background on the role of obsidian in Central Mexico during the Classic period.

THE IMPORTANCE OF OBSIDIAN AT TEOTIHUACAN

The site of La Sierra de las Navajas was the primary source of green obsidian for Teotihuacan, although a smaller, secondary source of greenish obsidian is also located at Tulancingo. It is comprised of dispersed mines, workshop locales, and households that reflect a wide range of production activities in all stages of reduction. The process began with deep and open pit mining and continued with preparing cores for making tools, weapons, and religious and status objects in various types of workshops. Teotihuacan developed an occupation with agricultural settlements in the peripheral areas of the source where the transport of obsidian products to trade destinations was organized (Pastrana and Domínguez 2009).

The importance of obsidian supply to the development of the Teotihuacan state has been a fundamental topic of archaeological research in Central Mexico since at least the 1960s. Archaeologists noted not only that there was a lot of obsidian material but also that it was used in agricultural and artisanal production; in military and religious activities; and in local, regional, and interregional trade. There is evidence of long-distance gift exchanges with the ruling elite of distant cultures as far away as the Maya and Zapotec areas. Obsidian studies have generated interesting hypotheses about Teotihuacan and the state control of obsidian economies, including source exploitation, production, distribution, and consumption. It has been argued that Teotihuacan's goal was to reap the rewards of an ever-increasing demand on the part of the growing and populous city, as well as peripheral and regional settlements (Carballo 2007; Santley 1984; Spence 1966). More recent studies suggest that part of the Teotihuacan expansion, albeit over a short period of time (Braswell 2004), involved strategies for creating similar systems of domination over Maya obsidian sources, which, in most cases, seem to have failed over the *longue durée*.

In addition to the green obsidian, Teotihuacan also organized the supply of the gray-black and reddish (*meca*) obsidian from the Otumba Estado de México Edo deposits and probably the Paredón, Puebla, source. The exploitation of green obsidian from La Sierra de las Navajas, Hidalgo, was carried out under different channels and forms of distribution to all sectors of society. The products exported from the source include: (1) raw material in the form of clastic blocks, cobbles, and slabs; (2) raw material in the process of production, such as bifacial preforms, macroflakes and blades; and (3) a wide variety of blade-core types. Other likely exports to Teotihuacan and its affiliates were finished instruments, as well as composite artifacts that integrated wood and armaments (e.g., atlatl darts) over approximately 500 years. In many cases the function and meaning of the different types of lithic artifacts are unknown, but we can divide them according to basic sociofunctional categories: instruments of work, weapons, objects of status attire, and religious-symbolic instruments.

Since 2003 we have been analyzing the debitage from Teotihuacan-related deposits from the Classic period to answer questions of political and economic control of green obsidian vis-à-vis the rise of the Teotihuacan state (Pastrana and

Domínguez 2010, 2011, 2012, 2013, 2014, 2016, 2018; Pastrana et al. 2018). Although it is a theme that has arisen in archaeological literature for many decades now (Spence 1967, 1981, 1996), previous theories have been developed from archaeological studies carried out in the city itself. The sources of information from Teotihuacan include the large number of "workshops" that were located on the surface and in the excavations (Carballo 2007), as well as the recovery of magnificent silhouettes from important ritual offerings of the main buildings, such as the Pyramid of the Sun, Pyramid of the Moon, and the Temple of the Feathered Serpent (i.e., Quetzalcoatl). Our study also examines the other side of the supply chain.

Teotihuacan was able to benefit economically from obsidian procurement, production, and exchange partially because of the nature of the material itself. The fracture of volcanic glass presents sharp and hard edges, but they are fragile and easily lost through cutting, scraping, friction, and impact. Hence, the constant supply of instruments for multiple practical functions in addition to religious functions was required. The multifunctionality of obsidian, thanks to its particular physical and aesthetic characteristics, made it of great value to all segments of society; it was the basic tool for commoners and specialized craftspeople alike and used in the production of both obsidian and nonobsidian goods throughout the city.

The common obsidian instruments used in housing complexes throughout Teotihuacan are prismatic knives, scrapers, and drills used for basic domestic activities. The many specialized crafts carried out with obsidian include the working of leather, skins, and bones; fibers of plant and animal origin, feathers, shells and snails, and soft woods; and minerals such as slate, volcanic tuffs, and sedimentary and other rocks with a low degree of metamorphism (Carballo 2007, 2011). Obsidian was also used as a component of more complex composite tools, weapons, and decorations, such as offensive and defensive weaponry, as well as clothing and showy headdresses of leaders, priests, the military, and deities. Its symbolic importance manifested in sumptuous offerings that included the bifacial silhouettes (i.e., eccentrics) found in the Pyramids of the Moon and Feathered Serpent. Some have been interpreted as rays of sunlight and lightning bolts associated with the sun and rain deities and also as anthropomorphic and feline silhouettes and effigies (Pastrana and Athie 2014; Sugiyama 2005; Sugiyama and Cabrera Castro 2018; chapter 1 here).

It should be noted that the techniques used to work Teotihuacan obsidian have been identified as direct and indirect percussion, pressure flaking, and pecking, but so far no evidence of polished and burnished objects has been identified. This pattern lies in stark contrast to Postclassic Mesoamerican traditions in Central Mexico, where the polishing of obsidian seems to begin with the Toltec culture and continue with the Aztecs. The manufacture of

Teotihuacan obsidians was masterful and efficient, favoring the natural aspect of the conchoidal fracture in percussion and pressure retouching. The eyes and ornaments of sculptures were manufactured taking advantage of the smooth and bright fracture planes that gave a dark mirror appearance without polish. An interesting topic for future investigation is why the polishing technique was used on a wide range of materials, such as shell, pyrite, and slate and mainly in green minerals such as jadeite, nephrite, and serpentines, but was not used for obsidian.

THE OBSIDIAN SILHOUETTES OF TEOTIHUACAN CULTURE

Based on the analysis of Teotihuacan workshops at La Sierra de las Navajas, the results of our replicative experimentation, as well as the identification and classification of ceramic material and camp types, Pastrana and team have defined the following types of workshop-camps in which obsidian production and domestic activities were mixed:

Family workshops: These were domestic family camps where the storage, preparation, and consumption of food occurred. There was also direct evidence of the accumulation of raw obsidian materials, their reduction by percussion, and the production of projectile points and prismatic blades. Indicators in the lithic materials suggest that these activities were done by family members or closely related individuals.

Specialized workshops: These were camps where the storage, preparation, and consumption of food was carried out, as well as the production of a single type of product, such as beads and sequins, made from prismatic blade segments that were also produced in the workshop. This labor was probably staged, involving multiple specialists, with an overall organization that superseded the family (Pastrana et al. 2018).

Specialized individual-size areas: Located in the vicinity of the camps and workshops, they consisted of chipping debris and artifacts in the process of production. Activities included the preparation of large nuclei and the production of macroflakes and blades used to supply blanks for the manufacture of bifacial and unifacial silhouettes (e.g., canids, felines, serpents, and perforated unifacial discs). In some of these individual size areas we have also found preforms of knives and dart points of the finest quality, which were likely carried out by master specialists.

At Teotihuacan itself, the production techniques and morphologies of obsidian artifacts (i.e., dart points, elliptical knives, and silhouettes) were very stable for long periods through the Tlamimilolpa and Xolalpan phases, 150 to 550 CE. The knapping techniques for obsidian included direct percussion, indirect percussion, pecking, and pressure. As mentioned, polishing in obsidian was absent during the Classic period in the Teotihuacan culture.

SILHOUETTE-MAKING TECHNIQUES

In the analysis of Teotihuacan silhouettes, we did not resort to the generic term *eccentric* as is traditional in Mesoamerican archaeology but rather used terms that emphasize technological and morphological characteristics, for example, *bifacial silhouettes*, *unifacial silhouettes*, and *modified flakes* and *blades*. Some of the silhouettes were also perforated. From a technological point of view, silhouettes were chipped by direct percussion, indirect percussion, pressure, and pecking from bifacial and unifacial directions. The contours of the silhouettes comprise various shapes: curvilinear, geometric, simple, or composite rectilinear as a combination of shapes.

Silhouettes were made from different types of blanks and preforms, such as *natural slabs* or *clasts*, *flakes* of various dimensions and types, and *primary* and *secondary macroblades* of different sizes. In several cases, it is clear that certain blanks and preforms were selected and modified specifically to make the silhouettes, while in others, some types of blades, flakes, and debitage were clearly not socially suitable for silhouettes. Thus, although the final shape may have ended up as irregular, and sometimes barely recognizable, it appears that since a piece was made with the proper blank it could still be deposited in important offering contexts, regardless of its imperfections (Kwoka and Shackley 2019). This pattern suggests that there may have been something inherent in the production process that was important to the symbolism and value of the final product.

Slabs: The tabular shape and size of many of the natural slabs, or clasts are suitable for biface blanks and the preforms for large silhouettes. Tabular slabs are usually limited by parallel cortex surfaces, which need to be carefully removed, without creating staggered or stepped fractures that impede consecutive flake removals.

Flakes: These macroflakes were detached by direct percussion from large blocks or cores. The flakes needed to be straight, flat, and thin and preferably of good-quality obsidian. They were detached by direct percussion through precise impacts by specialized knappers, since errors at this early stage caused the waste of considerable volumes of raw material. Much of this material was obtained at depth from a mine, which means an additional costly waste of several stages of work if unskilled errors occurred.

Knives and projectile points: Several of the silhouettes were made up of knives beautifully finished by series of consecutive pressure flakes, with the transverse-parallel flaking characteristic of the Teotihuacan tradition. For example, anthropomorphic silhouettes often seemed to combine the initial shape of a "laurel leaf" or elliptical knife from which the silhouette was later trimmed in the outline of a human figure.

Pressure blades (primary and secondary): These were used as preforms for certain types of smaller silhouettes, mainly snakes and small lightning bolts. Secondary blades (i.e., 2s and 3s) with parallel edges were usually chosen, since they were more suitable to reduce by means of series of consecutive pressure-flake detachments. The clefted contours of the silhouettes were created with a kind of wider notch that was knapped bifacially. Primary blades (i.e., 1s) with a central arris were also used as preforms, but in this case the silhouette was achieved simply by roughing out or retouching the margins of the blade rather than deep unifacial or bifacial reduction. This technique is marked by the limited extension of the flakes into the central mass of the blade, and probably reflects less skill required to produce. It should be noted, however, that esoteric knowledge may have been required to create proper elements and proportions of the piece—measurements that may not have been completely understood by the layperson.

In the analysis of the various types of finished silhouettes, it is sometimes possible to recognize evidence of previous stages of production, such as negative scars from direct percussion; certain irregularities in the general form that could be produced by production errors, such as flakes that plunged too deeply; or step fractures that occurred because a lack of force on impact or by pressure. Furthermore, irregularities present in the vitreous matrix of obsidian, such as internal fractures and incipient cones, the concentration of gas vesicles, or the presence of crystallization of spherulites, can lead to errors in the sequential creation of clean conchoidal fractures.

PRELIMINARY MORPHOLOGICAL CLASSIFICATION OF TEOTIHUACAN SILHOUETTES

In general, most of the Teotihuacan silhouettes are easy to identify thanks to their well-defined and often repetitive design elements. They contrast with the total body of Maya eccentrics, which feature a much greater variety of forms, even though they may have been founded on basic Teotihuacan forms in the Early Classic. We also find Teotihuacan forms in other materials, as well as in the iconographic and epigraphic record of Teotihuacan. For example, similar silhouettes are found in sheets of slate and segments of seashells, but also painted and incised on walls and ceramic vessels as priestly scepters and godly lightning bolts.

General Types of Silhouettes

Anthropomorphs: These forms often feature short arms or, perhaps more likely, arms bent at the elbow and behind the back; they may have bilateral symmetry from a longitudinal axis. They may represent captives with their arms bound behind their backs, or be obsidian representations of the "Charlie Chaplin" figurines made of *Spondylus* shell found throughout Mesoamerica but especially in Central Mexico and the Maya area (see figure 2.1; see also Johnson 2018). Hruby has recently identified gender in some figures, especially from the Pyramid of

FIGURE 2.2. *Example of a Teotihuacan serpentiform zoomorph. © Proyecto Pirámide de la Luna.*

FIGURE 2.3. *Example of a curved bifacial knife from the Pyramid of the Moon (opposite face shown in figure 1.2a). © Proyecto Pirámide de la Luna.*

the Moon, in which a pair of anthropomorphs feature a masculine effigy with a feminine one likely wearing a huipil. The duo may refer to some form of an Early Classic "creator couple" similar to the *Ometeotl* of the Aztecs.

Zoomorphs: These quadrupeds include felines and canids; serpents, including snakes with serrated ridges and crests, possibly plumed serpents, and snakes with smooth margins; and bivalves and gastropods reflecting shells and snails. The most common bifacial zoomorphs have the silhouette of a snake. Magnificent snake-shaped and lightning-shaped specimens have been found in the offerings of the Pyramids of the Feathered Serpent, Moon, and Sun (figure 2.2; Sugiyama 2005; Sugiyama and Cabrera Castro 2018; Sugiyama and López Luján 2006).

Weapons: These items are projectile points, dart points, and spearheads, as well as straight and curved knives (figure 2.3). However, because many are either miniature or hypertrophic, it is not clear if they were made to have a utilitarian function. It is possible that the silhouette of a weapon is a weapon, as well as an idealized or exaggerated form of that weapon. For example, a long laurel-leaf knife, curved knife, or serpentiform atlatl dart may or may not have been used as a typical weapon of war per se (Gazzola 2014). It is not known the extent to which silhouettes were used in blood sacrifice; however, it is clear that most of them were not used for utilitarian purposes. In other words, many of the

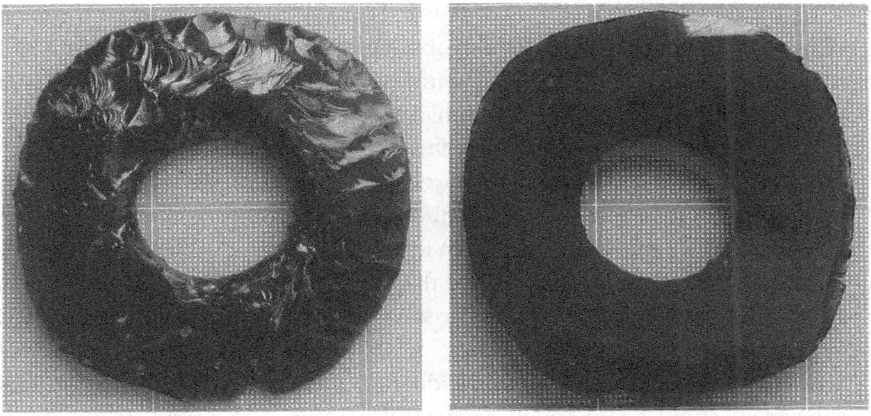

FIGURE 2.4. *Perforated-disc silhouette from excavations at Sierra de las Navajas.*

silhouettes in the form of weapons likely had the primary purpose of evoking the form of the weapon and the symbolic load that type of weapon carried, instead of functioning as such. At Teotihuacan they were used in exactly the same way as the more elaborate silhouettes in cache contexts.

Conflated silhouettes: Different forms are combined or conflated with one another (see next section). Some forms include serpent-projectile point and laurel-leaf anthropomorph. Saburo Sugiyama identified them as "figurines" and eccentrics, like the "Feathered Serpent and half projectile point" (Sugiyama 2005, 135–137). In the Maya area this process of conflation is quite common (see chapters 1 and 8). The Maya seemed to use this principle much more freely, which often led to new permanent forms that were spread form site to site.

Geometric silhouettes: These forms consist of ovals, large discs, medium discs, perforated discs (figure 2.4), rings, crescents, notched crescents, and other "U" forms. When the representation of the silhouette is yet unknown, this is a preliminary type that consists of geometric or stylistic silhouettes, which may also represent natural elements, such as lightning bolt, fire, sunbeam, stars, crescent moon, eyes, teeth, tusks, claws.

Next, the types are shown in the order of general classification elements that we propose for Teotihuacan silhouettes. The *various* category of silhouettes indicates shapes that have not been clearly identified as a known element. The remaining descriptions are at the geometric level and can be complemented by the characteristics of the finishing technique and the type of obsidian. Of course, each of these types can be further divided into varieties, and the different attributes can be described in terms of their production method. This kind of description requires more space than is available here.

To provide an example of a more detailed classification, we describe the *unifacial, perforated-disc* silhouette of green obsidian (see figure 2.4), which would be a piece made on a flat and wide flake with a relatively smooth, broad percussion-bulb. It was modified on the medial margins of the flake and is slightly curved and reflective on the ventral face, which was not worked. The dorsal face of the flake, due to its variable thickness, possibly had two or more edges formed by the negatives of the previous flakes. This flake was probably detached during the preparation of a core larger than 15 cm in length.

In the first phase of reduction, the flake was trimmed by direct percussion to create the circular perimeter. In the second phase the center was perforated by pecking to create microfractures, as well as gradually deepen a concavity to connect the ventral face to the dorsal. Once the initial indentation was achieved, it could be extended by careful pressure flaking from the center toward the periphery. In a third and final stage of reduction, the piece was retouched by pressure, with at least three series of semi-parallel pressure flakes with a concentric orientation (i.e., from the periphery toward the center), the consecutive sequence of the flakes like the pressure flakes of the Teotihuacan points and knives.

This piece was probably a component of a composite artifact and was combined with other raw materials such as shell or mica sheets. It could have been a part of an earspool or the circular outline of the eyes of an owl, feathered serpent, or perhaps Tlaloc, the god of storm and water. These types of elements, as far as we know, are not frequent in Teotihuacan offerings. There are some copies and fragments of the same type in the National Museum of Anthropology, but they are of different dimensions from the one found in the Pachuca workshops. The classification of this silhouette is presented in figures 2.5–2.9.

For Teotihuacan culture it seems that the silhouettes were preferably elaborated in the green obsidian of La Sierra de las Navajas, Hidalgo, and also the black-gray, reddish, coffee, or yellowish obsidian, commonly referred to as *meca* from the site of Otumba, State of Mexico (Glascock et al. 1994). It is likely that different types of obsidian were chosen for the manufacture of certain silhouettes, probably evoking different ideological meanings within the religion. The obsidian *meca* may have reddish, brown-yellow coloration, which was associated with blood, fire, and the sun (Sugiyama 2005). The green obsidian and its green varieties can be related to the color of jade, quetzal birds, and vegetation, while gold and golden green–yellow may have been associated with the sun. For Late Postclassic Central Mexico, green obsidian was directly related to the *chalchihuitl*, or preciousness; however, the most precious was *quetzalitzli*, which also had properties of the green obsidian but perhaps the bright chatoyant variety that evoked the same properties as quetzal plumes (Pastrana and Athie 2014; Thouvenot 1982).

FIGURE 2.5. *Teotihuacan anthropomorphic silhouettes (drawings by Pastrana).*

AN ITINERARY OF TEOTIHUACAN SILHOUETTE TYPES

Anthropomorphs

The *anthropomorphic silhouettes* probably represent men, and more infrequently women, from the ventral or dorsal view, with a wide variety in the dimensions and proportions of the body, which were probably limited by the morphology of the blank. The characteristics of the raw material and the technical dexterity of the carver almost certainly played into how elaborately the figure was chipped, since the opening of the legs, for example, is a technical feature of great difficulty. These silhouettes were made from blanks of different types and also on natural oblong clasts or slabs. The reduction strategies range from irregular pressure-trimmed macroflakes and slabs, to bifaces chipped by direct percussion, as well as some pieces finished by consecutive parallel flake removals by pressure. These also probably required indirect percussion to produce the openings or deeper notches (see chapter 5; Titmus and Woods 2003). The quality of the stone working for anthropomorphic silhouettes is variable, which is evident when comparing the work of occasional knappers on irregular flakes, to the work of master knappers on thin and fine bifaces. Generally speaking, the largest anthropomorphic silhouettes would require the highest skill level to produce. The different varieties of this type are provided here (see figure 2.5).

a. Unifacial asymmetric anthropomorph, on irregular flake created through pressure flaking or raking.
b. Unifacial symmetric anthropomorph, on regular flake created through direct percussion and pressure notching.
c. Unifacial symmetric anthropomorph, on regular flake created through direct percussion, pressure notching, and pressure flaking.

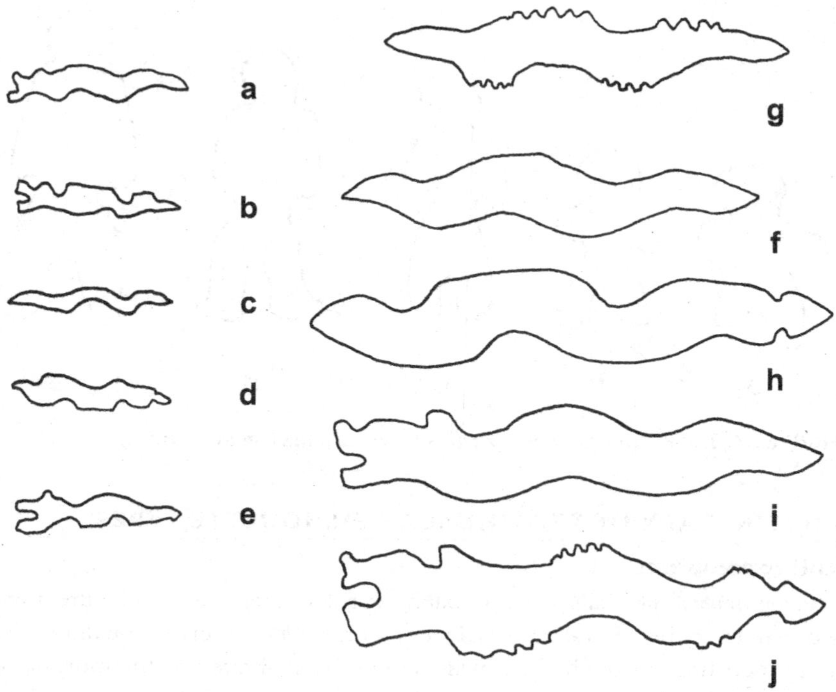

FIGURE 2.6. *Teotihuacan serpentiform zoomorphic silhouettes (drawings by Pastrana).*

d. Bifacial symmetric anthropomorph, on regular flake elaborated by direct percussion, pressure notching, and pressure flaking.
e. Bifacial symmetric anthropomorph, on regular flake elaborated by direct percussion, pressure notching and pressure flaking, and possible indirect percussion.

Zoomorphic Serpentiform Silhouettes (see figure 2.6)

a. Unifacial serpent with open mouth, on flake elaborated by pressure flaking or raking.
b. Unifacial serpent with open mouth, on blade elaborated by pressure flaking or raking.
c. Unifacial serpentiform with pointed ends, on prismatic blade elaborated by pressure notching and pressure flaking—possible lightning or solar ray.
d. Unifacial serpentiform with one notched pointed end, on prismatic blade elaborated by pressure notching and pressure flaking—possible serpent with crotale.

e. Unifacial serpent with open mouth, on prismatic blade elaborated by pressure notching and pressure flaking.
f. Bifacial serpentiform with pointed ends, on biface created by direct percussion, pression notching, and pressure flaking—possible lightning or solar ray.
g. Bifacial feathered serpentiform with pointed ends, on biface created by direct percussion, pressure notching, and pressure flaking—possible lightning or solar ray.
h. Bifacial serpentiform with one notched pointed end, on biface created by direct percussion, pressure notching, and pressure flaking—possible serpent with crotale.
i. Bifacial serpent with open mouth, on biface created by direct percussion, pressure notching, and pressure flaking.
j. Bifacial feathered serpent with open mouth and notched end, on biface created by direct percussion, pressure notching, and pressure flaking—possible serpent with crotale.

Zoomorphic Quadruped Silhouettes (see figure 2.7)

a. Unifacial quadruped on flake, created by direct percussion and pressure notching—possible feline.
b. Unifacial quadruped on flake, created by direct percussion and pressure notching—possible canid.
c. Unifacial quadruped on flake, created by direct percussion, pressure notching, indirect percussion notching, and pressure flaking—possible feline.
d. Unifacial quadruped on flake, created by direct percussion, pressure notching, and pressure flaking—possible canid.

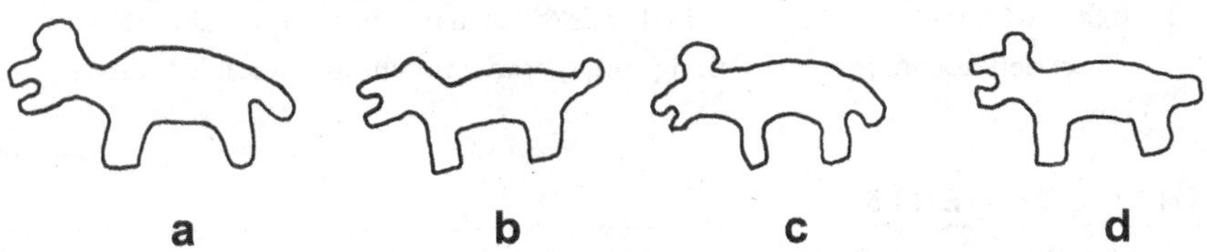

FIGURE 2.7. *Teotihuacan quadrupedal zoomorphic silhouettes (drawings by Pastrana).*

Conflated Silhouettes (see figure 2.8)

a. Bifacial serpent with open mouth combined with projectile point on biface, elaborated by direct percussion, pressure flaking, and pressure notching.
b. Unifacial projectile point serpentiform on flake, created by direct percussion and pression flaking.
c. Bifacial double projectile point on bifacial, created by direct percussion, pressure flaking, and pressure notching.

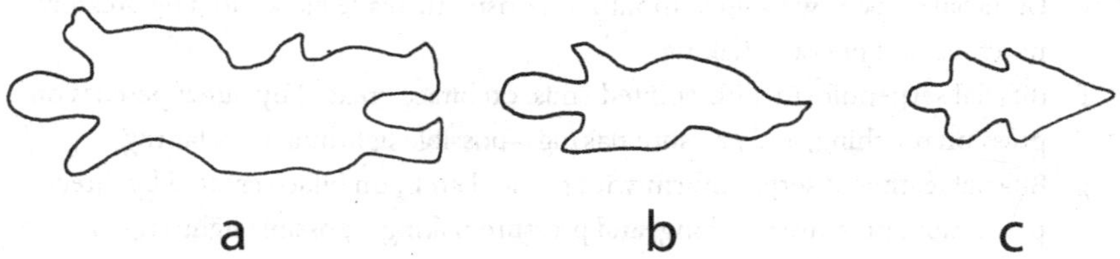

FIGURE 2.8. *Teotihuacan conflated silhouettes (drawings by Pastrana).*

Geometric Silhouettes (see figure 2.9)

a. Unifacial perforated-disc circular with central perforation, created by direct percussion, pecking, and pressure flaking.
b. Unifacial irregular oval with side or eccentric perforation, created by direct percussion, pressure flaking—possible valve of shell.
c. Biface with thin irregular curve and straight section with pointed ends, created by direct percussion and pressure flaking—possible thin-pointed curve knife.
d. Biface with wide irregular curve, created by direct percussion—possible sacrificial knife.
e. Biface with wide spiral section and straight section, created by direct percussion and pressure flaking (bifacial spiral).
f. Bifacial trefoil, created by direct percussion, pressure flaking, and pressure notching—possible drops of blood or water (Stocker and Spence 1973).
g. Bifacial half circle (bifacial lunar crescent).
h. Biface with thin curved section, created by possible direct percussion and pressure flaking.
i. Biface with wide "U" with notched ends (bifacial notched "U"), elaborated by direct percussion, pressure flaking, and pressure notching—possible human jaw.

FINAL COMMENTS

The objective of this chapter is to promote the systematic classification of silhouettes in Mesoamerican cultures beginning with those from Teotihuacan. Our goal is to shift from the misleading terminology of the *eccentric*, to the use of more common terminology and attribute definitions that focus on both technology and morphology. For Teotihuacan silhouettes, we define them as "lithic objects chipped from laminar or tabular obsidian with a perimeter or contour that is figurative." The shape of the silhouette had a certain meaning for the Teotihuacan culture that created it; that is, they are cultural products. The classification system we propose for Teotihuacan silhouettes is in its initial phase and is open to modifications based on new information that undoubtedly will be uncovered during the development of research and new objectives over time.

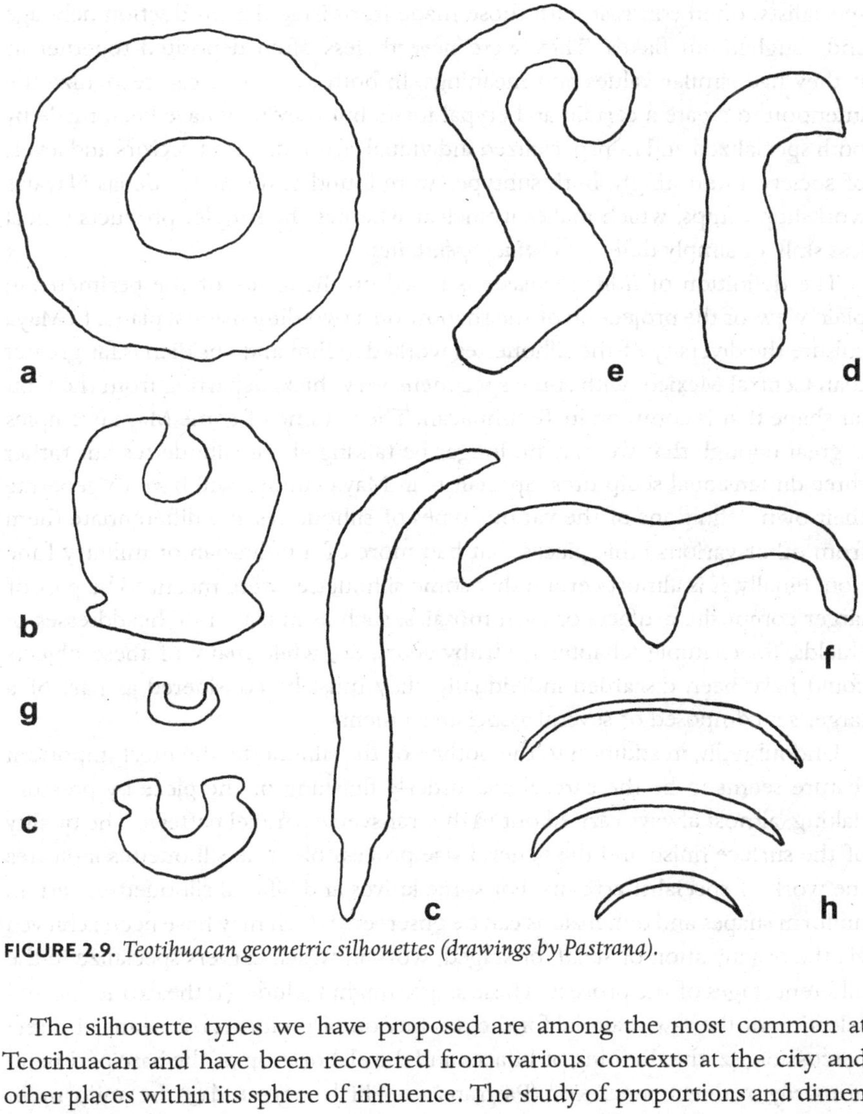

FIGURE 2.9. *Teotihuacan geometric silhouettes (drawings by Pastrana).*

The silhouette types we have proposed are among the most common at Teotihuacan and have been recovered from various contexts at the city and other places within its sphere of influence. The study of proportions and dimensions will allow for the definition of subtypes and variations elaborated from the base shape or archetype. For example, serpentiform silhouettes range from miniature specimens to large examples larger than 1:50 scale, the limits of which were determined by the physical characteristics of the initial raw material and knowledge of knapping techniques. It is also notable that the silhouettes varied according to the blanks and preforms from which they were made but also the skill level of the knapper and complexity of the technique used to manufacture it. Magnificent specimens, for which one can appreciate the work of master specialists, often contrast with those made from irregular production debitage and roughed-out flakes. They were nevertheless often deposited together as if they had similar values and meanings. In both cases, one can recognize the intention to create a certain archetypal form, but they may have been made by both specialized and nonspecialized individuals from different sectors and levels of society. Interestingly, both subtypes were found at the Sierra de las Navajas workshop-camps, which makes it unclear whether the simpler products reflect less skill, or simply different *chaînes opératoires*.

The definition of *lithic silhouette* is based on the shape of the perimeter in plain view, or the projection of the shadow on a two-dimensional plane. In Maya culture the diversity of the silhouettes worked in flint and obsidian is far greater than Central Mexico, with some specimens very thick, departing from the tabular shape that is common in Teotihuacan. The volume of some Maya examples is great enough that we may no longer be talking about silhouettes but rather three-dimensional sculptures. Specialists in Maya culture will have to generate their own definitions of the various types of silhouettes and differentiate them from other various lithic pieces that had more of a utilitarian or military function. Finally, it is almost certain that some silhouettes were meant to be part of larger composite artifacts or even mosaics, such as in elaborate headdresses or shields, for example (chapter 1; Hruby 2007). So, while many of these objects could have been discarded individually, they must be considered as part of a larger set composed of several associate elements.

Undoubtedly, in addition to the outline of the silhouette, the most important feature seems to be the careful and orderly finishing of the piece by pressure flaking, almost always carried out in the transverse parallel pattern. The quality of the surface finish and the general size process of some silhouettes indicates the work of specialist artisans. For some knives and bifacial silhouettes, certain uniform shapes and dimensions can be observed, which may have been achieved via the organization of *serial*, or staged, work in which carvers specialize in the different stages of the process. These stages might include: (1) the extraction and selection of the raw material for the production of preforms by means of direct percussion; (2) the thinning of bifaces and the elaboration of the base perimeter primarily by direct percussion (Pastrana and Hirth 2003); and (3) the uniform finishing of the working surfaces and detailed final contouring of the silhouette by means of pressure and possibly also by indirect percussion as illustrated by the experimental works of Gene Titmus, James Woods, and John Clark (Titmus and Woods 2003; chapters 4 and 5 here). We can hypothesize that the elaboration of the larger Teotihuacan silhouettes, as well as those with more detailed designs, was carried out by elite specialist knappers linked to the state.

The relative uniformity of the designs of the Teotihuacan silhouettes suggests to us that they were originally conceived and dictated by the ruling institutions as part of religion and official politics. These forms may have later been assumed by the sectors of the global population, as an essential part of their culture (Carballo 2007, 2011; Sugiyama 2005). The study of the symbolism and meaning of Teotihuacan designs can bring us closer to knowing the structure of the official ideology under which this great Mesoamerican state developed and maintained itself for over 500 years.

REFERENCES

Blom, Frans, and Oliver La Farge. 1928. *Tribes and Temples*. New Orleans, LA: Tulane University.

Braswell, Geoffrey E. 2004. *The Maya and Teotihuacan: Reinterpreting Early Classic Interaction*. Austin: University of Texas Press.

Carballo, David M. 2007. "Implements of State Power: Themed Obsidian Production near the Moon Pyramid, Teotihuacan." *Ancient Mesoamerica* 18:173–190.

Carballo, David M. 2011. *Obsidian and the Teotihuacan State: Weaponry and Ritual Production at the Moon Pyramid*. University of Pittsburgh Memoirs in Latin American Archaeology 21. Center for Comparative Archaeology. Pittsburgh and Mexico City: University of Pittsburgh, and Instituto de Investigaciones Antropológicas, Universidad Nacional Autónoma de México.

Childe, V. Gordon. 1958. *Rebuilding the Past*. Translated by María Teresa Rabiela de Rojas. Mexico City: UNAM.

Coe, William R. 1959. *Piedras Negras Archaeology: Artifacts, Caches, and Burials*. Museum Monographs. Philadelphia: University Museum, University of Pennsylvania.

"effigy (*n.*)." 2020a. *Merriam-Webster Dictionary*, s.v. Accessed January 2, 2020. https://www.merriam-webster.com/dictionary/effigy.

"eccentric (*adj.*)." 2019. *Merriam-Webster Dictionary*, s.v. Accessed December 31, 2019. https://www.merriam-webster.com/dictionary/eccentric.

"excéntrico." 2014. *Diccionario de la Real Academia Española*, s.v, 23rd ed. https://dle.rae.es/exc%C3%A9ntrico.

Eliade, Mircea. 1983. *Blacksmiths and Alchemists*. Humanities Section (553). Mexico City: Editorial Alliance.

Gann, Thomas W. F. 1918. *The Maya Indians of Southern Yucatan and Northern British Honduras*. Bureau of American Ethnology Bulletin 64:655–692. Washington, DC: Smithsonian Institution.

Gann, Thomas W. F. 1925. *Mystery Cities*. London: Camelot Press.

Gann, Thomas W. F. 1929. *Discoveries and Adventures in Central America*. New York: Scribner.

Gann, Thomas W. F., and Mary Gann. 1939. *Archeological Investigations in the Corozal District of British Honduras*. Bureau of American Ethnology, Anthropological Papers, 7. Washington, DC: Smithsonian Institution.

Gazzola, Julie. 2014. "Estudio y resultados preliminares de los artefactos de obsidiana localizados en un conjunto de fases tempranas en Teotihuacan." En *Estudio de la lítica arqueológica en Mesoamérica*, coordinated by Lorena Mirambell and Leticia González Arratia, 213–246. Colección Arqueológica. Serie Logos. Mexico City: INAH.

Glascock, Michael D., Héctor Neff, Joaquín García Barcena, and Alejandro Pastrana. 1994. La obsidiana meca del centro de México: Análisis químico y petrográfico. *TRACE* 25:66–73.

Gruning, E. L. 1930. "Report on the British Museum Expedition to British Honduras, 1930." *Journal of the Royal Anthropological Institute* 60:477–483.

Holmes, W. H. 1917. *Handbook of American Antiquities, Part I: Introductory, Bulletin 60 of the Bureau of American Ethnology*. Washington, DC: Smithsonian Institution.

Hruby, Zachary. 2007. "Ritualized Chipped-Stone Production at Piedras Negras, Guatemala." In *Rethinking Craft Specialization in Complex Societies: Archaeological Analysis of the Social Meaning of Production*, edited by Zachary Hruby and Rowan K. Flad, 68–87. Archaeological Papers of the American Anthropological Association 17. Arlington, VA: American Anthropological Association.

Johnson, Lisa M. 2018. "Little Helpers, Companions, and Ancestors: Miniature Figures in the Maya Lowland Region During the Late Preclassic and Classic Periods (AD 15–800)." *Latin American Antiquity* 29(3):616–622.

Joyce, Thomas A., Thomas Gann, E. L. Gruning, and R. C. E. Long. 1928. "Report on the British Museum Expedition to British Honduras, 1928." *Journal of the Royal Anthropological Institute of Great Britain and Ireland* 258:323–350.

Joyce, Thomas A. 1932. "Presidential Address: The 'Eccentric Flints' of Central America." *Journal of the Royal Anthropological Institute of Great Britain and Ireland* 62:xvii–xxvi.

Kidder, Alfred V. 1947. *Artifact of Uaxactun, Guatemala*. Washington, DC: Carnegie Institution of Washington, Publication 576.

Kwoka, J. Joshua, and M. Steven Shackley. 2019. "Technological Analysis and Source Provenance of Obsidian Artifacts from a Sun Pyramid Substructure Cache, Teotihuacan, Mexico." *Latin American Antiquity* 30:205–210.

Lane Fox, A. H. 1871. "A Flint Implement Brought from Honduras." *Proceedings of the Society of Antiquaries of London* 5:93–95.

Linné, Sigvald. 1934. *Archaeological Researches at Teotihuacan, Mexico*. Publication 1. Stockholm: Ethnographical Museum.

López Luján, Leonardo, and Saburo Sugiyama. 2018. "Ritual Deposits in the Moon Pyramid at Teotihuacan." In *Teotihuacan: City of Water, City of Fire*, edited by Matthew Robb, 82–89. Oakland: Fine Arts Museums of San Francisco, de Young and University of California Press.

Maler, Teobert. 1908. *Explorations in the Department of Petén, Guatemala, and Adjacent Regions*. Memoirs of the Peabody Museum 4(2). Cambridge, MA: Peabody Museum, Harvard University.

Mason, J. Aldon. 1935. "Preserving America's Finest Sculptures." *National Geographic Magazine* 68(5):537–570.

Moholy-Nagy, Hattula. 1999. "Mexican Obsidian at Tikal, Guatemala." *Latin American Antiquity* 10(3):300–313.

Moholy-Nagy, Hattula, and William R. Coe. 2008. "Appendix 11: Typology and Description of Flint and Obsidian Eccentrics and Incised Obsidians." In *The Artifacts of Tikal: Ornamental and Ceremonial Artifacts and Unworked Material, Tikal Report 27A*. Philadelphia: University of Pennsylvania Museum of Archaeology and Anthropology. Accessed February 1, 2018. https://doi.org/10.6067/XCV8CJ8D43.

Pastrana, Alejandro, and Ivonne Athie. 2014. "The Symbolism of Obsidian in Postclassic Central Mexico." In *Obsidian Reflections: Symbolic Dimensions of Obsidian in Mesoamerica*, edited by Marc Levine and David Carballo, 75–110. Boulder: University Press of Colorado.

Pastrana, Alejandro, and Kenneth G. Hirth. 2003. "Biface Production and Craft Specialization: A View from Sierra de las Navajas, Hgo." In *Mesoamerican Lithic Technology*, edited by Kenneth G. Hirth, 197–207. Salt Lake City: University of Utah Press.

Pastrana, Alejandro, and Silvia Domínguez. 2009. "Changes in the Strategy of the Exploitation of the Obsidian of Pachuca: Teotihuacan, Tula and the Triple Alliance." *Ancient Mesoamerica* 20(1):129–148. https://doi.org/10.1017/S0956536109000133.

Pastrana, Alejandro, and Silvia Domínguez. 2010. *Proyecto Yacimientos de Obsidiana, Temporada de Campo 2009*. Informe técnico al Archivo del Consejo de Arqueología, INAH. Archival manuscript, Mexico City.

Pastrana, Alejandro, and Silvia Domínguez. 2011. *Proyecto Yacimientos de Obsidiana, Temporada de Campo 2010*. Informe técnico al Archivo del Consejo de Arqueología, INAH. Archival manuscript, Mexico City.

Pastrana, Alejandro, and Silvia Domínguez. 2012. *Proyecto Yacimientos de Obsidiana, Temporada de Campo 2011*. Informe técnico al Archivo del Consejo de Arqueología, INAH. Archival manuscript, Mexico City.

Pastrana, Alejandro, and Silvia Domínguez. 2013. *Proyecto Yacimientos de Obsidiana, Temporada de Campo 2012*. Informe técnico al Archivo del Consejo de Arqueología, INAH. Archival manuscript, Mexico City.

Pastrana, Alejandro, and Silvia Domínguez. 2014. *Proyecto Yacimientos de Obsidiana, Temporada de Campo 2013*. Informe técnico al Archivo del Consejo de Arqueología, INAH. Archival manuscript, Mexico City.

Pastrana, Alejandro, and Silvia Domínguez. 2016. *Proyecto Yacimientos de Obsidiana, Temporada de Campo 2015*. Informe técnico al Archivo del Consejo de Arqueología, INAH. Archival manuscript, Mexico City.

Pastrana, Alejandro, and Silvia Domínguez. 2018. *Proyecto Yacimientos de Obsidiana, Temporada de Campo 2016–2017*. Informe técnico al Archivo del Consejo de Arqueología, INAH. Archival manuscript, Mexico City.

Pastrana, Alejandro, Silvia Domínguez, and Linda R. Manzanilla. 2018. "La producción, distribución y uso de las cuentas y lentejuelas de obsidiana teotihuacanas." In *Teopancazco como centro de barrio multiétnico de Teotihuacan: Los sectores funcionales y el intercambio a larga distancia*, Edited by Linda R. Manzanilla, 497–528. Mexico City: UNAM.

Price, W. H. 1897–1899. "Excavations on Sittee River, British Honduras." *Proceedings, Society of Antiquarians, London* 17:339–344.

Rice, R. Garraway. 1909. "Notes on a Flint in Human Shape Found in the Thames." *Proceedings of the Society of Antiquaries of London* 22:359–360.

Ricketson, Oliver G. 1929. "Excavations at Baking Pot, British Honduras." In *Contributions to American Archaeology* 1(1):1–28. Washington, DC: Carnegie Institute of Washington.

Ricketson, Oliver G. 1931. "The Present Status of Archaeological Investigation in Guatemala." *Year Book* (30):112–114. Washington, DC: Carnegie Institution of Washington.

Ricketson, Oliver G., and Edith B. Ricketson. 1937. *Uaxactun, Guatemala: Group E-1926–1931*. Carnegie Institution of Washington, Publication 477. Washington, DC: Carnegie Institution of Washington.

Russell, Y. Virgil. 1951–1957. *Indian Artifacts*. Boulder, CO: Johnson Publishing Company.

Santley, Robert. 1984. "Obsidian Exchange, Economic Stratification and the Evolution of Complex Society in the Basin of Mexico." In *Trade and Exchange in Early Mesoamerica*, edited by Kenneth Hirth, 43–86. Albuquerque: University of New Mexico Press.

"silhouette (*n*.)." 2020b. *Merriam-Webster Dictionary*, s.v. Accessed January 2, 2020. https://www.merriam-webster.com/dictionary/silhouette.

Smith, A. L. 1934. "Two Recent Ceramic Finds at Uaxactun." Carnegie Institute of Washington Publication 608, Contributions to American Archaeology No. 5, Washington, DC.

Spence, Michael W. 1996. "Commodity or Gift: Teotihuacan Obsidian in the Maya Region." *Latin American Antiquity* 1(1):21–39.

Spence, Michael W. 1967. "The Obsidian Industry of Teotihuacan." *American Antiquity* 32(4):507–514.

Spence, Michael W. 1981. "Obsidian Production and the State in Teotihuacan." *American Antiquity* 46(4):769–788.

Stephens, John Lloyd. 1860. *Incidents of Travel in Yucatan*. New York: Harper and Bros.

Stocker, Terrance, and Michael Spence. 1973. "Trilobal Eccentrics at Teotihuacan and Tula." *American Antiquity* 38(2):195–199.

Stuart, David. 2000. " 'The Arrival of Strangers': Teotihuacan and Tollan in Classic Maya History." In *Mesoamerica's Classic Heritage: From Teotihuacan to the Aztecs*, edited by Davíd Carrasco, Lindsay Jones, and Scott Sessions, 465–513. Boulder: University Press of Colorado.

Stuart, David. 2024. Spearthrower *Owl: A Teotihuacan Ruler in Maya History*. Studies in Pre-Columbian Art and Archaeology. Washington, DC: Dumbarton Oaks, Trustees for Harvard University.

Sugiyama, Saburo. 2005. *Human Sacrifice, Militarism, and Rulership: Materialization of State Ideology at the Feathered Serpent Pyramid, Teotihuacan*. Cambridge: Cambridge University Press.

Sugiyama, Saburo, and Ruben Cabrera Castro. 2018. "The Moon Pyramid and the Ancient State of Teotihuacan." In *Teotihuacan: City of Water, City of Fire*, edited by Matthew Robb, 74–81. Oakland: Fine Arts Museums of San Francisco, de Young and University of California Press.

Sugiyama, Saburo, and Leonardo López Luján, eds. 2006. *Sacrifices of Consecration in the Pyramid of the Moon*. Mexico City: INAH.

Taube, Karl. 1999. "The Writing System of Ancient Teotihuacan." *Ancient America* 1(1).

Taube, Karl. 2011. "Teotihuacan and the Development of Writing in Early Classic Central Mexico." In *Their Way of Writing Scripts, Signs and Pictographies in Pre-Columbian America*, edited by Elizabeth Hill Boone and Gary Urton, 77–109. Washington, DC: Dumbarton Oaks.

Thompson, J. Eric S. 1939. *Excavations at San Jose, British Honduras*. Carnegie Institution of Washington, Publication 506. Washington, DC: Carnegie Institution of Washington.

Thompson, J. Eric S. 1963. *Maya Archaeologist*. Norman: University of Oklahoma Press.

Thouvenot, Marc. 1982. *Chalchihuitl-Le Jade Chez les Aztèques*. Mémoires l'Institute d'Éthologie XXI. Paris: Muséum National d'Histoire Naturelle, Institute Paris.

Titmus, Gene L., and James C. Woods. 2003. "The Maya Eccentric: Evidence for the Use of the Indirect Percussion Technique in Mesoamerica from Preliminary Experiments Concerning Their Manufacture." In *Mesoamerican Lithic Technology: Experimentation and Interpretation*, edited by Kenneth G. Hirth, 132–146. Salt Lake City: University of Utah Press.

Willey, R. Gordon. 1972. *The Artifacts of Altar of Sacrifices, Papers of the Peabody Museum*64(1). Cambridge: Peabody Museum.

3

An Apologia for the Punch

Ethnographic, Ethnohistoric, and Archaeological Evidence of Indirect Percussion Techniques in North and Central America

BENJAMIN EBLE AND ZACHARY HRUBY

> *Fracture is accomplished by two classes of processes. Distinguished by the manner in which the shaping forces are applied; these are known as percussion and pressure. The percussive method implies: (1) the use of a hard and heavy implement with which the stone to be shaped is struck, producing direct fracture in its simplest form; and (2) the use of two tools, a punch-like implement, set upon the stone at the point to be fractured, and a heavy tool with which this implement is struck, producing fracture indirectly. This process has been mentioned by a number of observers, but is apparently not well understood by anyone.*
>
> William Henry Holmes, *Natural History of Flaked Stone Instruments* (1894)

Since the turn of the century, many prominent nineteenth- and twentieth-century American researchers recognized two fundamental flaking processes classed as *percussion flaking* and *pressure flaking*. These two classes of flaking were understood to be carried out via three methods. The first class, percussion flaking, utilized both *direct percussion*, which involved a percussor and the mass being struck and *indirect percussion*, which employed a percussor and some manner of

https://doi.org/10.5876/9781646427819.c003

intermediary device, such as a punch. The second consisted of various *pressure* techniques whereby the knapper uses a device to press a flake or blade off of a prepared core. This triad of flaking methods was attested by early explorers, ethnographers, and archaeologists, such as Frank Hamilton Cushing, Lieutenant P. H. Ray, George Byrd Grinnell, and A. F. Bonney, as well as by archaeologists such as William H. Holmes, Otis Tufton Mason, Henry C. Mercer, William Martin Beauchamp, Robert Alexander Stewart Macalister, Arthur Caswell Parker, Mark Harrington, Mark F. Seeman, Fredrick Williams, Harlan Ingersoll Smith, Clark Wissler, Frederick Hodge, Ernest A. Hooten, Gerard Fowke, W. H. Collins, William Snyder Webb, William Delbert Funkhouser, and William Ritchie, and even by linguists such as Edward Sapir, to name a few. Also, it is possible that some past observers recognized more sophisticated flaking technologies, such as flaker aided by blow, which would have enjoined aspects of both direct percussion and indirect percussion into a single process. In this chapter, we explore these researchers' accounts of chipped-stone technologies to provide analogies for use in lithic analysis, replication, and interpretation in Mesoamerica and beyond. In particular, we focus on archaeological, ethnohistoric, and ethnographic evidence for the widespread use of indirect percussion techniques in the Americas.

In a lecture originally published in 1879, Frank Hamilton Cushing mentioned this triad of flaking techniques:

> Surface-flaking, which is the thinning of the unfinished arrow by the detachment of flakes running from the edge to the center, is the most difficult part of the whole process. Arrows upon which no signs of it appear were always the work of beginners. It may be produced either by direct blows with a hammer-stone [i.e., *direct percussion*], by pressure with a flaker [i.e., *pressure*], or by a combination of the two methods, the hammer being used with the flaker as if the latter were a stone chisel [i.e., *indirect percussion*]. Each of these methods leaves it unmistakable mark upon the finished implement, so that it is easy to determine by simple inspection of the chipped article to what degree of perfection the art had come, at the time when it was made. (Cushing 1879a, 57; emphasis and additional notes ours)

While citing an array of historical accounts, Henry Mercer (1897, 69) commented on the flaking triad and their subvariants, stating: "So much for the accounts, which I believe comprise all of importance thus far published in America, by eyewitnesses. We learn from them . . . of flaking: (a) by direct percussion; (b) by indirect or hammering on punches; (c) by direct pressure; (d) by impulsive pressure, or pressure aided by a blow." Holmes (1894, 126) similarly attested to these three general flaking methods, while stating that indirect percussion is still not well understood by *anyone*. We argue that Holmes, and his predecessors, were on the correct evidentiary path, and we provide additional evidence to demonstrate that indirect percussion was not used simply for the most elaborate work but as a core element of most New World lithic technology. We also provide historical context for how and why archaeologists virtually abandoned this avenue of research in the 1950s and up to the present day.

This chapter reviews ethnohistorical, ethnographic, and archaeological evidence for various forms of indirect percussion in the Americas but with a special nod to Maya flintknappers and the clearest examples from Belize recorded by Harry Shafer (2023). In many ways the study of Maya *eccentrics* or *silhouettes* was the catalyst for a reconsideration of indirect percussion in the Americas, since it was so clearly used to create the elaborate forms they are famous for (Clark et al. 2012; Titmus and Woods 2003; chapters 1, chapter 2, and chapter 4 here). In other words, silhouettes represent the most widely accepted example of the use of indirect percussion in the New World. We argue that previously undocumented flintknapping technologies, which likely had origins as early as Paleoindian times, persisted in the Maya world and deserve recognition as evidence of long-standing, Native American culture traits. However, we take a broader view and present ethnohistorical, ethnographic, and archaeological accounts from throughout the Americas, which first show that indirect percussion in the New World existed in various forms, and second that it was actually quite common throughout the Archaic, Late Prehistoric, and Proto-Historic eras. There is additional evidence to suggest that indirect percussion was carried out in other media, such as shell, wood, and other stones, indicating a possible conceptual, technological, and historical link between different kinds of crafting, including the production of the famous stone monuments of Mesoamerica.

If we allow for the possibility that manufacturing techniques were not as simple as presently reconstructed, and that the role of indirect percussion is much greater than is assumed by most archaeologists, then a number of issues and implications arise. The first has to do with the differences between academic Western ontologies and Native American ontologies associated with chipped-stone technology. Most modern knappers approach the process from the perspective of the hyperrational individualist: the lone knapper. The main goal is to not only replicate a technology but also to be able to do it alone. Once this feat is accomplished, then the speediest method is usually selected as the most appropriate analogy for interpreting the archaeological record. While Native knappers undoubtedly were interested in efficiency, especially in terms of resource exhaustion and waste, it must be considered that their practices were passed down through generations, centuries, and even millennia. These practices were not picked up in a grad seminar or a yearly knap-in but were a lifelong endeavor and fully situated embodied practices that were no doubt informed by worldview, ethnophilosophy, and religion. As we discuss in this chapter, many knapping activities were not done by a single person, could involve rituals and chanting during the knapping process, and involved different levels of socially recognized specialization and expertise. We argue that early- and middle-stage knapping was a social activity among many New World peoples—something done in groups or in tandem. Thus, while resource exhaustion may have been of the greatest importance to Native knappers, the aspect of time so treasured in Western ideologies may not have been a priority. Occam's razor is culturally constructed based on who is doing the cutting. "Waxaklajun Ubah K'awiil's Taaj" may be a more appropriate principle here.

A related issue is staging and portability. Many knappers today and in the ancient world treasured their favorite hammerstones. A good one, particularly of a softer stone, allows a crafter to produce virtually every kind of biface, and an array of different-sized hammerstones is even better. Hammerstones are an absolute necessity for early-stage reduction, especially when working harder materials. The downside to hammerstones is that if they break, then you have to find another one. In many areas good hammerstones are few and far between. At many sources, particularly obsidian sources, bringing in percussors from the surrounding countryside must have been part of the planned activity, since that is where blanks, macroflakes, and cores were made to take elsewhere. The benefit of indirect percussion is that antler sizes and densities are relatively consistent—much more so than stones from different geological zones. Furthermore, dense hardwoods, bones, and even other stones can be used as punches, which can expand resource options if antler is scarce. Knappers could use any kind of rock as a percussor; as long as they had their light portable punch they did not need to tote a hammerstone, or stones, with them. Even if many forms of indirect percussion may require more people to bring off properly, the tools are more portable, are easier to master, yield a lower failure rate than direct percussion, and result in a product that was culturally acceptable. We provide evidence of this kind of staging later in the chapter but first provide an array of probable indirect percussion flakers as an example from Kentucky and Michigan in figure 3.1.

The reader may wonder how archaeologists reconstructed the history of chipped-stone technology in the New World and then subsequently came to see the matter as "settled." The simple answer is that anthropologists and experimenters since the end of the nineteenth century have been presented with evidence that has never been well understood and subsequently avoided, or rationalized away. Evidence of direct percussion and pressure flaking is generally well recognized in the archaeological record, but evidence of indirect percussion is less so. We show that many archaeologists actually advocated for a much larger role for indirect percussion in Native American technologies up until the 1950s. More recently, one of the few places where indirect percussion techniques are widely embraced but little understood is in Maya archaeology, in which researchers have, at least in theory, embraced indirect percussion for the manufacture of blade types and effigy flints (see Clark et al. 2012; Titmus and Woods 2003).

The goal of this chapter is not to propose a specific method of indirect percussion used by all Maya cultures over time but rather to show that indirect percussion was used and what it might have looked like in various formats. We not only examine Maya knapping tools, of which there are few recorded; we also show that indirect percussion was a Pan-American phenomenon to which the Maya also contributed. This ethnological study is just a first step that will require the replication of both tools and possible tool practices, carried out along with a concomitant study of debitage and

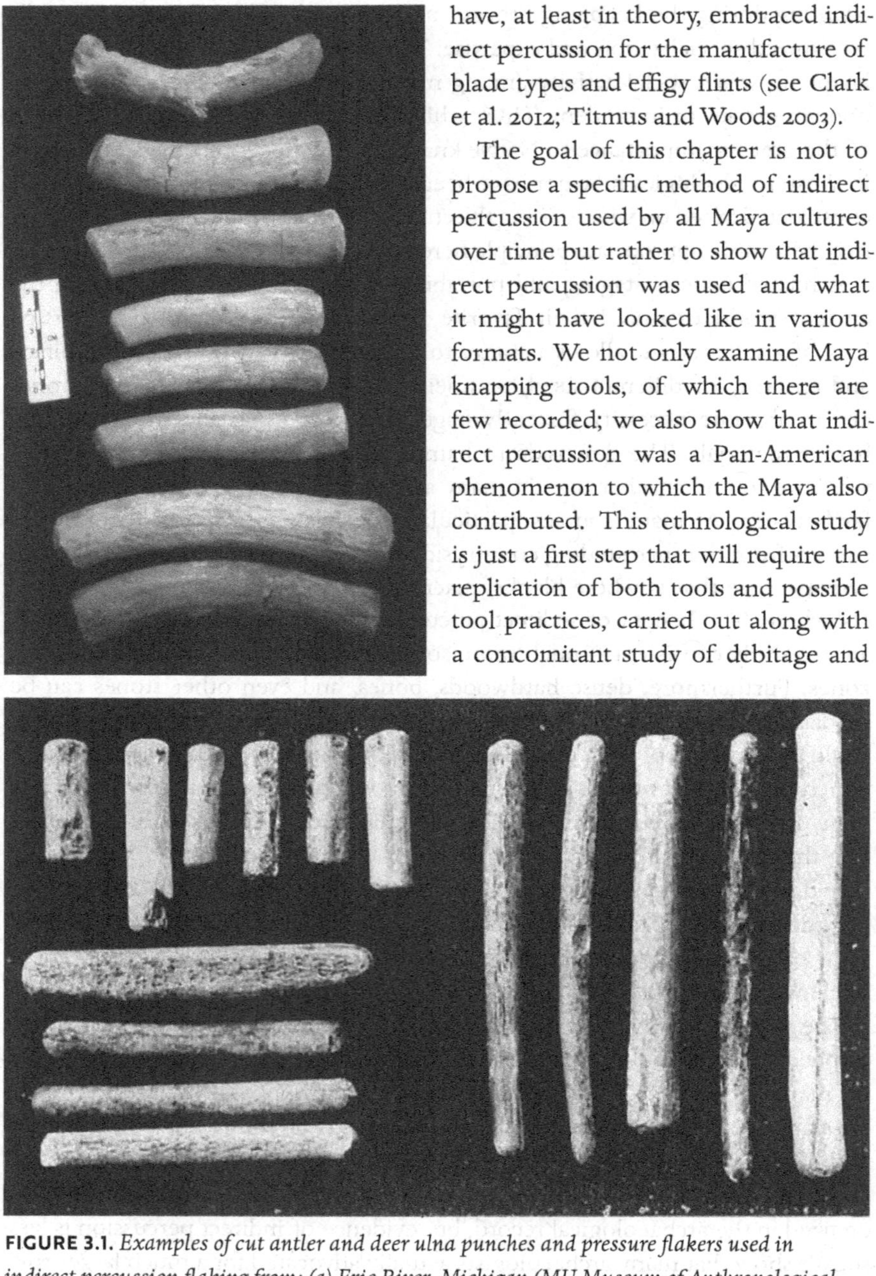

FIGURE 3.1. *Examples of cut antler and deer ulna punches and pressure flakers used in indirect percussion flaking from: (a) Erie River, Michigan (MU Museum of Anthropological Archaeology); and (b) various sites in Kentucky (Webb and Funkhouser 1928, fig. 88).*

formal instruments. It might well not be possible to recognize the evidence of such technologies without replication and a more detailed analysis of nonlithic artifacts as definitive cultural traits, much less authentically replicate those techniques. In North American archaeology, many researchers continue to wish this problem away and avoid systematic approaches to studying indirect percussion. One of the field's leading scholars, Harry Shafer (2023), has correctly identified some of the key issues and material correlates in the Maya area, as well as Texas, that implicate the employment of little-understood, indirect percussion processes—processes that we believe are as essential to Native American chipped-stone traditions as direct percussion or pressure techniques.

HARRY SHAFER AND THE WORKSHOPS AT COLHA, BELIZE

The Colha site is famous among Mesoamerican sites for its incredible flake density and evidence of large-scale workshops (Shafer and Hester 1983, 1991). Along with lithic samples at Copan, it was used to address a debate topic in the 1980s and 1990s on the role of lithic industries in the creation of wealth and state power (Aoyama 1999, 2001). Given the modest size of the site, most interpretations concluded that chipped-stone production was not a pathway to great state expansion and power (King and Potter 1994; Potter and King 1995). It was also one of the hotbeds for nailing down the different tool forms and technologies employed by the ancient Maya from Preclassic to Postclassic times (Hester 1985). While many of these tool forms were replicated by project members, including Shafer, not all of the forms could be properly replicated, especially the *eccentric* or *silhouette* bifaces from some of the deposits.

In an unpublished field report, Shafer (2023, 23–24) offered the following observations on punches and toolkits:

> Few archeologists in Texas are aware of the Maya punches and the large thin bifaces made using them. If the Maya could make large thin bifaces using punches, then it is reasonable to assume the central Texas flintknappers could as well. Missing in bone assemblages across Central Texas are antler billets and the bone debitage related to their manufacture; antler debitage from punch manufacture was present at Colha. If such tools and debitage do occur, they are not being reported. The most common flintknapping tools recovered archeologically are antler punches (often misidentified), antler tine pressure flakers, and deer ulna pressure flakers [figure 3.2 here]. Examples of presumed antler flakers from Archaic sites are not convincingly shown to have been billets.

The use of indirect percussion using an antler punch is foreign to most modern flintknappers, but it was a method widely used by prehistoric knappers from North America to Central America. Punches made of deer antler bases have

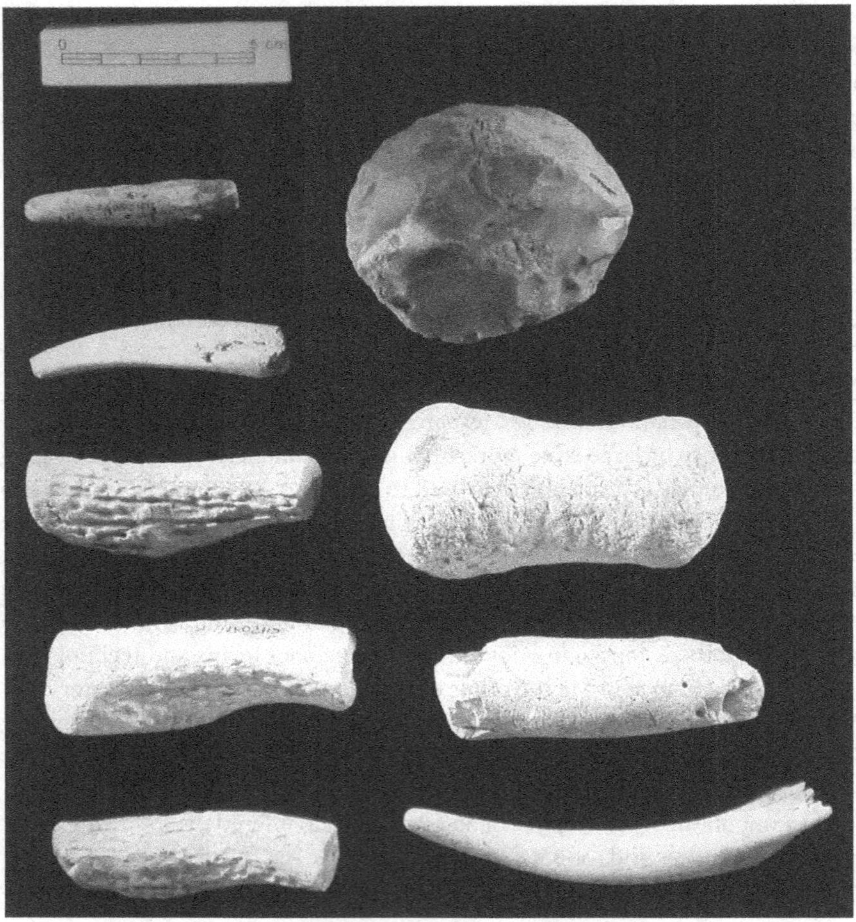

FIGURE 3.2. *Examples of antler punches and cut tine flakers from Colha, described as tools of indirect percussion by Shafer (2023).*

large contact areas, which produce wide flake initiations consistent with the manufacture of large, biface-thinning flakes (Geib 2004). The rare finds of flintknappers' toolkits provide a glimpse into the tool set used for chipped stone manufacture. Flintknapper kits from Horseshoe Ranch Caves (Shafer 1986, 105), Burial 119 at Morhiss (Dockall and Dockall 1999), Feature 9 at the Crestmont site (Hall 2002), Lemens Rockshelter (Smith 1994), and the Sand Dune Cave cache in Utah (Geib 2004) are cases in point. The toolkit discovered at Horseshoe Ranch Caves clearly provisioned one for the hunt:

> It contained not only biface blanks, spare flakes, antler punches, sinew, and an edge abrader, but also a scarifier, jackrabbit mandibles, and buckeye and mountain laurel seeds for hunting rituals; all components of the technological system supporting his technological style of hunting and associated ritual behavior . . . The Lemens Rockshelter (Smith 1994) kit contained only nonperishable items, but here, too, the burial assemblage consisted of seven antler tools, two of which are clearly punches; possibly three others are punches as well, although Smith describes four as flakers. One (Smith 1994, Fig. 7a) is identical to Postclassic antler punches from Colha, Belize. These antler base tools at Colha were originally described as billets (Shafer 1985) but were later examined microscopically by the author and John Dockall. We identified them as punches based on wear patterns; they were used in the manufacture of very large thin bifaces. (Shafer 1986, 105)

More than a century after William Holmes (1894) touched upon the lack of knowledge regarding indirect percussion flaking methods, Shafer (2023) clearly notes that antler punches known to the archaeological record are often "misidentified" and go unrecognized (see also Titmus and Woods 2003). Shafer also points out that an experimental flintknapping tool hypothetically proposed by a European researcher during the 1940s, referred to as an antler billet, is largely absent from faunal assemblages, including assemblages from Colha. He suggests that the use of indirect percussion is largely unknown to modern knappers. Shafer (2023) offers clear physical evidence of the flintknapping punches recovered from Colha. Also, in at least one instance, he amended his own publication. After carrying out a microscopic study of a tool that he had originally interpreted as a billet, he reclassified it as a tool of indirect percussion (Buttles 2002).

Over the last ten years or so, modern knappers such Patrick Blank, also known as Jack Crafty, have embraced indirect percussion using what is termed as a "rocker punch." The rocker punch involves holding a flaking agent in the same parallel as the stone being flaked, and then striking down upon the upper broadside of the core, driving the flake downward, and away from the plane that the biface or preform is held in. In its mechanical form, the rocker punch is a welded contraption that allows a single knapper to carry out indirect percussion easily without the help of an assistant. While many of the products have been astounding, and verify the utility of one form of indirect percussion, the single-person-operated rocker punch, which is essentially a modern shortcut, has actually stymied exploration into how indirect percussion was likely carried out anciently. That many forms of indirect percussion were likely carried out in groups or teams is an important factor in reconstructing prehistoric *chaînes opératoires* and the implications of more complex social and craft organization.

THE BILLET AND THE ORIGIN OF DIRECT PERCUSSION AS AN EXPLANATORY MODEL

In order to evaluate Shafer's comments, one must have a clear and precise understanding of what is meant by the term *antler billet*, how it is defined archaeologically, and how it might be recognized as a significant part of the Native American *chaînes opératoires* over the millennia and across the continents. We use the term *chaîne opératoire* to describe the acts and behaviors associated with the procurement of resources for production and use of, in this case, flake cores, blade cores, and bifacial cores. The concept goes beyond the "life cycle" of a tool, and includes all the people, places, and things needed to make something. Here we are most concerned with the tools that were used to make chipped-stone items. If there is no archaeological evidence for an essential stage in a proposed *chaîne opératoire* (e.g., the use of an antler billet), then hypotheses based on those reconstructed stages should be rejected. The reason this staging is so important to us, is that we argue indirect percussion was an essential aspect of Native American lithic technology going back at least to the Early Archaic. Based on archaeological evidence, which we present in the text that follows, we argue that various indirect percussion techniques were an essential part of *chaînes opératoires* passed down through generations and across continents—to such an extent that indirect percussion could be considered a core cultural trait of most ancient American cultures.

The term *billet* was originally a word used to describe a small stick, such as a stick of firewood. The "billet technique," or the "wood technique" as it was described during the 1940s, is an experimental flaking technique that was developed at the Pitt River Museum under Professor Alfred Barnes, between 1935 and 1943. In essence, the billet technique involves swinging a stick or stick-like percussor and striking the edge of a stone, in order to detach a flake. The billet technique was developed in order to answer an enduring question regarding hypothetical "intermediary flakes" that academia had been unable to resolve during the preceding seven decades, from the 1860s until the 1930s.

During the nineteenth century, people who puzzled over chipped-stone tools common to Europe began to recognize the associations between debitage and bruised "pebble hammers," known from archaeological contexts. Sometimes, these associations were inadvertently discovered during excavation projects, when hammerstones and chipped flint were discovered together. As a result, these associations between hammerstones and the evidence of stone chipping led to speculation that chipped-stone tools were simply fashioned using hammerstones via direct percussion. Conjecture regarding the mode of manufacture continued until the middle of the nineteenth century, when it was revealed that Edward Simpson, the notable relic forger aka "Flint Jack," was clandestinely creating forgeries with a soft iron hammer while employing a method not unlike the English gunflint knappers of his day (figure 3.3a). This revelation in Europe led to an intense study of the English gunflint knappers and their direct percussion techniques, which continued into the late nineteenth century. The similarity will be noted between the English gunflint knappers (figure 3.3a) and the forger (figure 3.3b). Both are seated in an upright position, holding the stone on their left leg, and are swinging hammers to chip the stone. The forger may have used a softer iron hammer than the gunflint makers, but this is unclear.

Ironically, these studies led to the conclusion that the direct percussion fracture created by hammerstone would preclude creating the finishes found on the most exquisite European bifaces. The inherent bulbing and edge crushing creating by hard hammer percussors failed, because the edges became far too thin with each blow to maintain the great width-to-thickness ratios observed in the archaeological record. In essence, the thinner the stone became, the less effective knappers found hard hammer percussion to be in replicating the thinnest bifaces. This realization led to the hypothesis of a "secondary" or "intermediate" flaking process, which would yield flakes comparable in width to those made with direct percussion, but also with a thinness comparable to that of pressure flakes. Such a process would yield flakes with a very diffuse bulb of percussion required to replicate the finest finishes on European artifacts. Since they were unable to experimentally reproduce such flaking, by the 1920s some researchers began to lament that stone chipping was truly a "lost art."

In 1943, nearly a century after Flint Jack's methods were first revealed, Barnes and Leon Coutier (figure 3.3c) carried out a series of experiments at the Pitt River Museum, Oxford, England. They employed what was termed *soft hammer percussion* flaking, via the use of an antler baton, a wooden baton, and an ivory hammer. These experiments were published in a paper titled "Alfred S. Barnes Technique Series." In his experiments, Barnes documented two types of direct percussion bulbs: the salient bulb and the diffuse bulb (figure 3.3d). The salient bulb originates from a single Hertzian cone-like point and indicates that a hard percussor such as steel, quartzite, or some other hard percussor was most likely used. The diffuse bulb of percussion does not show a distinctive hard hammer cone. Instead, the diffuse bulb of percussion is generally flatter and may even be lipped, indicating that a soft hammer such as wood or antler was most likely used.

On account of these experiments Sir Francis Knowles (1953, 63) wrote: "The experimental work of Professor A. S. Barnes in this country, of Dr. L. S. B. Leakey in Kenya, and of M. Leon Coutier and M. Francois Bordes in France, have shown that flaked implements can be made by means of wood, horn, bone, and antler batons and hammers." Interestingly enough, while even Ice Age antler artifacts were already known to researchers, from the European archaeological record, neither Barnes nor Knowles cited any direct evidence from Europe, which would suggest that such an antler tool had indeed been employed during prehistoric times. Instead they cited the recent use of the wooden baton by Australian

FIGURE 3.3. *Early experimenters: (a) English flintknappers at Brandon, England (Skertchly 1879); (b) Edward Simpson "Flint Jack" using his mallet as a predecessor to baton technique (photo by F. Treble, 1863); (c) Leon Coutier's wooden billet technique and subsequent experimentation with indirect percussion and pressure (Lacaille 1954, fig. 1); (d) illustration of the "salient" bulb and the "diffuse" bulb (Barnes 1943).*

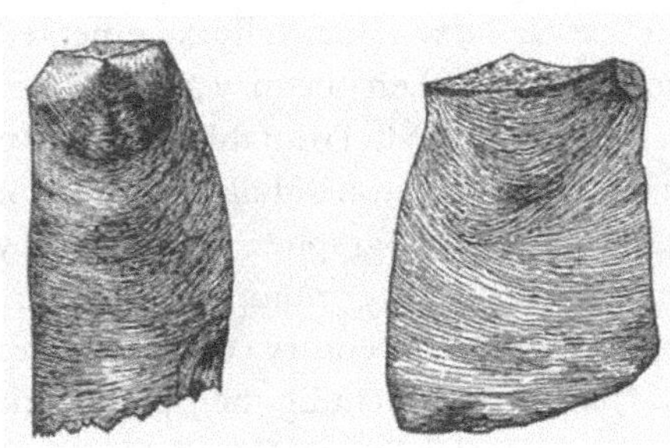

aborigines for reducing glass tektites. Additionally, Knowles noted some small antler tools from a Pre-Colombian quarry site, which potentially appear to be punches (1953, 64). Although he did not provide any direct evidence for the soft hammer baton in Europe, Barnes (1943) proposed that it could be inferred from experimentally created flakes. In essence, Knowles credits Barnes, Leakey, Coutier, and Bordes for experimentally demonstrating that soft impactors can be used to chip stone tools. Thus, the *billet technique* was born.

In 1957, while writing about lithic technologies of the "American Indians," Ellis also confirmed that the billet technique was developed in recent times: "The so-called 'billet technique,' the use of a wooden club in chipping flint [figure 3.3c], was developed by N. Leon Coutier working under the able direction of L'Abbe Breuil. Coutier demonstrated this process to H. C. Shetrone of the Ohio State Museum, in 1937. Although there is as yet no satisfactory documentary proof that this method was employed by the aborigines of North America" (Ellis 1957, 11–13). To the credit of European academia, many researchers were careful to simply suggest that such a tool *could have been used*, while not asserting that there was direct evidence of its use. This unfolding scenario begs the question: Can one infer a flaking method from flakes and flake scars alone, apart from a knowledge of the tools that were used? Such an inference cannot be positively demonstrated as long as the possibility of unknown technologies, and tools, continues to exist. It is also worth noting that Barnes's original soft hammer experiments, his hammer and boxwood punch experiments, and even his ivory hammer experiments all produced characteristics similar to the soft hammer technique. Instead of producing salient bulbs of percussion, the ivory hammer produced a lipped flake with no cone of percussion, while the hammer and punch method also produced diffused bulbs rather than salient bulbs.

To recognize a flintknapping tool as an essential part of *chaînes opératoires* based on evidence from the archaeological record, we argue that one must demonstrate and document the existence of a flintknapping tool, demonstrate evidence of its use, demonstrate evidence of its refurbishment during use, and finally provide evidence of its final discard in a cultural context, whether in a midden or burial. If all of these facets of a proposed tool's life cycle can be systematically documented in association with a reconstructed *chaînes opératoires*, then one may propose that the tool is part of a cultural tradition or a core culture trait. From this perspective, there is no evidence for the hypothesized soft hammer baton or billet (stick)—not in the archaeological record, documented by any ethnographer, or attested by Native Americans in any New World culture. Thankfully, multiple researchers dating back to the late nineteenth century and twentieth century consistently collected evidence for flintknapping punches, while documenting their prehistoric creation, their refurbishment, their apparent use, and their final disposal either in middens, or with the deceased prehistoric flintknapper.

ACCOUNTS FROM THE NEW WORLD THAT DIRECTLY OR INDIRECTLY ATTEST TO THE EMPLOYMENT OF INDIRECT PERCUSSION

The introduction of the horse in the fifteenth century led to the distribution of both steel and glass over the continent, which in turn initiated the partial abandonment of Native stoneworking practices. For cutting tools, steel quickly replaced stone, and for piercing tools both steel and glass replaced stone. The difference between the two is that glass chipping still retained some aboriginal techniques. Thus, glass chipping can be considered a continuation of Native technology, but with a new format and materials. Furthermore, the rapid modernization of the North and South American continents, thanks in large part to the introduction of cross-continental rail, quickly brought more isolated Native American populations into contact with Westernized peoples. In some cases, these encounters resulted in people inadvertently becoming primary or secondary witnesses to lithic reduction practices that in other parts of the world were assumed to have been "lost arts."

There is also evidence that the original lithic reduction practices were to a degree retained in some areas even after the introduction of glass and steel, often as a fallback for when Western materials were not available. As we discuss in this section, some crafters who continued lithic reduction arts became trinket makers during the nineteenth century, possibly plying their wares to early tourists who traveled the western states by rail. Some chipped-stone technologies continued into the early 1900s, when trading posts proliferated and arrowheads were offered for sale en masse by mail-order catalogue. By the early 1900s some state legislatures began to pass legislation that criminalized the forgery of artifacts, as it had become a growing problem. Forgers such as Marvin McCormick may have manufactured thousands of fluted Folsom points that still might be in museums today. McCormick used an entire set of lithic reduction techniques that he learned as a boy, including indirect percussion for fluting in a hollow log anvil.

In 1933, California gemologist Fred S. Young wrote about the loss of the original Native lithic practices and the transition to modern forgery:

> The fashioning of arrow and spear points and other blades was an art with the early Indian, confined to a few expert members of the tribe. Squaws were never permitted to do this work . . . The tools used by the early Indian for working obsidian and agate into points, were, bone, hardwood tools. Beaver and porcupine teeth were often used as tools for the purpose of working the very small bird points. A piece of bone about five inches in length and thicker than a pencil was the common tool for working obsidian. The work was held between the knees, padded with buckskin, the smaller points were worked by holding with the hands, padded with buckskin . . . In the making of the large blades some white men have found it advantageous to use a clamp or vise to hold the work. The fashioning of a blade from obsidian of even twenty to thirty inches in length was a very difficult task for a number of reasons . . . It was a matter of considerable pride and honor to the Indian exhibiting the largest and best made blade, and he was signally honored with the rare privilege of dancing last, at the ceremonial . . . The skill of a point maker was generally judged by his ability at producing fine long blades. The

> largest blade of obsidian of which the writer has any authentic record, was made by an aged Indian with some fifty years' experience. This blade was made from a block of Lake County, Oregon obsidian, which would weigh approximately 900 pounds in the rough. The finished blade measured 42 inches in length, 9 inches wide and nearly 2 inches thick, and was of very excellent workmanship. It was sold to a collector for a rather substantial sum. (129)

While the person with "fifty years' experience" Young describes is probably Native knapper Ted Orcutt (see later in chapter), sadly it seems that the forgers were responsible for carrying on most of the Native American stoneworking techniques. In the following section we review a sample of ethnographic and ethnohistoric accounts of stoneworking collected by Eble but with a special focus on those that describe indirect percussion. These accounts are presented primarily in a chronological order and include observations of practices carried out by peoples with little contact with the West, as well as those by practitioners who had been integrated into the modern world.

During the 1850s, Sir Edward Belcher visited Cape Lisburne, Alaska, where he witnessed Native people extracting chert from a seam at the base of some cliffs and subsequently fashioning it into arrowheads and harpoon heads. Belcher noted the employment of three distinct methods: direct percussion, indirect percussion, and pressure flaking. John Evans (1872, 39) later commented on what Belcher witnessed:

> Sir Edward Belcher some years ago kindly explained the process to me, and showed me both the implements used, and the objects manufactured. It appears that the flake from which the arrow-head is to be made is sometimes fixed by means of a cord in a split piece of wood so as to hold it firmly, and that all the large surface flaking is produced either by blows direct from the hammer, or through an intermediate punch or set formed of reindeer horn. The arrow- or harpoon-head thus roughly chipped out is afterwards finished by means of the arrow-flaker.

Clearly, Belcher took note of the triad described in the preceding text.

In October 1866, at the meeting of the International Congress of Paleoethnology in Switzerland, Mortillet (1866, 517–518) records coastal surveyor Mr. Frank de Pourtalès as giving the following account of stone tool manufacture by Native Americans:

> M. de Pourtalès gives details on the way the Indians shape flints. After having detached by means of a shock a suitable blade, they make use of an instrument in compact bone, bearing at the extremity a slight notch. They insert the edge of the blade into this notch, and with a sideways blow, they knock out a small splinter . . . It is mainly arrowheads that they manufacture in this way. They sell the arrows very easily, but it is very difficult to obtain the instrument which is used to make them. Moreover, it is an industry that is being lost.

Around 1853, lawyer, artist, and amateur ethnographer George Catlin (1868, 187–189) traveled eastward from the Pacific Coast, crossed what is now modern-day California and Arizona, and then traveled through Apache territory, heading from the Gila River to Santa Fe. Catlin visited multiple Apache villages. While commenting on their weaponry, he provided the following insights regarding stone tool manufacture:

> Like most of the tribes west of and in the Rocky Mountains, they manufacture the blades of their spears and points for their arrows of flints, and also of obsidian, which is scattered over those volcanic regions west of the mountains; and, like the other tribes, they guard as a profound secret the mode by which the flints and obsidian are broken into the shapes they require.
>
> Every tribe has its factory, in which these arrow-heads are made, and in those, only certain adepts are able or allowed to make them, for the use of the tribe. Erratic boulders of flint are collected (and sometimes brought an immense distance), and broken with a sort of sledge-hammer, made of a rounded pebble of horn-stone, set in a twisted withe, holding the stone, and forming a handle.
>
> The flint, at the indiscriminate blows of the sledge, is broken into a hundred pieces, and such flakes selected as, from the angles of their fracture and thickness, will answer as the basis of an arrow-head.
>
> The master workman, seated on the ground, lays one of these flakes on the palm of his left hand, holding it firmly down with two or more fingers of the same hand, and with his right hand, between the thumb and two fore-fingers, places his chisel (or punch) on the point that is to be broken off; and a cooperator (a striker) sitting in front of him, with a mallet of very hard wood, strikes the chisel (or punch) on the upper end, flaking the flint off on the under side, below each projecting point that is struck. The flint is then turned and chipped in the same manner from the opposite side, and so turned and chipped until the required shape and dimensions are obtained, all the fractures being made on the palm of the hand.
>
> In selecting a flake for the arrowhead, a nice judgment must be used, or the attempt will fail: a flake with two opposite parallel, or nearly parallel, planes is found, and of the thickness required for the centre of the arrow-point. The first chipping reaches near to the centre of these planes, but without quite breaking it away, and each chipping is shorter and shorter, until the shape and the edge of the arrow-head are formed.
>
> The yielding elasticity of the palm of the hand enables the chip to come off without breaking the body of the flint, which would be the case if they were broken on a hard substance. These people have no metallic instruments to work with, and the instrument (punch) which they use, I was told, was a piece of bone; but on examining it, I found it to be a substance much harder, made of the tooth (incisor) of the sperm-whale, which cetaceans are often stranded on the coast of the Pacific. This punch is about six or seven inches in length, and one inch in diameter,

with one rounded side and two plane sides; therefore presenting one acute and two obtuse angles, to suit the points to be broken. This operation is very curious, both the holder and the striker singing, and the strokes of the mallet given exactly in time with the music, and with a sharp and rebounding blow, in which, the Indians tell us, is the great medicine (or mystery) of the operation.

Catlin painted a scene showing the first flintworker wielding the stone maul, and two other flintworkers operating the punch and mallet (figure 3.4). Catlin provides his readers and viewers with many interesting details in this account, which do not appear to be spurious and that concur with other lines of study. Here we isolate each important facet.

Geographic isolation: He points out that the chipping and flaking of stone persisted in areas that are geographically isolated, to the west of the Rockies, where people do not yet have an adequate supply of metal. Subsequent surveys of historical knappers revealed that people in more remote areas preserved some flintknapping practices, even until the early to middle part of the twentieth century.

Dedicated workstation: Catlin describes the flintworking site, or "flint factory," as he terms it, which is not unlike stage reduction sites, found varying distances away from quarries, where knappers worked during prehistoric times. It is worth pointing out that Catlin chose the word *factory*, as factories were known to house assembly lines. In other passages, Catlin spoke of the division of labor among flintworkers. By referring to the flint workshop as a "factory," he implies that it was also the site of an assembly-line operation.

Mobility: Catlin notes that sometimes boulders of tool stone were carried far and wide. This is actually possible, since the introduction of the horse in the Americas could have allowed for long-distance transportation of large stones.

Specialists: Catlin notes that certain adepts were chiefly responsible for chipped-stone manufacture. This observation also concurs with what is seen in the prehistoric record, where some but not all men were interred with flintknapping tools, including small antler punches and pressure tools, hammerstones, and partially worked or finished points. If knapping was a skill mastered by all, then there would be a much wider distribution of knapping tools in burial contexts.

Use of maul or hafted hammer: Like Belcher before him, Catlin (1868, 138) describes the initial stage of lithic reduction as hard hammer spalling that is carried out with a stone maul (cf. Belcher's hafted nephrite hammer). This tool use concurs with other accounts and lines of information.

Palm-based support: Importantly, Catlin notes the process of laying a flake on the palm of the left hand while holding the flake down firmly with two or more fingers. This technique is identical to the palm-based support method utilized in conjunction with pressure flaking recorded in other descriptions and accounts. In fact, this description of flake stabilization appears to be identical to the way

FIGURE 3.4. *Painting of a flintknapping scene by George Catlin (drawing by Hruby after Smithsonian Museum archive).*

Ishi reduced obsidian with a pressure flaker. Only, in Catlin's account, two men wield a punch and mallet.

Shock absorption: Unlike Belcher's account (in Evans 1872, 39), which refers to the Alaskan chert being held in a sturdy wooden vise, Catlin notes that the Apache knapper deliberately avoided placing the stone on anything hard, to avoid breakage, while chipping with a punch. Furthermore, Catlin explains that the soft portion of the palm acts as a sort of "shock absorber" to prevent a hard blow from breaking the stone being chipped.

Possible penis bone or antler flaker: Catlin speculates that the punch tool was fashioned from a sperm whale's tooth. It is possible, however, that Catlin misidentified a flaker that was actually made from a sea mammal's penis bone, possibly from the not-so-distant Pacific Coast. If so, this would concur with other lines of evidence, even from the arctic, which speak of penis bone flakers and penis bone flaking tool tips being employed, sometimes even during prehistoric times. A smoothed antler may have also been mistaken for such a tooth.

Rebounding blow: Catlin describes the blow of the mallet as being carried out with a peculiar rebounding strike. This may sound strange, but if the stone that Catlin witnessed being worked was already thin, then the peculiar "rebounding blow" could have been intended to prevent the hard punch blow from overstriking the edge, which would cause a margin collapse on the already-thin edge, or snapping the point altogether.

Isolated platforms: Catlin also seems to identify what modern flintknappers refer to as an "isolated platform." He calls them "projecting points," and he explains that the punch blow removes flakes from the underside of the point, after each blow, upon the palm. Catlin's contemporary Charles Rau (1873, 143) commented on Catlin's account, and referred to them as "protuberances." The concept of an isolated platform being used to remove flakes from the surface of a biface is a detail that would probably be recognizable by some of the best modern knappers.

Punch description: Since Catlin states that he physically inspected the hard bone punch being used, we have reason to believe that he was offering more than a hypothesis regarding how flintknapping could have been carried out.

Ritualized production: Although we have been focusing on the technology, it is significant that singing and knapping in time with the music (i.e., medicine or mystery) appears to be an essential aspect of the production process. Rather than an oddity, this facet of production is one shared by other Native American cultures such as the Aztecs and Maya (Hruby 2007). In other words, it may be fruitful to escape Western norms of production practice (i.e., quiet, efficient labor) when reconstructing or considering how other cultures may have done it. There is clearly a religious facet or ritualization of production recorded here.

Recognition of a problem in the academic world: In prefacing his well-detailed flintknapping description, Catlin, who was also a resident of Europe during the

1860s, indicates that he was aware of the general state of understanding, among academia, when he penned the following: "Like the other tribes, they (Apachees [*sic*]) guard as a profound secret the mode by which the flints and obsidian are broken into the shapes they require. Their mode is very simple, and evidently the only mode by which those peculiar shapes and delicacy of fracture can possibly be produced; for civilized artizans [*sic*] have tried in various parts of the world, and with the best of tools, without success in copying them." At the time Catlin (1868, 187–189) wrote this, such attempts were already being met with failure in Europe. He concludes that the only way a certain type of delicate facture can be created is via indirect percussion, which he witnessed among the Apache.

It appears that Catlin's reports may have been met with more than a dose of healthy skepticism prior to the publication of *Last Rambles*. In fact, after witnessing a paper read in 1864 to the Geological and Polytechnic Society in England, one researcher, P. O'Callaghan (1864, 312), publicly stated the following:

> The method of making these flint weapons was for a long time a mechanical puzzle. Mr. Catlin, the celebrated American traveler, gave me a wonderful account of his visit to the workshop of an Indian, who made arrowheads, in the rocky mountains. It was quite a romantic story. He told me that the secret of the manufacture was concealed with the most religious care, and that the door of the wigwam in which this trade was carried on, was vigilantly guarded by an armed sentry. He told me also, that this mysterious profession was held in the highest estimation, and hereditary, and that these cunning workmen were looked upon as priests. Sir William Wylde, in writing upon this subject, says, "The trade of the gun-flint-maker gives but a faint idea of the process by which such specimens of exquisite workmanship could have been fashioned."
>
> However, Sir Edward Belcher, in his last voyage to the arctic regions, not only saw the whole process of making flint arrow-heads, but easily acquired the art himself, and shewed it to us all, at the meeting of the British Association at Oxford. The simple tool employed for this purpose, is nothing more than a small bone, about three inches long, with a little piece of horn, or chert, fastened into it, near the end. With this rude instrument resting upon a wooden block, or bench, the flint-flake was readily converted into an arrow or spearhead, by nipping little chips off it. If Mr. Catlin or Sir William Wylde could have seen the beautiful series of flint arrow-heads which I have the pleasure to exhibit on this occasion, and which have been all made at Bridlington, in your own county, by the celebrated "Flint Jack," their speculations as to their mysterious fabrication would have been brought to a ludicrous conclusion.

O'Callaghan was aware of the direct percussion methods employed by Flint Jack, and he had witnessed Belcher reproduce the pressure-flaking methods of the Eskimo, in London. As a result, in light of what he knew, and on account of his own experience, he viewed Catlin's accounts as "ludicrous." What O'Callaghan did not know is that: (1) Evans would go on to document that indirect percussion *was involved* in Belcher's account; and (2) the study of gunflint knappers would eventually reveal that hard hammerstone percussion alone *was not enough* to explain the finishes found on some of the finest stone tools.

In a manner reminiscent of Catlin's description and illustration, historian, publisher, and author Paul A. Jones (1937, 111), born 1882 in Lyons, Kansas, on the Santa Fe Trail, wrote the following in "Coronado and Quivira":

> From our observation on village sites we are ready to agree that there were certain shops or places where arrowheads were made in great quantities. Tradition also has it that two men worked at the heavier shaping of the flint implements. One held a flint flake on the palm of his hand, grasped with a finger or two folded back. He shifted the bone punch with his right hand to the point desired. When the punch was at the proper place and angle the helper, on command, would hit it with a stone hammer. To avoid a long and tiresome series of commands the twain would sing a song in unison and at certain cadences the helper would hit the punch. It was the business of the holder to have the punch at the right place and angle when the proper cadence was reached else the implement might be ruined. As Indians became more effete a skin pad on one knee was substituted for the calloused palm.

While Jones does not explain from where he learned of this "tradition," the details are complementary to Catlin's account.

During the 1880s, and 1890s, Edwin and Therese Deming traveled the West, visiting various tribes including the Apache, Blackfoot, Sioux, Umatilla, Crow, Puebloan, and others. The husband-and-wife duo dedicated their lives to painting scenes from Native American life and writing stories for young readers to show what life was like among various tribes. These stories were written from notes kept in diaries (Deming 1932, 84–86), as they attempted to document the remaining vestiges of Native American life.

In a children's story describing stone tool manufacture, we are told that the knapper used a lapstone as an anvil. A flake of flint was struck off by the knapper, while the core was rested on the anvil in his lap. (A similar account of a lap anvil used by a Shasta knapper, also with implications for indirect percussion, was related by Caleb Lyon in 1860 [recorded in Deming 1932].) The flake that was to be worked into a stone tool was then held very carefully in the hands of a helper. The knapper used a hammerstone and a punch to strike at the edge, while the knapper held the stone very carefully. After each flake removal, the helper turned the piece over to check and make sure that the flakes were coming off correctly from the underside. Once the piece was sufficiently thinned, the knapper placed a small handstone in the palm of his hand. He rested the point on top of the handstone, and used a bone tool to further chip a sharp edge. Deming (1932) furnishes the reader with the following painting of the scene (figure 3.5).

FIGURE 3.5. *Edwin Deming's 1932 illustration of a flintknapping scene (drawing by Hruby).*

Larry Dean Olsen (1997, 151), the late founder of the Anasazi Foundation, reported that he learned three forms of indirect percussion from his uncle around 1950. His uncle, in turn, learned from an old man in Burns, Oregon. Burns, Oregon, is home to the Burns Paiute Tribe. Burns is also the site of a large obsidian outcrop. The three forms of indirect percussion that Olsen learned from his uncle involve: (1) holding a small biface on the palm, and having an assistant chip it with hammer and punch; (2) resting the preform or biface on a small padded anvil-stone, and chipping it with hammer and punch; and (3) holding the stone between the knees, and chipping it with hammerstone and punch.

> If a blank has a difficult knob or thick edge in such a place as to make direct percussion impractical, it can be worked down individually with the aid of a punch made from deer antler or other antler of similar density. This process requires the assistance of a second person. The blank is held in the left hand, which is protected with a leather pad, and the punch is placed against the edge of the blank and steadied with the other hand. The assistant then strikes the punch sharply with a hammerstone. In this way the force is applied to the exact spot intended.
>
> Percussion flaking with the blank resting on another stone (anvil) works but is rarely necessary except for removing troublesome knobs and for striking off larger flakes to thin down a thick blank. For this it is best to use the indirect percussion method with the blank resting on a padded anvil.
>
> A variation of the indirect percussion method calls for the blank to be placed between the knees with the edge to be flaked facing up. One person can then use the punch and hammerstone without an assistant. There is no danger that the chips will cut the knees if the blank is placed in the fold of a thick piece of leather.

Regarding the use of the knees by aboriginal American obsidian workers, in 1933, Fred Young reports the following:

> The fashioning of arrow and spear points and other blades was an art with the early Indian, confined to a few expert members of the tribe. Squaws were never permitted to do this work. It is sometimes thought that heat and water were used in the chipping of the obsidian and agate but this method was not used. The tools used by the early Indian for working obsidian and agate into points, were, bone, hardwood tools. Beaver and porcupine teeth were often used as tools for the purpose of working the very small bird points. A piece of bone about five inches in length and thicker than a pencil was the common tool for working obsidian. The work was held between the knees, padded with buckskin, the smaller points were worked by holding with the hands, padded with buckskin.

Beckwith also depicts the use of the knees, in conjunction with indirect percussion, in 1911 (figure 3.6).

Around the same time, Aurther Caswell Parker (1917, 161) explained that a few of the Iroquois still remembered the basic lithic reduction practices that had previously been employed. He affirmed that some flaking methods were positively known to have been employed. And he authored a paper outlining his experiments under Putnam, with known processes, that was never published. Parker (161) also cites the use of indirect percussion presumably on a padded anvil, with a "pad of skins" and a "pitching tool":

> In the description which follows I have combined previously known facts regarding the chipping of flintlike stones with other facts gleaned from a series of experiments conducted by myself under the direction of Prof. F. W. Putnam, in

FIGURE 3.6. *Thomas Beckwith's 1911 photograph of indirect percussion using the knees and legs.*

> the American Museum of Natural History. These results were embodied in a paper which has never been published. Much of the description which follows later is taken from this paper. In the description of the various processes the reader must understand that where positive statements of methods are made that these methods were those used in experiments and are in accord with methods known to have been used. The tools used in shaping arrowheads were few and simple, consisting merely of a stone hammer and a flaker. For larger implements a stone anvil, a pad of skins, and a pitching tool were used in addition. The flaker was one of the most important tools in the process and with it the most delicate work was done.

George Byrd Grinnell (1923, 2008 [1872]), an American soldier, who spent a considerable amount of time among the Cheyenne on the Great Plains during the 1870s, recorded the following details regarding direct percussion, indirect percussion, and pressure flaking:

> The Cheyennes in ancient times used to a considerable extent arrowheads and knives of stone. Black Moccasin, who died about 1884, remembered when such implements were still often employed. Persons recently living were told by their grandfathers that in their time, say early in the nineteenth century, such implements were in use. Some knives were made of the ribs of elk or moose. For their cutting and piercing implements they employed the hardest stones they could secure, usually flint. When a suitable stone had been found for making a celt, it was broken off to the right length by a sharp blow of a small stone hammer[;] it was then held in the left hand and rested on a block of wood, and repeated blows were given it with the hammer, it being turned and chipped by these blows until the desired edge and point were produced. When a knife—or an arrowpoint—had been worked down quite thin, but had not yet received a satisfactory cutting edge, the piece was held in the left hand, between the thumb and forefinger, while a small stone punch was held between the fore and middle fingers. The punch was pressed against the edge of the blade, and was struck [using] sharp downward blows with a hammer, each blow taking off a small flake, and this process was continued until the edge was finished. A better cutting edge was finally given by the flaking off of small chips from near the margin. The flint was held in the palm protected by a wad of hair or piece of tanned hide and a small point of antler or bone suddenly applied with force against the stone at the required point. This pressure cracked off a small chip and the operation was repeated as needed. (Grinnell 1923, 185)

Edward Curtis, a companion of Grinnell's, took over 40,000 photos of eighty tribes and recorded the following regarding the Piegan: "Arrow-points were of bone from the buffalo shoulder-blade, and of flint, which was shaped by holding between the fingers and chipping under the edge with the tip of a piece of elk-horn; the other end of the horn was struck sharply with a heavy stick and a flake was thus chipped off" (1911, 156).

In 1878, Edwin Tappan Adney (1964, 18) described the manufacture of stone tools used in North American boatbuilding. He noted that direct percussion, pressure flaking, and indirect percussion were all employed by northern tribes in this work:

> Stone tools were formed by chipping flint, jasper, or other forms of quartz, such as chalcedony, into flakes with sharp edges. This was done by striking the nodule of stone a sharp blow with another stone held in the hand or mounted in a handle of hide or wood to form a stone hammer. The flakes were then shaped by pressing the edges with a horn point say, part of a deer antler—to force a chip from the flake.
>
> The chipping tool was sometimes fitted with a hide or wood handle set at right angles to the tool, so that its head could be hit with a stone or horn hammer.

FIGURE 3.7. *Hupa or Wintu punch, part of the Ray Collection (Smithsonian Museum Archive, also depicted as part of a knapping toolkit in Otis [1886: Plate XXI]).*

> The flake being worked upon, if small, was often held in the hand, which was protected from the slipping of a chipping tool by a pad of rawhide. Heat was not used in chipping, and some Indians took care to keep the flake damp while working it, occasionally burying the flake for a while in moist soil. The cutting edge of a stone tool could be ground by abrasion on a hard piece of granite or on sandstone, but the final degree of sharpness depended upon the qualities of the stone being used as a tool. Slate could be used in tools in spite of its brittleness. In general, stone tools were unsuitable for chopping or whittling wood.

Travis Hudson and Thomas Blackburn (1982, 104) mention an account of stone-on-stone indirect percussion being used by Channel Islanders from 1884:

> In 1884 Henshaw collected this description of their manufacture from Juan Pico: "To chip a stone knife or arrowhead the flint was placed in the cleft of a large stone. A very hard transparent slightly pointed pebble of agate was held against it and struck a light quick tap with a second pebble of quartz held in the right hand. Pico states that the Islanders were more skillful in the production of stone implements[—]these formed their stock in trade." As a kind of footnote to the above description, Heizer (1955:151) added that "this method of flint flaking is not the usual one practiced by California Indians, but something like it is known in the northern part of the state."

In 1885, while researching the Hupa, Ray (Mason 1886, 227) documented an arrow-making kit and noted that men over the age of forty still knew the art. Among the items recorded, Ray illustrated a hard column of antler or bone, approximately six inches long and one inch thick (figure 3.7), which was reserved for the finest stone in the production of "flakes," "blades," and "spalls."

Borrowing language from stoneworkers, Ray referred to the tool as a "pitching tool," or a "cold chisel," thus denoting it as a tool of indirect percussion. The notes displayed with the Hupa punch at the Columbian Historical Exposition (Madrid, Spain, 1895) stated: "The chisel of hard antler struck by a cooperator as in Catlin's description." This brief note indicates that there was some consensus on how these tools were used, and similarities were noted between accounts from Catlin, Jones, Deming, Olsen, and Ray.

In 1895, in "From the Indian Trail to the Railroad," Grinnell (1895, 147) wrote again regarding aboriginal American flintknapping, with a possible allusion to indirect percussion:

> Near a lodge, small and weather-beaten, two men seated under a shade are hard at work. Each holds between his knees a block of stone, from which, by light sharp blows of a small stone hammer, he is chipping off triangular flakes of flint for making arrowheads. The material used by one of the men is a black obsidian obtained by trade from the Crows to the south, while the other has a piece of milky chalcedony picked up in the mountains to the west. Each of these blocks has been sweated by being buried in wet earth, over which a fire has been built, the object of this treatment being to bring to light all the cracks and checks in the stone, so that no unnecessary labor need be performed on a piece too badly cracked to be profitably worked. As the workmen knock off the chips, they turn the blocks, so that after a little they become roughly cylindrical, always growing smaller and smaller, until at length each is too small to furnish more flakes. They are then put aside.
>
> Each man now collects all the flakes he had knocked off, and, piling them together on one corner of his robe, carefully examines each one. Some are rejected at a glance, some put in a pile together as satisfactory, while over others the arrow-maker ponders for a while, as if in doubt. Presently he seems to have satisfied himself, and prepares for his second operation. For this he takes in his left palm a pad of buckskin large enough to cover and protect it while holding the sharp flake, while over his right hand he slips another piece of tanned hide something like a sailmaker's "palm," and used for the same purpose. Against his "palm," the arrow-maker places the head of a small tool[,] a straight piece of deer or antelope horn or of bone about four inches long, and pressing its point against the side of the piece of flint held in the other hand, he flakes off one little chip of the stone and then another close to it, thus passing along the edge of the unformed flint until one side of it is straight, and then along the other. He works quickly and apparently without much care, except when he is near the point, for this is a delicate place, and carelessness or haste here may endanger the arrowhead for, if its point should be broken, it is good for nothing. Sometimes an unseen check will cause the head to break across without warning, and the labor expended on this particular piece is thus wasted. But usually the arrow-maker works rapidly and

spoils but few points. After the head is shaped, there are often left some thin projecting edges which mar its symmetry and add nothing to its effectiveness. These are broken off either by pressure or by a sharp blow with some light instrument, such as a bit of bone or of hard wood.

George Escol Sellers (1886, 874) later published Catlin's descriptions of blade manufacture both by pressure and pressure aided by blow, a form of indirect percussion. He provided an illustration of the tools employed, as well as a description of core immobilization techniques:

In some cases the stone operated on was secured between two pieces or strips of wood like the jaws of a vise, bound together by cords or thongs of rawhide; on these strips the operator would stand as he applied the pressure of his weight by impulse. The best flakes, outside of the home-made, were a subject of commerce, and came from certain localities where the chert of the best quality was quarried in sheets or blocks, as it occurs in almost continuous seams in the intercalated limestones of the Coal Measures. These seams are mostly cracked or broken into blocks, that show the nature of the cross-fracture, which is taken advantage of by the operators, who seem to have reduced the art of flaking to almost an absolute science, with division of labor; one set of men being expert in quarrying and selecting the stone, others in preparing the blocks for the flaker. This was done when the blocks were nearly right-angled at the corners, by striking off the corner where the flaking was to commence, and, with a properly directed blow with a hard pebble stone, knock off of the upper edge a small flake, making a seat for the point of the flaking-tool. Sometimes these blows were carried entirely across the front upper edge of the block, making a groove entirely across the edge, when the first row of flakes have been thrown off. It is the work of this operator to prepare seats for a second row, and so on. What was meant by almost absolute science was a knowledge and skill that would give the proper direction to the pressure to throw off the kind of flake required. The staffs of these flaking-tools were selected from young hard-wood saplings of vigorous growth. A lower branch was utilized to form the crotch in which the blow was struck (see Fig. 16.). Another branch on the opposite side was used to secure a heavy stone to give weight and increase the pressure. When the stone to be flaked was firmly held, the point adjusted to give the pressure in the required direction, the staff firmly grasped, the upper end against the chest of the operator, he would throw his weight on it in successive thrusts, and if the flake did not fly off, a man standing opposite would simultaneously with the thrust give a sharp blow with a heavy club, it being so shaped that its force is downward close in the crotch. It has been represented to me that a single blow rarely failed to throw off the flake, frequently the entire depth of the block of stone, sometimes as much as ten or twelve inches. The tooth or tusk of the walrus was highly prized for tips of the flakers.

In 1910, in translating a Yana myth, Edward Sapir (1911, 43) recorded the mythological story of a clan that was entrapped in a sweat lodge made from flint rock. According to the myth, one man wielded a stone maul, while the second man held a "pitching tool" against the flint wall. The blows of the stone maul against the "pitching tool" eventually caused the flint wall to give way and shatter, allowing the clan to escape.

> The South people entered, they were chuck full in the sweathouse. "Let us go outside," said the chief, talking (to his own people). "Let us make a fire outside in the night time." The sweat-house was crowded, the people filled it entirely. Those who belonged to that house all moved outside. There were the Geese people inside, all by themselves, sitting inside the sweat-house. Flint Rock Chief shut the door of the sweat-house, so the sweathouse was totally dark; there was no daylight whatever. The people outside, the owners of the sweat-house, were making much noise, having a good time. Three times it was day and three times it was night, and there was no daylight inside; it was always night. The people outside were having a good time pounding acorns and hunting deer.
>
> "This looks bad. Daylight must have appeared long ago." The people inside felt around with their hands. There was no fire there and they were hungry and thirsty. "He has shut the door on us, he is angry at us," said they inside. "Four days and four nights have passed and there is no daylight yet. What shall we do? We are all going to be killed."
>
> "Hehe'?! Would that I could get outside again! Have not any of you perchance a *flint flaker*? Have not any of you perchance a *flaking maul*?" "Yes," said Ma'ldama. "I have a *pitching tool*." "I also have a *pitching tool*," said Bopdidjū'si. "Yes," said the chief, "it is you that always say that you have supernatural power." The *pitching tool* was like this here, the *flaking maul* was like this here. Those two men, little Bopdidjū'si and Ma'ldama, had *pitching tools*. They arose in the sweat-house in the night that surrounded them all. The sweat-house was made entirely of flint rock, thick was the flint rock. They put out their hands inside and felt around all over. They were all like blind men. "Now! pound away!" This is how they did, pounding away at the flint rock to test for a thin spot. Now he pushed his pitching tool against the flint rock and pounded on it with his maul. This is how Ma'ldama did it (Accompanied by tapping ruler on knife against window).
>
> "S*s*" said the chips of flint as they fell to the ground. The flakes made a noise as they were thrown to the ground. Thus he kept it up all day, and little Bopdidjū'si worked too. Every little while they pounded around with their hands to see flow thick it was. Now it became thin and they pounded away at that spot. "S*!" said the flint chips as they fell splintered off to the ground. They pounded with their hands to see how thick it was. "S*!" said the flakes falling down on the ground. Again they pounded with their *pitching tools*. Thus they did and burst

right through the wall. Now they had made a hole right through. The light of day streamed in, it became daylight as soon as the hole had been burst through. The sweat-house was lit up. Now the people returned. They all came out again, returned out of the sweat-house. (emphasis ours)

After living with the Zuni during the 1880s, Cushing (in Putnam 1896, 209) gave a lecture in which he outlined a series of reductive steps employed in creating arrowheads, first by spalling and chipping with hammerstones, then by further flaking via indirect percussion, and pressure. With regard to indirect percussion, Cushing recorded the following:

> In finally forming arrow-points from these trimmed blanks, the smallest of them only were chosen. The first care in fashioning one of these was to remove protuberant points from its edge and sides and to thin it down by means of a "pitching tool" of buck-horn. This was effected in several ways, usually by clamping it in a folded pad of buckskin under the knee against a hammerstone or notched wooden block, so that the projecting edge rested over the margin or else over the pit of the stone, or notch if a block or log were used, and with one hand holding the point of the pitching-tool very lightly and slantingly and at a wide angle, against or just over the points to be chipped, sharply tapping the tool with a maul or with a knapping hammer. Thus the blade was quickly thinned down and made almost even edged.

Professor A. F. Bonney (1918, 340) wrote an article for *Scientific American* describing an elderly Apache man, in Arizona, who was found using a punch made from a tenpenny nail, a mallet, a leather pad, and a piece of wood with a crack in it, to create projectile points to sell to tourists as trinkets:

> While in Arizona a number of years ago, the writer saw an old Apache indian in Arizona chipping arrowheads of quartzite. He might have been an expert, sixty or eighty years previously, but his eyes had grown dim and his hands unsteady, and the pieces he produced were rude, but good enough to be used in making arrows which he sold to tourists—after the relics had been well smoked to give them age. The old man's knowledge must have come from the Old Time, for finding a flake to suit him, a rather flat one which was not too thick in the center, and sloped to an edge in all directions from the middle, he roughed it into a leaf shape by pounding the flat of the edge with a piece of stone. Then with a 10d nail for a punch, and a billet of wood for a mallet he began working the edge, first flaking from one side and then the other, holding the punch at right angles to the surface of the stone. Then he thinned the edge by holding the punch at an obtuse angle to the surface, finally finishing it by putting notches into the base. He worked against a thick piece of leather laid on his thigh, but had a piece of wood with cracks in it which he sometimes used to hold the flake when the edge was too

thick to yield by the first method, which suggested a couple of pieces of wood in a vise for holding the work.

In 1916, Grace Nicholson photographed an aging Karuk knapper named Theodore Orcutt, carrying out finishing flintknapping processes at his cabin with a set of antler pressure flakers, and antler punches. Nicholson sent twenty-one plates and notes outlining the process, to Holmes, at the US National Museum. According to the step-by-step notes, Orcutt held the obsidian biface between his thighs, while chipping it with a set of small antler punches and a pestle-like hammerstone used as a mallet. The antler punches and pressure flakers used by Orcutt and collected by Nicholson are depicted here in figure 3.8.

While Holmes initially wrote, in 1894, that indirect percussion was not well understood, by "anyone," he afterward did respond more glowingly twenty years later, when presented with the methods employed by Orcutt:

> You have given me a great surprise in sending the photographs of Theodore Orcutt illustrating his obsidian work. Every step of the manipulation is clearly shown and the series must always remain one of very marked scientific value. It tells not only of the manipulation of the stone in every detail, but it makes it clear to archaeologists and to the public that the art is *not* a dead art and that many of the implements in our collection are of modern manufacture. This latter fact, however, is not so serious since the work is done by an Indian who merely carries on the work of the fathers. (Holmes 1919; emphasis ours)

Holmes (1919) also stated that when Ishi was queried as to how he would make a larger implement, such as a spearhead, Ishi explained, presumably through his translator Sam Batwi, that what was needed was a short, stout piece of bone or wood with a blunt to be used as an intermediate tool, as a punch or lever—not to divide the mass, but to trim down the mass. Regrettably, no one ever bothered to ask Ishi to create this tool, much less demonstrate its use in fashioning a spearpoint or larger implement. It is possible this method was the same one employed by Orcutt, especially considering that they were both in the same general geographic area and both worked obsidian.

In 1941, Californian Journalist Paul Wilhelm wrote an article for *Desert Lore*, titled "Arrowheads Easy to Make—If You Happen to Be an Indian." Wilhelm traveled to the Martinez Reservation, in Southern California, and met with a certain arrowhead maker named "Roxie."

> "To make an arrowhead," Roxie told me, "you need just a few tools—like these." He held out a small hammerstone, a deer bone awl, and a deer antler. Roxie tied a piece of buckskin around his left hand. "It is all in knowing the particular stone that you are working with and the method of pressure, percussion, and pitching with the tools." . . . Roxie explained that a spearhead takes larger tools. "A

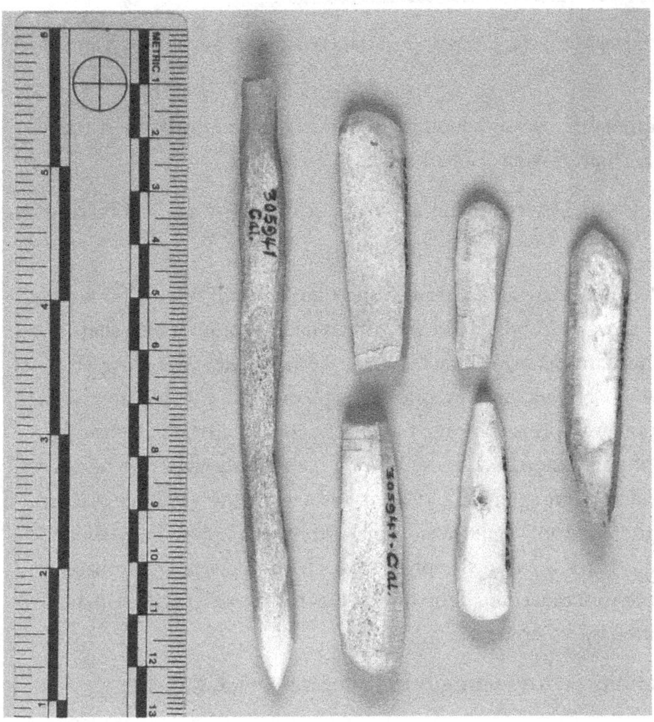

FIGURE 3.8. *Photographs of Theodore Orcutt's knapping toolkit including an array of antler punches and flakers (Smithsonian Museum Archive).*

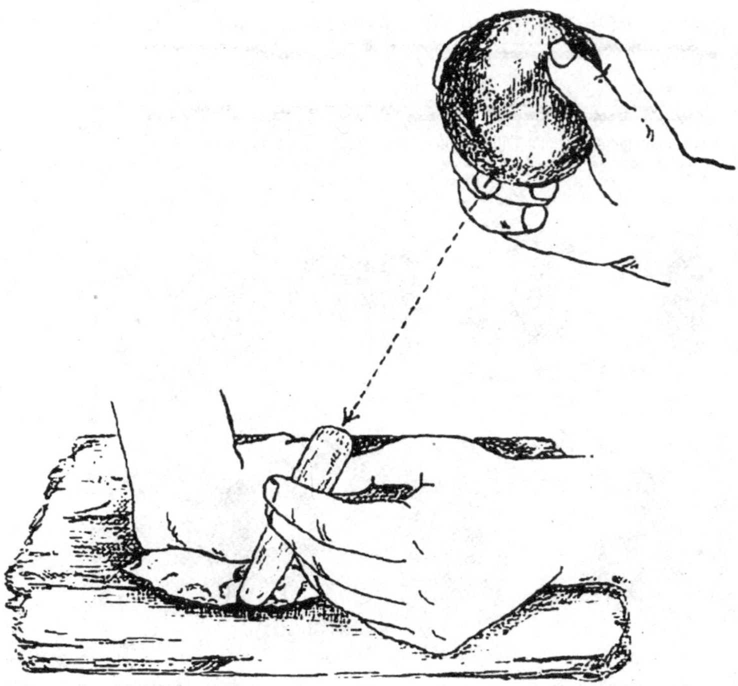

FIGURE 3.9. *An illustration of Mark Harrington's (1941) knapping process with punch and anvil.*

deerhorn and a hammerstone act as a pitching tool and a mallet. And, the method would be the same no matter what tribe of indians you visited."

In 1941, in an article for *Desert Magazine*, archaeologist Mark R. Harrington (1941, 11–13) wrote:

> Every Indian I have seen making arrowheads—and there are still some who know the art—employ the same methods . . . So the Indian had to rough it into shape by percussion with his hammerstone or with a deer antler mallet, knocking off large chips. Or, he tied his bone flaker to a stout long handle that would give him more leverage and power to press off longer and larger flakes. Another method was a combination of percussion and pressure, for which the services of a helper were needed. He held the flint against a block of wood while the workman set an antler cylinder or "pitching-tool" against the edge at the point where he wished to remove the flake, and struck the tool sharply with his hammerstone. Of course, as in all flint-work, the tool must touch the flint at just the proper angle, which can only be determined by experience.

There is an artistic rendition of Harrington's process provided (figure 3.9).

In 1943, in *Notes on the Culture of the Yana*, Sapir and Leslie Spier (1943, 268) wrote:

> The artisan fabricating stone arrowheads or larger blades had at least four pieces of equipment: a bone rod (bo'pdiyauna), 1 1/2 inches long, used as a flaker; a slim flat stone (bo'badjayauiaina) with which to strike the flaker against the flint; a deerhorn or bone chipper or notcher (do's, ityauna), about 2 inches long; and, to protect the palm, a small piece of buckskin (o'niyauna; buckskin, p'adata'igi). The supplementary notes with this statement describe two separate stages of manufacture: (1) the flint was chipped around roughly with the flaking rod and hammerstone; and (2) was then put on the buckskin pad in the palm while rough protuberances were pressed down upon and broken off with the horn chipper.

In *The Upper Stalo Indians of the Fraser Valley, British Columbia*, Wilson Duff (1953, 59) wrote: "To make a chipped 'flint' point, the blank was set on its edge in a groove in a piece of wood and shaped by indirect percussion with a pencil of nephrite and a hammer. One side shaped, it was turned over and the other side completed." He (60) went on to implicate the use of indirect percussion in the manufacture of slate projectile points, as well: "They were able to, however, give vague descriptions of ground-slate blades used prior to the advent of steel. These were said to have been about 3 inches long and 1 inch wide, and to have protruded almost 2 inches beyond the valves. Their exact shape was not made clear, but they were shaped by *indirect percussion*, using a 6-inch stone pencil and hammer, then ground on a whetstone" (emphasis ours).

During the 1950s, a group of intrepid explorers hiked for many days into Chiapas and filmed the lives of the modern-day Lacandon (Scherer et al. 2021). It was found that they still did not possess guns. In manufacturing their arrows it was found that they carried out a peculiar form of handheld blade manufacturing with tools that harken back to prehistoric Mayan culture (figure 3.10 from Clark 1991).

Various accounts of indirect percussion such as these, made by historical and ethnographic observers of Native American flintknappers, led many researchers to eventually conclude that they employed a triad of flintknapping methods, including various forms of direct percussion, indirect percussion, and pressure flaking. Those who ascribed to this view include Cushing (1879a, 1879b), Williams (1906), Holmes (1894), Gerard Fowke (1896), Otis Tufton Mason (1895), Mercer (1897), William Martin Beauchamp (1897), William James McKnight (1905), Félix Outes and Florentino Ameghino (2012 [1905]), Robert Macalister (1921), Arthur Caswell Parker (1922), Harrington (1941), Mak Seeman (1984), Shafer (1985, 2023), and others.

There is evidence indicating that indirect percussion was carried out in both bifacial and blade-core reduction using various surfaces of the punch. There is also evidence to suggest that indirect percussion processes were carried out in a

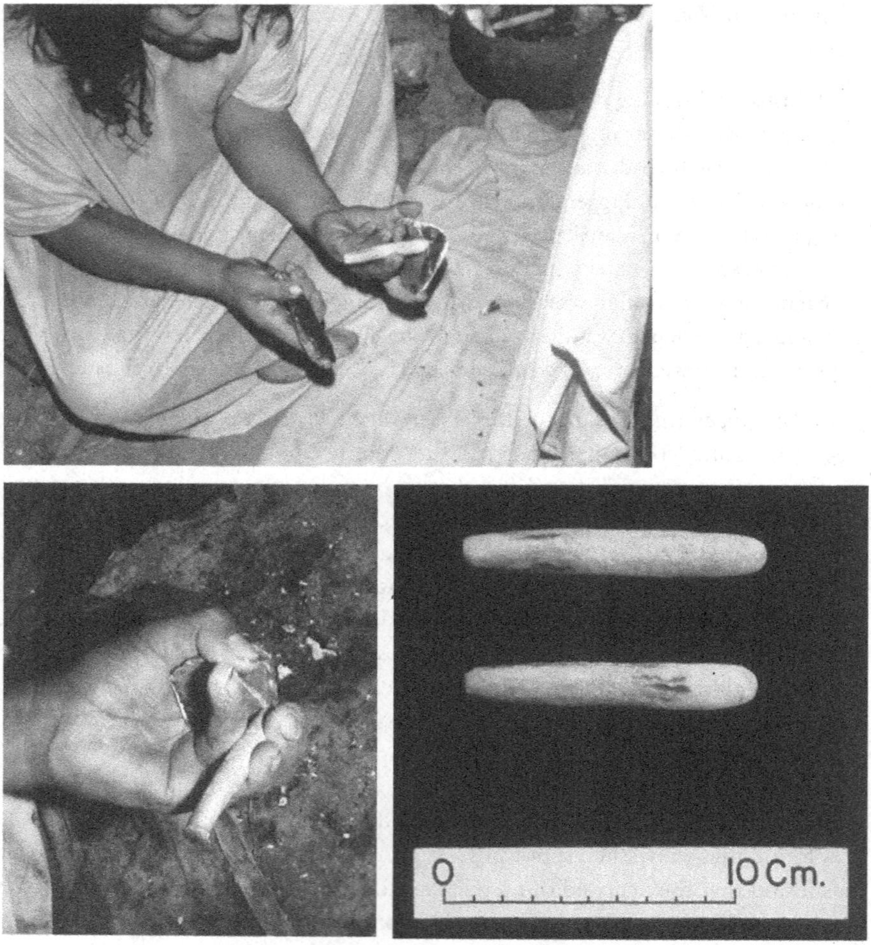

FIGURE 3.10. *Lacandon punch technique: (a) in process; (b) hand positions for punch, core, and stone mallet; (c) examples of the antler punches employed (photos by John Clark).*

vastly different scales, ranging from finger-held punches operated with one hand, to staff flaking punches, operated by two men. We also find evidence of a variety of punch tool materials, such as antler, ivory, stone, and possibly even wood. To add to this complexity, some modes of indirect percussion could generate fractures that resemble those of direct percussion.

From a mathematical perspective, if one takes the two basic forms of reduction (i.e., bifacial and blade-core reduction), multiplied by four or five possible scales of operation, multiplied by broadside striking versus butt-end striking, multiplied by a range of possibly three or four punch and mallet materials, then what becomes apparent is that there could be many different possible forms and applications of indirect percussion that can be thoughtfully created according to the basic principles of such knowledge. Further, of these dozens of hypothetical forms of indirect percussion, it is possible that some may be highly advantageous for reducing certain types of materials, and at certain stages of reduction. On account of all of these various factors, there could be an entire range of artifactual correlates still going unrecognized in the archaeological record simply for lack of knowledge, or an undue emphasis on the hypothetical billet technique developed by European researchers, such as Coutier, Bordes, and Leakey (see Knowles 1953).

SOME ARCHAEOLOGICAL ACCOUNTS FROM NORTH AMERICA

While previous nineteenth-century researchers, like Holmes, only had access to prior historical and ethnographic records, we now have a vast dataset that encompasses over a century's worth of archaeological data that spans the Historic period all the way back to the Paleoindian. The question that must rightfully be answered is whether there is evidence from this dataset to exonerate the views of past researchers like Holmes who reported evidence of the use of indirect percussion. We submit the following archaeological accounts that emphasize one particular tool known variously as antler drift, antler cylinder, cylindrical antler punch, flintknapping punch, and others, which were recognized by early North American archaeologists. The term *drift* was borrowed from the industrial machine shops of the day. A "drift pin" was a small steel rod or cylinder that was used in conjunction with a hammer to localize force. In a cursory survey, Eble collected over thirty explicit references to, or illustrations of, these antler or bone instruments. For the sake of space, we directly quote only a sample of these sources and provide references for the remainder.

We see the first allusions to the archaeologically noted presence of such tools in North America beginning in the end of the nineteenth century. In 1906, Williams (1906, 114) wrote:

> Stone chipping is now believed to be a lost art. The ethnologists of the Smithsonian Institute have never found an artisan who, even when supplied with all the tools of modern art, was able to imitate some of the leaf-shaped implements of prehistoric man. And the most skillful of the flint knappers of Brandon, England, men whose occupation is making gun flints[,] also failed after months of effort to produce the forms made by a savage whose only tools were stones and bones. It is not certainly how the Indian made these arrow points, working such a brittle material as white quartz into the exquisite forms . . . The flakes were gradually chipped down into shape with the little knockers. When the stone had thus been partly outlined, it was finished by another process. Either some hard object as stone, bone or horn was used as a chisel driven by a hammer to force off

little flakes from either side alternately, or the so-called flakers were used to push suddenly against the arrow, being worked from alternate sides, each impulsion of the tool taking off a little splinter opposite the point of impact. Various arrow flakers have been found among the surviving savages. The only tool resembling these from this section that we have seen . . . resembles the alleged bone flakers from the prehistoric cemetery of Madisonville, Ohio. We are unable to conceive no other use for the above implement.

In 1906, in writing *The Explorations of the Baum Prehistoric Village Site*, William Mills (1906, 65–66) penned a section in a chapter entitled "Implements used in Flaking Knives."

> The specimens shown in Fig. 53 were no doubt used in flaking the large flint knives so common in the village. The flaking tools were always made of deer and elk horn, and varied in length from one and one-half inches to four inches, and in diameter from one-half inch to three-fourths inch. One end is square, while the other end is oval, the longest point being in the middle of the specimen. The square end on almost all the specimens shows a splintered and battered condition, indicating that the implement had been struck with a heavy blow. The oval end also frequently shows a splintered condition, caused by improperly placing the tool against the flint, and striking the blow. The flaking tools are found everywhere in the village, especially in the sites of the teepees and the graves.

In 1917, in writing, "The Feurt Mounds and Village Sites," Mills (1917, 126–128) explained:

> The flaking tools found at the Feurt site were of two kinds, both being found in great numbers, and for the most part all were made of deer and elk horn. Fine examples of one kind are shown in Fig. 79. All are cylindrical in form and vary in length from one and one-half inches to five and one-half inches. One end is usually cut at right angles, while the other end is made oval. Many of the specimens in Fig. 79 show use as flaking tools, one end being usually battered, from being struck by the stone hammer in making the large flakes, and in the manufacture of the long knives and spear points.

Around the same time Alanson Skinner (1919, 99; Throggs Neck) wrote:

> A fragment of a cylinder made from a deer's horn was found. This is no doubt a piece of one of the common tools used by all the Eastern tribes for flaking arrowheads from suitable stone . . . They have been found by the writer in Erie graves at Ripley, Chautauqua county, New York, accompanying flint raw material and finished points, as though deposited in a bag with the other petty paraphernalia of the owner. Houghton has similarly obtained them from Seneca graves, and the writer found one with several hundred chips of flint, jasper, and quartz, and a perfect

> arrowhead, in a small deposit beneath the Inwood shellheap. They have likewise been recorded from all tidewater sites of the late prehistoric to the historic period.

Eble collected ten other references to similar antler tools from 1900 to 1920 (Heye 1915; Holmes 1919; Hooten and Willoughby 1920; Houghton 1909, 1916; Parker 1907; Skinner 1915, 1921; Smith 1907; Smith et al. 1910).

In the accounts of aboriginal flint chipping there are few references to the punch and hammer. It is apparent, however, that their use was widespread as a preliminary process to the final pressure flaking by the ordinary antler tools with special working ends. In the collections of the Illinois Museum there are examples of these punches from the village sites in other sections of Ohio, from the Iroquoian sites in New York, one good specimen from a Maine shell-heap, and a number from various places in the Mississippi Valley.

> The relatively large number of the short antler punches found in comparison with the long pressure flakers (the ratio being roughly about one hundred to one) would seem to indicate that the smaller punches may have been used in the final flaking of at least a considerable portion of the coarser chipped implements. Figures l, m, plate 6, represent examples of a certain group of stones, showing abrasions and other marks of use, which probably served in preparing sinew and perhaps as hammers for antler punches in arrow-making. Stones of like character were found with three arrow-making outfits in Iroquoian graves during the Peabody Museum explorations in western New York. (Hooten and Willoughby 1920, 49)

In 1929, in writing "Some Shell-Heaps in Nova Scotia," Harlan Ingersoll Smith and W. J. Wintemberg wrote: "The pin shaped piece of antler found in heap A (pl. 16, fig. 6) was brought to nearly a cylindrical form, with one end nearly flat and the *cells there compressed* from being POUNDED. Slivers were broken off at the edge. The other end is slightly smaller and so bluntly pointed as to be nearly dome shaped. It shows *longitudinal compression of the cells*, as if from being held against something while the other end was *pounded*" (emphasis ours). Eble collected twelve more references to indirect percussion from 1920 to 1940 (Collins 1928; Harrington 1922, 1924; Guernsey 1931; Houghton 1922; Langford 1927; Lily 1937; Parker 1920; Webb 1939; W. S. Webb and Funkhouser 1928; Webb and Haag 1939; Willoughby 1935; Wintemberg 1936).

In his 1942 article "An Archaeological Survey of the Pickwick Basin in the adjacent portions of the states of Alabama, Mississippi, and Tennessee," Webb (1942, 85) wrote:

> Burial No. 92.—This type-2a burial was only 1.4 feet below the surface in square 85L15. With this burial, shown in plate 83, figure 2, were 43 objects listed as field specimens. Among them were six flint points, six flint knives, three bone awls, a slate whetstone, an antler spear point, and a *series of antler drifts in various sizes*.

> Two of these were drilled horizontally in such a way as to suggest they were used as shaft straighteners. This collection of tools might at once suggest that this man in life had been specially skilled in the working of flint. Some of these flint points with this burial were especially well made, and are shown in plate 92, figure 2. These associations, made of bone and antler, are shown in plate 97. (emphasis ours)

Later, when writing "The Adena People," Webb (Webb and Snow 1945, 169) pondered cut antler segments: "This trait is fairly frequent in Adena villages. Sections of antler some three inches long seem to have been used as drifts. The rounded ends show battering as if they were interposed in percussion between the hammer and the object struck."

In 1946, the State of Illinois Museum displayed a flintknapping technology exhibition showing the triad of flaking practices, known from archaeological study (figure 3.11). In the same year Frederick Johnson (1946, 135) wrote *Man in Northeastern North America*, stating:

> Dorset tools for working chert include the bone pressure flaker and the bone or antler cylindrical piece used for indirect percussion. As far as I know, the latter type is not found in other Eskimo cultures, and must, therefore, have been derived from the Indians. In the northern Dorset area, primary flaking was done with a hafted hammer of bone or antler; in Newfoundland, cylindrical hammerstones were used, some of which were also hafted. I believe that the production of large stone blades, a characteristic of southern Dorset sites, is related to the use of the hammerstone. Any comparison of Dorset and Indian chipped implements should include a consideration of the methods by which they were shaped—by direct or indirect percussion, or by pressure flaking. Although I have examined many Eskimo and Indian stone specimens, unfortunately I did not do it with this question in mind, and furthermore I do not feel qualified to distinguish the results of the different techniques. I have not been assisted by the habit of so many archaeologists of lumping together under the same vague term, "flaker, punch, knapping tool," etc., quite different types of bone tools, without specifying whether the tool was simply pressed against the stone, or whether its butt end was struck with a hammer. I can only compare gross features of Dorset stone types with those of the Indians, conscious that this is inadequate.

Johnson goes on to list eight citations to back up the claim that the antler cylinder was used as a tool of indirect percussion.

In 1949, in writing *An Archaeological Survey of the Trent Waterway in Ontario*, Ritchie describes: "Antler tools presumably utilized in indirect percussion flint chipping, apparently the prevailing method among the western Iroquoian tribes, who also employed more stone projectile points than the Onondaga, Oneida and Mohawk, and extending back, with typological variations into Laurentian

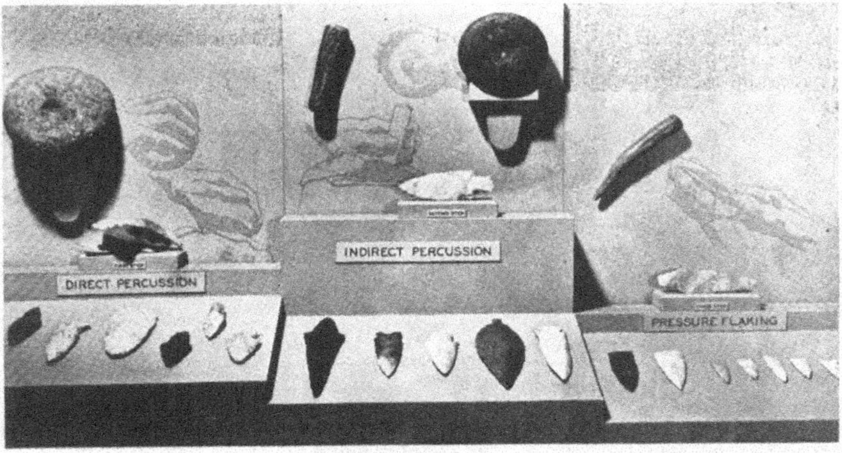

FIGURE 3.11. *Display at the Illinois State Museum showing three phases of knapping.*

times in the same general area, were present in the village debris (total five) and in the graves."

In 1962, in writing "Wedding of The Waters Cave, 48 Ho 301, A Stratified Site in the Big Horn Basin of Northern Wyoming," George Frison stated: "A section of deer antler made from a cylinder of antler, split lengthwise by cutting into the cancellous part of the antler from opposite sides, may have been some sort of punch. One of these half cylinders was then rounded on both ends. One end has no cancellous bone, being the base of the antler where it joined the skull; and this end has a number of indentations which had fragments of flint and chert embedded in them (figure 4.1)." Eble collected nine more accounts from 1938 to 1975 (Fowler 1951; Frison 1968; J. Griffen 1974; W. Griffen 1969; Mayer-Oakes 1955; Ritchie 1938, 1959, 1961, 1969; Webb 1946; Webb and Snow 1945; Wedel 1943; Wood and Pangborn 1968).

Phil Geib (2004) analyzed a set of prehistoric sheep horn punches and bifaces, which were recovered in an animal skin pouch from a Basketmaker II site. Geib speculated that the sheep horn punches were used in indirect percussion to fashion the bifaces (figure 3.12), given that the flake scars are relatively large starting from the initiation. He goes on to further point out that while the Basketmaker II bifaces appear to be made via indirect percussion, older Archaic bifaces appear to have been made by direct percussion with deer tines. To test his theory regarding the use of the sheep horn punches in indirect percussion, Geib carried out an experimental study, involving replicating one side of a biface with each technology. Also, Geib carried out a study with a scanning microscope, and shows that there are indeed embedded silicate particles found in the ends of the punch tools.

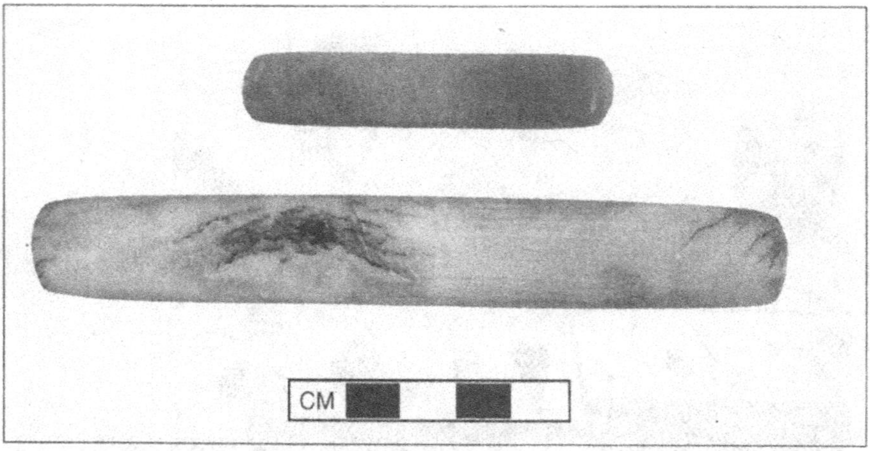

FIGURE 3.12. *Sheep horn punches described by Phil Geib (2004).*

Most recently, David Hally (2008) notes:

> Polhemus (1987, Fig. 11.19, personal communication 1999) identifies antler cylinders as flaking tools on the basis of the existence of one specimen from Toqua with a chert flake embedded in its end and the occurrence of wear facets on the rounded ends of some other specimens. Polhemus also argues that the different lengths of antler cylinders at Toqua reflect differences in tool use life—shorter specimens having been worn down and reground more frequently. Illustrated specimens (Polhemus 1987, figure 11.19) vary between 10 mm and 180 mm in length. In support of Polhemus's identification is an antler cylinder from the Little Egypt site that had a faceted end and was physically associated with a large concentration of flint debitage in Structure 3 (Hally 1980, 354–355). Also supportive is the physical location of antler cylinders adjacent to flintknapper kits in two King site burials.

This array of evidence deserves some discussion. First, there is a wide range of terminology that was coined and employed by archaeologists referring to this particular tool type (see quotations), especially in instances when it was formally recognized as a potential flintknapping tool of indirect percussion. By the 1960s, some researchers realized that many past scholars had been independently describing essentially the same tool with a plethora of different, unstandardized names. In instances where associations with flintknapping were not noted, the tool was sometimes referred to as a "rubbing tool," "game piece," "hair roller," "object of unknown use," "antler plug," and others. Sometimes it is no longer possible to know whether it is indeed the same type of artifact that is being referred to, or even whether the characteristics of the artifact conform to other similar examples known to the archaeological record. On the other hand, when the artifacts are well documented with photographs, then the trends become much more apparent, in spite of the diversity of terminology that was employed.

It is important to note that the antler drift phenomena has been documented for the Early Archaic to the Late Prehistoric era, and into the Proto-Historic, with one possible Paleoindian cylinder from Blackwater Draw (Haynes 1991). While many projectile point types changed, this particular flintknapping tool appears to have been consistently used throughout every millennium right into the Proto-Historic era. However, in terms of the geographic distribution of cylindrically shaped antler drifts in North America, the tradition of employing antler cylinders seems to be mostly confined to the chert-bearing areas in the eastern portion of the continent. In the western portion of the continent, there is evidence that very similar punches, such as those used by the Basketmaker Culture and by Puebloan cultures, were not cylindrical in shape but, rather, lozenge shaped, or round ended, yet were used in a similar manner in bifacial reduction.

It is also noteworthy that the sizes and shapes of the tools are not universally uniform according to ethnographic-ethnohistoric and archaeological evidence of indirect percussion flakers. For example, in eastern North America we find long-standing evidence of "cylinder" style punches being employed, yet in the western portion of the continent we find evidence of both tine flakers and round-ended punches used as tools of indirect percussion. On account of these discrepancies it is important to take the evidence as a whole, extrapolate the overall evidence of indirect percussion use, and then use an interpretation of the overall evidence to interpret newer evidence. This way, even if there are minor discrepancies between tools and processes in different lines of evidence, it still should be possible to recognize new evidence of indirect percussion based upon an overall interpretation.

Importantly, indirect percussion is not fully technologically distinct from direct percussion, based on flake morphology. Many of the flake attributes that result from direct percussion can also be seen in some indirect percussion flakes. For example, the bulb of percussion is frequently created by indirect percussion fracture. Only, the bulb may be more diffuse or almost negligible. Further, among the ethno-historical records, there is some evidence for the stabilization of the mass, or core, before direct percussion, and sometimes even pressure flaking, was carried out. For example, some accounts note that stone was rested upon padded anvils while the hammerstone process was being carried out. To be consistent, the fundamental question is whether the same process was also used for indirect percussion, and whether there is archaeological evidence to support it. We argue that all evidence should be considered when trying to reconstruct ancient technologies and *chaînes opératoires*.

CONCLUDING THOUGHTS AND A RETURN TO COLHA

So far, our study has documented some of the historical and ethnographic evidence for the use of indirect percussion chipped-stone technology in the Americas. We have demonstrated how such evidence resulted in a number of early researchers positively concluding that indirect percussion was one facet of a cultural triad of flaking methods, which also included direct percussion and pressure flaking. This triad was understood to cover the entire range of chipping and flaking, from roughing out the stone, to creating a finer surface finish, to a final regularization of the edge of the stone via pressure flaking. Also, we have demonstrated that in their wisdom, aboriginal Native American knappers recognized that not all stone is "equal" and various processes could have been thoughtfully employed at various stages, depending on either the shape or the inherent nature of the stone being worked. Such thinking is superior to a "one-method-fits-all approach," since it takes into account variations of the nature and shape of the stone itself and not just angle and location of platform. Also, we have shown that, by honest admission, the evidence of indirect percussion still was not well understood at the end of the nineteenth century (Holmes 1894). We have further demonstrated how a subsequent generation of twentieth-century archaeologists similarly recognized evidence of indirect percussion in the archaeological record, including evidence from both the prehistoric and Proto-Historic times. Armed with this information and these interpretations, we turn our attention to Colha, and the chert-bearing regions of Belize, which in some respects bear similarity to the chert-bearing regions of centarl and eastern North America.

The prehistoric flintknapping tools arguably most similar to the chipping tools known from Colha are the basal antler chipping tools recovered from Mill Creek quarries of the Ohio Valley. At Mill Creek quarries prehistoric knappers obtained large, flattish chert nodules that were fashioned into chipped bifaces, hoes, and celts. These end products were trades throughout the region, even to what are nearby modern states. Holmes (1894, figs. 75 and 76) offers two possible interpretations of the use of the antler chipping tools: use as a direct hammer in the hand, and as a hammer hafted to a hard handle like a modern hammer. We present three other possible interpretations regarding the use of the antler tools from Mill Creek.

In accordance with the data presented in this chapter, the first possibility is that the tools could have been "punches" used to create indirect percussion fractures. Another possible interpretation is that the tools could have been a type of "sling mallet" or sling hammer that was held with a leather thong rather than a formal handle. Such a tool could have been swung simply while being held with a "strap handle"—a piece of leather held around the center. In fact, we believe that this type of direct percussion was an incredibly important part of the Native American toolkit, but these arguments must be reserved for a separate publication. A third possible interpretation for the use of the tool involves a very advanced hybridization of both indirect percussion and direct percussion in a single process. In this case, the chipping tool would have been used much like a traditional punch, but the tool actually could have been held some distance away from the platform, and the sharp blow to the antler tool would cause the antler tool to collide with the platform at a very high rate of speed.

Was indirect percussion employed at Colha? According to Shafer, it was and the tools are going unrecognized, not just at Colha but also in other places like Texas. Shafer (2023) also amended his own previous reports, after carrying out a microscopic study of the antler chipping tools used at Colha, and concluded that the tools had been used in indirect percussion. George Herbert Michaels (1987) noted the following antler tools associated with evidence of flintknapping. The antlers were cut from the basal section of whitetail deer antler, as well as intersecting portions of the antler. The tools are on average 7.59 cm in length with a mean variation of 1 cm, and an average width of 2.11 cm. The tools are described as being "slightly curved, and cylindrical." Michaels also notes a lack of antler tine pressure flakers (see figure 3.2 for examples).

In 2002, Palma Jean Buttles also wrote of the punches recovered at the Colha site, used by prehistoric Mayan knappers:

> The antler punch is a soft hammer indirect percussion instrument utilized in the production of stone tools (figure 9.3; Hester and Shafer 1979, 1991a; Michaels 1994; Shafer 1979). At Colha, hard hammer percussion prevailed during the Preclassic and Late Classic eventually being replaced by soft hammer technique in the Postclassic (Hester 1982; Hester and Shafer 1991; Shafer 1979). Their use during the Postclassic at Colha is confirmed by the presence of lipped biface thinning flakes (Shafer 1979:52). Within this form category are found two subforms based predominantly on the section of antler it was produced from. (see figure 3.2 for examples)

Shafer (1979, 2023) documented punches used at Colha (see figure 3.2), which he compared to those found in Texas. Clearly, like other aboriginal American flintknappers in prehistoric North America, the Maya of Colha were employing indirect percussion. Whether indirect percussion existed outside of Colha is a different issue. Most projects in the Maya area have specialists that focus either on the lithic artifacts or the faunal remains, and this division of labor actually results in a lack of communication. While Hruby (2006) notes that cut tine flakers are present at most of his study sites (see also Coe 1959, fig. 57f for an example of one used for indirect percussion), these objects have not been studied in a technological manner. This issue must be resolved in future studies.

In very recent history, indirect percussion has gained some foothold in the literature. In fact, many of the scholars in this volume (chapters 2, 4, and 5; Clark 1989; Clark et al. 2012; Titmus and Woods 2003) have argued now for years that indirect percussion was indispensable in creating Mesoamerica's most elaborate bifaces and effigies. Combinations of notching by pressure and indirect percussion were the mainstays of this art. We argue, however, that indirect percussion was not reserved just for the fancy stuff. Instead, a wide array of indirect percussion techniques were a common, and even essential, aspect of chipped-stone production going back to the initial peopling of the Americas. It seems likely, then, that indirect percussion was not something controlled only by elite knappers who shared the "eccentric" traditions across Mesoamerica; rather, the traditions were already common to stoneworkers across the continent. For us, and especially for the lead author, who has been working on this thesis for many years, indirect percussion is a deep-rooted American tradition that was part of the core aspects of chipped-stone production in North, South, and Mesoamerica. In other words, this is not a novel "trick move" used in rare cases, or esoterica for vocational knappers to have fun with, but rather, in our view, one of the main forms of cultural transmission in the New World.

Although we have spent a great deal of time here on North America, and accounts from outside of Mesoamerica, we did so because we see Mesoamerica sharing in the same Pan-American tradition. It is important to revisit John Clark's (1991) breakthrough ethnoarchaeological account of Lacandon flake production here. Instead of a strange post-Conquest application of indirect percussion, disconnected from earlier modes of chipped-stone production, we understand Lacandon techniques as a continuation in *practice* of one of the most common methods in the New World. Interestingly, Clark notes that the Lacandones would chant during the production process, as is described for Aztec blade makers (Clark 1989; Hruby 2007; Hruby and Rich 2014), as well as noted earlier in some of the North American accounts (see Catlin). Thus, we should understand chipped-stone production not necessarily in terms of Western efficiency but rather as deeply rooted and culturally situated practices that may have been ritualized for both religious and functional mnemonic reason. Now that we have provided the historical, ethnographic, and archaeological foundations for understanding indirect percussion, the next stage in this endeavor is to describe our replication experiments of these techniques to provide correlates and analogies for better interpreting the archaeological record.

REFERENCES

Adney, Edwin Tappan. 1964. *The Bark Canoes and Skin Boats of North America*. Washington, DC: US Government Printing Office.

Aoyama, Kazuo. 1999. *Ancient Maya State, Urbanism, Exchange, and Craft Specialization: Chipped Stone Evidence of the Copán Valley and the La Entrada Region, Honduras*. Memoirs in Latin American Archaeology 12. Pittsburgh, PA: University of Pittsburgh Press.

Aoyama, Kazuo. 2001. "Classic Maya State, Urbanism, and Exchange: Chipped Stone Evidence from the Copan Valley and Its Hinterlands." *American Anthropologist* 103:346–360.

Barnes, Alfred S. 1943. "Letter to the Editor: High-Angle Edge Flaking of Flint." *Nature* 152 (October 23):477.

Beauchamp, William Martin. 1897. *Aboriginal Chipped Stone Implements of New York*. Bulletin of the New York State Museum 4(16). Albany: University of the State of New York.

Beckwith, Thomas. 1911. *The Indian, Or Mound Builder: The Indians, Mode of Living, Manners, Customs, Dress, Ornaments, etc., Before the White Man Came to the Country, Together with a List of Relics Gathered by the Author. Geology, Ethnology and Archæology of this Country and the Pacific Tribes Treated to a Limited Extent*. Cape Girardeau, MO: Naeter Brothers Publishers.

Bonney, A. F. 1918. "The Stone Age in America." *Scientific American Supplement* 2213:340.

Buttles, Palma Jean. 2002. "Material and Meaning: A Contextual Examination of Select Portable Material Culture from Colha, Belize." PhD diss., University of Texas, Austin.

Catlin, George. 1868. *Last Rambles Amongst the Indians of the Rocky Mountains and the Andes*. London: Sampson Low, Marston and Company.

Clark, John E. 1989. "Obsidian: The Primary Mesoamerican Sources." In *La obsidiana en Mesoamérica*, edited by Margarita Gaxiola G. and John E. Clark, 299–319. Mexico City: Instituto Nacional de Antropología e Historia.

Clark, John E. 1991. "Modern Lacandon Lithic Technology and Blade Workshops." In *Maya Stone Tools: Selected Papers from the Second Maya Lithic Conference*, edited by Thomas Hester and Harry J. Shafer, 251–265. Madison, WI: Prehistory Press.

Clark, John E., Fred W. Nelson, and Gene L. Titmus. 2012. "Flint Effigy Eccentrics." In *Ancient Maya Art at Dumbarton Oaks*, edited by Joanne Pillsbury, Mariam Doutriaux, Reiko Ishihara-Brito, and Alexandre Tokovinine, 273–287. Pre-Columbian Art at Dumbarton Oaks (4). Washington, DC: Dumbarton Oaks.

Coe, William R. 1959. *Piedras Negras Archaeology: Artifacts, Caches, and Burials*. Museum Monographs. Philadelphia: University Museum, University of Pennsylvania.

Collins, W. H. 1928. "National Museum of Canada." In *National Museum of Canada Annual Report for 1926*, Bulletin (50):32–70. Ottawa: King's Printer.

Columbia Historical Exposition. 1895. Commission of the United States for the Columbian Historical Exposition at Madrid. Report with Special Papers. Washington, DC: Government Printing Office.

Curtis, Edward S. 1911. *The North American Indian: Being a Series of Volumes Picturing and Describing the Indians of the United States, and Alaska (Piegan, Cheyenne, Arapaho)*. Vol. 6. Cambridge, MA: University Press.

Cushing, Frank Hamilton. 1879a. "Curious Discoveries in Regard to the Manner of Making Flint Implements by the Aborigines and Prehistoric Inhabitants of America." *Scientific American* 41:57.

Cushing, Franklin Hamilton. 1879b. "Curious Discoveries in Regard to the Manner of Making Flint Implements by the Aborigines and Prehistoric Inhabitants of America." *Engineering and Mining Journal* 28(6):91.

Deming, Therese O. 1932. *Red People of the Wooded Country: A Story of Indian Life*. New York: Laidlaw Brothers.

Dockall, J. E., and H. D. Dockall. 1999. "Knapping Implements in Mortuary Context: A Case from Morhiss (41VT1) with Comparative Data from Texas and the Midcontinent." *Bulletin of the Texas Archeological Society* 70:547–562.

Duff, Wilson. 1953. *The Upper Stalo Indians of the Fraser Valley, British Columbia*. Anthropology in British Columbia, Memoir (1). Victoria: British Columbia Provincial Museum, Department of Education.

Ellis, H. Holmes. 1957. *Flint-Working Techniques of the American Indians: An Experimental Study*. Columbus: Ohio Historical Society.

Evans, John. 1872. *Ancient Stone Implements, Weapons, and Ornaments, of Great Britain*. London, England: J. S. Virtue and Co.

Fowke, Gerard. 1896. *Stone Art Stone Art: Thirteenth Annual Report of the Bureau of American Ethnology to the Secretary of the Smithsonian Institution, 1891–1892*, 47–178. Washington, DC: Government Printing Office.

Fowler, Melvin L. 1951. "The Ethel R. Wilson Site: A Preliminary Report." *Journal of the Illinois State Archaeological Society* 1(3):90–94.

Frison, George C. 1962. "Wedding of the Waters Cave, 48 HO 301: A Stratified Site in the Big Horn Basin of Northern Wyoming." *Plains Anthropologist* 7(18):246–265.

Frison, George C. 1968. "A Functional Analysis of Certain Chipped Stone Tools." *American Antiquity* 33(2):149–150.

Geib, Phil. 2004. "AMS Dating of a Basket Maker II's Hunter's Bag (Cache 1) from Sand Dune Cave, Utah." *Kiva* 69(3):271–282.

Griffen, John W. 1974. *Investigations of Russel Cave: Russel Cave National Monument, Alabama*. Publications in Archaeology (13). Washington, DC: National Park Service, Department of the Interior.

Griffen, William B. 1969. *Culture Change and Shifting Populations in Central Northern Mexico*. Anthropological Papers of the University of Arizona 13. Tucson: University of Arizona Press.

Grinnell, George Byrd. 1895. *The Story of the Indian*. D. Appleton & Company: New York.

Grinnell, George Byrd. 1923. *The Cheyenne Indians: Their History and Ways of Life*. Vol. 1. New Haven, CT: Yale University Press.

Grinnell, George Byrd. 2008 (1872). *The Cheyenne Indians: Their History and Lifeways.* Bloomington, IN: World Wisdom.

Guernsey, Samuel James. 1931. *Explorations in Northeastern Arizona: Report on the Archaeological Fieldwork of 1920–1923.* Papers of the Peabody Museum of American Archaeology and Ethnology, Harvard University 12(1). Cambridge, MA.

Hall, G. D. 2002. *Archaeology at the Crestmont Site.* General Investigations Report 1. Lubbock: Archaeology Laboratory, Texas Tech University.

Hally, David J. 1980. "Archaeological Investigation of the Little Egypt Site (9MU102) Murray County, Georgia, 1970–1972 Seasons." Report submitted to the National Park Service, United States Department of the Interior.

Hally, David J. 2008. *King: The Social Archaeology of a Late Mississippian Town in Northwestern Georgia.* Tuscaloosa: University of Alabama Press.

Harrington, Mark R. 1922. *Cherokee and Earlier Remains on Upper Tennessee River.* New York: Museum of the American Indian, Heye Foundation.

Harrington, Mark R. 1924. *An Ancient Village Site of the Shinnecock Indians.* Anthropological Papers of the American Museum of Natural History, vol. 22, pt. 5. New York: The American Museum of Natural History.

Harrington, Mark R. 1941. "There's No Mystery About Arrowheads." *Desert Magazine* (December):10–13.

Haynes, C. V., Jr. 1991. "Geoarchaeological and Paleohydrological Evidence for a Clovis Age Drought in North America and Its Bearing on Extinction." *Quaternary Research* 35:438–450.

Heizer, Robert F. 1955. *California Indian Linguistic Records: The Mission Indian Vocabularies of H. W. Henshaw.* Anthropological Records 15(2). Berkeley and Los Angeles: University of California Press.

Hester, Thomas R. 1985. "The Maya Lithic Sequence in Northern Belize." In *Stone Tool Analysis: Essays in Honor of Don E. Crabtree*, edited by M. G. Plew, J. C. Woods, and M. G. Pavesic, 187–210. Albuquerque: University of New Mexico.

Heye, George G. 1915. *Contributions from Museum of the American Indian.* Vols. 2–3. New York: Heye Foundation.

Holmes, William H. 1894. "Natural History of Flaked Stone Implements." In *Memoirs of the International Congress of Anthropology*, edited by C. Staniland Wake, 120–139. Chicago: Schulte Publishing.

Holmes, William H. 1919. *Handbook of Aboriginal American Antiquities. Part I, The Lithic Industries, Bulletin 60.* Washington, DC: Smithsonian Institution Bureau of American Ethnology.

Hooten, Ernest A., and Charles C. Willoughby. 1920. "Indian Village Site and Cemetery Near Madisonville, Ohio." *Papers of the Peabody Museum of American Archaeology and Ethnology* 8(1). Cambridge, MA: Harvard University.

Houghton, Fredrick. 1909. *Indian Village Camp and Burial Sites on the Niagara Frontier.* Bulletin of the Buffalo Society of Natural Sciences 9(2). Buffalo, New York.

Houghton, Fredrick. 1916. "The Characteristics of Iroquoian Village Sites of Western New York." *American Anthropologist* 18(4):508–520.

Houghton, Fredrick. 1922. *The Archaeology of Genesee County.* Researches and Transactions of the New York State Archaeological Association 3(2). Rochester: Louis H. Morgen Chapter.

Hruby, Zachary X. 2006. "The Organization of Chipped-Stone Economies at Piedras Negras, Guatemala." PhD diss., University of California, Riverside.

Hruby, Zachary X. 2007. "Ritualized Chipped-Stone Production at Piedras Negras, Guatemala." In *Rethinking Craft Specialization in Complex Societies: Archaeological Analysis of the Social Meaning of Production*, edited by Zachary Hruby and Rowan K. Flad, 68–87. Archaeological Papers of the American Anthropological Association 17. Arlington, VA: American Anthropological Association.

Hruby, Zachary X., and Michelle Rich. 2014. "Flint for the Dead: Ritual Deposition of Production Debitage from El Perú-Waka', Burial 39." In *Archaeology at El Perú-Waka': Ancient Maya Performances of Ritual, Memory, and Power*, edited by O. C. Navarro-Far and M. Rich. Tucson: University of Arizona Press.

Hudson, Travis, and Thomas C. Blackburn. 1982. *The Material Culture of the Chumash Interaction Sphere: Food Procurement and Transportation.* Menlo Park, CA: Ballena Press.

Johnson, Frederick. 1946. *Man in Northeastern North America.* Papers of the Robert S. Peabody Foundation for Archaeology 3. Andover, MA: Phillips Academy Foundation.

Jones, Paul A. 1937. *Coronado and Quivira.* Lyons, KS: Lyons Publishing Company.

King, Eleanor, and Daniel Potter. 1994. "Small Sites in Prehistoric Maya Socioeconomic Organization: A Perspective from Colha, Belize." In *Archaeological Views from the Countryside: Village Communities in Early Complex Societies*, edited by G. Schwartz and S. Falconer, 64–90. Washington, DC: Smithsonian Press.

Knowles, Sir Francis H. S. 1953. *Stone-Worker's Progress: A Study of Stone Implements in the Pitts River Museum.* Occasional Paper on Technology 6(63). Great Britain: Oxford University Press.

Lacaille, A. D. 1954. "Stone Age Tools." *Transactions of the Glasgow Archaeological Society*, n.s., 13:17–32.

Langford, George. 1927. "The Fisher Mound Group, Successive Aboriginal Occupations near the Mouth of the Illinois River." *American Anthropologist* 29(3):153.

Lily, Eli. 1937. *Prehistoric Antiquities of Indiana: A Description of the More Notable Earthworks, Mounds, Implements and Ceremonial Objects Left in Indiana by Our Predecessors, Together with Some Information as to Their Origin and Antiquity, and the Prehistory of Indiana.* Indianapolis: Indiana Historical Society.

Mason, Otis Tufton. 1886. "The Ray Collection from the Hupa Reservation." In *Annual Report of the Board of Regents of the Smithsonian Institution, Showing the Operations, Expenditures, and Condition of the Institution*. Washington, DC: Smithsonian Institution.

Mason, Otis Tufton. 1895. *The Origins of Invention*. London: Walter Scott.

Mayer-Oakes, Willam J. 1955. *Excavations at the Globe Hill Shell Heap (46Hk34-1): Hancock County, West Virginia*. Moundsville: West Virginia Archaeological Society.

Macalister, Robert Alexander Stewart. 1921. *A Text-Book of European Archaeology*. Vol. 1, *The Palaeolithic Period*. Cambridge: Cambridge University Press.

McKnight, William James. 1905. *A Pioneer Outline History of Northwestern Pennsylvania, Embracing the Counties of Tioga, Potter, McKean, Warren, Crawford, Venango, Forest, Clarion, Elk, Jefferson, Cameron, Butler, Lawrence, and Mercer: Also a Pioneer Sketch of the Cities of Allegheny, Beaver, Du Bois, and Towanda: My First Recollections of Brookville, Pennsylvania, 1840–1843, When My Feet Were Bare and My Cheeks Were Brown*. Philadelphia: J. B. Lippincott Company.

Mercer, Henry C. 1897. "Researches upon the Antiquity of Man in the Delaware Valley and the Eastern United States." *Nature* 55:459.

Michaels, George Hubert. 1987. "A Description and Analysis of Early Postclassic Lithic Technology at Colha, Belize." MA thesis, Texas A&M University.

Mills, William C. 1906. *The Explorations of the Baum Prehistoric Village Site*. Columbus, OH: F. J. Heer.

Mills, William C. 1917. "The Feurt Mounds and Village Sites." In *Ohio Archaeological and Historical Publications* 26:305–449. Columbus: Museum Illinois.

Mortillet, Gabriel de. 1866. *Matériaux pour l'histoire positive et philosophique de l'homme, bulletin des travaux et découvertes concernant l'anthropologie, es temps anté-historiques, l'époque quaternaire, les questions de l'espèce et de la génération spontanée*. Paris.

Nicholson, Grace. 1916. *Grace Nicholson Papers and Addenda*. San Marino, CA: Huntington Library.

O'Callaghan, P. 1864. *Report of the Proceedings of the Geological and Polytechnic Society of the West-Riding of Yorkshire*. Vol. 4.

Olsen, Larry Dean. 1997. *Outdoor Survival Skills*. Chicago: Chicago Review Press.

Outes, Félix F., and Florentino Ameghino. 2012 (1905). *La faceta articular inferior única del astrágalo de algunos mamíferos no es un carácter primitivo*. Charleston, SC: Nabu Press.

Parker, Arthur Caswell. 1907. *An Erie Indian Village and Burial Site at Ripley, Chautauqua County, New York: Being the Record of the State Museum Archaeological Expedition of 1906*. Albany: New York State Education Department.

Parker, Arthur Caswell. 1917. "How Flint Arrowheads Are Made." *American Indian Magazine* 5:160–165.

Parker, Arthur Caswell. 1920. "The Archaeological History of New York." *New York State Museum Bulletins* 235, 236, 237, 238 (July–October)[2 vols.]

Parker, Arthur Caswell. 1922. *The Archeological History of New York*. Albany: University of the State of New York.

Polhemus, Richard. 1987. *The Toqua Site: A Late Mississippian Phase Town*. University of Tennessee, Department of Anthropology, Report of Investigations (41), Tennessee Valley Authority, Publications in Anthropology (44). Knoxville: Tennessee Valley Authority.

Potter, Daniel, and Eleanor King. 1995. "A Heterarchical Approach to Lowland Maya Socioeconomies." In *Heterarchy and the Analysis of Complex Societies*, edited by R. M. Ehrenreich, C. L. Crumley, and J. E. Levy, 17–32. Archaeological Papers of the American Anthropological Association 6(1). Arlington, VA: American Anthropological Association.

Putnam, Fedrick. 1896. *Proceedings of the American Association for the Advancement of Science* 45. Massachusetts: The Salem Press.

Rau, Charles. 1873. "Indian Netsinkers and Hammerstones." *American Naturalist* 7(3):139–146.

Ritchie, William Augustus. 1938. "Certain Recently Explored New York Mounds and Their Probable Relation to the Hopewell Culture." *Research Records* (4):1–42. Rochester, NY: Rochester Museum of Arts and Sciences.

Ritchie, William Augustus. 1949. *An Archaeological Survey of the Trent Waterway in Ontario, Canada: And Its Significance for New York State Prehistory*. Rochester, NY: Lewis Henry Morgan Chapter.

Ritchie, William Augustus. 1959. "The Stony Brook Site and Its Relation to Archaic and Transitional Cultures on Long Island." *New York State Museum and Science Service, Bulletin 372*. Albany: University of the State of New York, the State Education Department.

Ritchie, William Augustus. 1961. "A Typology and Nomenclature for New York Projectile Points." *New York State Museum and Science Service, Bulletin 384*. Albany: University of the State of New York, the State Education Department.

Ritchie, William Augustus. 1969. *The Archaeology of New York State*. 2nd rev. ed. Garden City, NY: Natural History Press.

Sapir, Edward. 1911. *Yana Texts*. Cambridge, MA: University Press.

Sapir, Edward, and Leslie Spier. 1943. *Notes on the Culture of the Yana*. University of California Anthropological Records 3(3). Berkeley: University of California Press.

Scherer, Andrew, Charles Golden, and Stephen Houston. 2021. "Lamb's Journey and the 'Lost City.'" In *A Maya Universe in Stone*, edited by Stephen Houston, 11–36. Los Angeles: The Getty Research Institute.

Seeman, Mark F. 1984. "Craft Specialization and Tool Kit Structure: A Systemic Perspective on the Midcontinental Flint Knapper." In *Lithic Resource Procurement: Proceedings from the Second Conference on Prehistoric Chert Exploitation*, edited by Susan C. Vehik,

7–36. Occasional Paper, Center for Archaeological Investigations, Southern Illinois University at Carbondale 4. Carbondale: Southern Illinois University.

Sellers, George Escol. 1886. *Observations on Stone-Chipping, United States Congressional Serial Set*. Vol. 2431. Washington, DC: Smithsonian Institution.

Shafer, Harry J. 1979. "A Technological Study of Two Maya Workshops at Colha, Belize." In *The Colha Project, 1979: A Collection of Interim Papers*, edited by Thomas R. Hester, 22–78. San Antonio: Center for Archaeological Research, University of Texas at San Antonio.

Shafer, Harry J. 1985. "A Technological Study of Two Maya Workshops at Colha, Belize." In *Stone Tool Analysis: Essays in Honor of Don E. Crabtree*, edited by M. G. Plew, J. C. Woods, and M. G. Pavesic, 277–316. Albuquerque: University of New Mexico Press.

Shafer, Harry J. 1986. *Ancient Texans: Rock Art and Lifeways Along the Lower Pecos*. Austin: Texas Monthly Press.

Shafer, Harry J. 2023. "Legacy of the Ancient Maya Tranchet Flake: Or How Two Texas Archaeologists Ended Up in the Maya World." *Journal of Texas Archaeology and History* 7(1):1–24.

Shafer, Harry J., and Thomas R. Hester. 1983. "Ancient Maya Chert Workshops in Northern Belize, Central America." *American Antiquity* 48(3):519–543.

Shafer, Harry J., and Thomas R. Hester. 1991. "Lithic Craft Specialization and Product Distribution at the Maya Site of Colha, Belize." *World Archaeology* 23:79–97.

Skertchly, Sydney B. J. 1879. *On The Manufacture of Gun-Flints, The Methods of Excavating for Flint, The Age of Palaeolithic Man, and The Connexion Between Neolithic Art and the Gun-Flint Trade*. Memoirs of the Geological Survey, London.

Skinner, Alanson. 1915. *The Indians of Newark Before the White Men Came*. Newark, NJ: Newark Museum Association.

Skinner, Alanson. 1919. *The Pre-Iroquoian Algonkian Indians of Central and Western New York*. New York: Museum of the American Indian, Heye Foundation.

Skinner, Alanson. 1921. *Notes on Iroquois Archeology*. New York: Museum of the American Indian, Heye Foundation.

Smith, Harlan Ingersoll. 1907. *Archaeology of the Gulf of Georgia and Puget Sound*. Memoirs of the American Museum of Natural History, Vol. 4, Pt. 6 of Publications, Jesup North Pacific Expedition. New York: E. J. Brill Limited.

Smith, Harlan Ingersoll, Aleš Hrdlička, and Clark Wissler. 1910. *The Prehistoric Ethnology of a Kentucky Site*. Anthropological Papers of the American Museum of Natural History. Vol. 5, pt. 1. New York.

Smith, Harlan Ingersoll, and W. J. Wintemberg. 1929. *Some Shell-Heaps in Nova Scotia*. Bulletin of the National Museum of Canada (47), Anthropological Series (9). Ottawa: F. A. Acland.

Smith, J. E., II. 1994. "Lemens Rockshelter (41SV60), a Late Prehistoric Site in Somervell County, Texas." *Bulletin of the Texas Archeological Society* 65:207–256.

Titmus, Gene L., and James C. Woods. 2003. "The Maya Eccentric: Evidence for the Use of the Indirect Percussion Technique in Mesoamerica from Preliminary Experiments Concerning Their Manufacture." In *Mesoamerican Lithic Technology: Experimentation and Interpretation*, edited by Kenneth G. Hirth, 132–146. Salt Lake City: University of Utah Press.

Webb, William Snyder. 1939. *An Archaeological Survey of Wheeler Basin on the Tennessee River in Northern Alabama*. Washington, DC: US Government Printing Office.

Webb, William Snyder. 1942. *An Archeological Survey of Pickwick Basin in the Adjacent Portions of the States of Alabama, Mississippi and Tennessee*. Smithsonian Bureau of American Ethnology, Bulletin 129. Washington, DC: Government Printing Office.

Webb, William Snyder. 1946. *Indian Knoll, Site Oh 2, Ohio County, Kentucky* 4(3), pts. 1–2. Lexington: Publications of the Department of Anthropology and Archaeology, University of Kentucky.

Webb, William Snyder, and William Delbert Funkhouser. 1928. *Ancient Life in Kentucky: A Brief Presentation of the Paleontological Succession in Kentucky Coupled with a Systematic Outline of the Archaeology of the Commonwealth*. Frankfort: Kentucky Geological Survey.

Webb, William Snyder, and William George Haag. 1939. *The Chiggerville Site: Site 1, Ohio County, Kentucky*. Lexington: University of Kentucky, Department of Anthropology and Archaeology.

Webb, William Snyder, and Charles Earnest Snow. 1945. *The Adena People*, 6. Lexington: Publications of the Department of Anthropology and Archaeology, University of Kentucky.

Wedel, Waldo R. 1943. *Archeological Investigations in Platte and Clay Counties, Missouri*. United State National Museum Bulletin 183. Washington, DC: United States Government.

Wilhelm, Paul. 1941. "Desert Lore: Arrowheads Easy to Make—If You Happen to Be an Indian." https://www.yumpu.com/en/document/view/17282982/paul-wilhelm-collection-calisphere#google_vignette.

Williams, Fredrick H. 1906. "Prehistoric Remains of the Tunxis Valley." In *Bristol, Connecticut, In the Old Time New Cambridge, Which Includes Forestville*, edited by Eddy N. Smith, George B. Smith, and Allen A. J. Dates, 79–126. Hartford, CT: City Printing Co.

Willoughby, Charles C. 1935. *Antiquities of the New England Indians, with Notes on the Ancient Cultures of the Adjacent Territory*. Cambridge, MA: Peabody Museum of American archaeology and ethnology, Harvard University.

Wintemberg, W. J. 1936. *Roebuck Prehistoric Village Site, Grenville County, Ontario*. Ottawa: National Museum of Canada Department of Mines.

Wood, W. R., and R. E. Pangborn. 1968. "The Eureka and Comstock Mounds, Southwestern Missouri." *Plains Anthropologist* 13:1–17.

Young, Fred S. 1933. "Oregon Obsidian Arrowpoints." *Hobbies* 38(8):129.

PART TWO

Studies of Chert and Flint Artifacts

4

Exploring Maya Flint Mastery

An Autopsy of a Ceremonial Scepter

JOHN E. CLARK

The object is as impressive as hell. Perhaps it would be more appropriate to say it is as impressive as heaven. It still scares me to handle it. Even in its decapitated state, the ceremonial lancehead weighs 493 g and spans 36.3 cm. It was buried in the eighth century and exhumed near the end of the twentieth in Buenos Aires, a coastal village of southernmost Chiapas, Mexico. It lost its tip and parts of its edges during discovery. The main body was given to me in 1992 by a man I do not know and on a day I failed to record.

Such are the humble beginnings for this object's new lease on life. Here I relate the story of the Buenos Aires Biface (BAB), or as much as I have pieced together. The flint biface still has much to teach me, and I hope I find time to listen. I explore here the awe of its ancient craftsmanship, a topic not handled competently in science reportage. Facts are critical to the story, but I arrange them differently here than I would in a treatise. I begin with an imaginary obituary of the scepter and then delve into the real autopsy behind it conducted in August 2019. I end with an exploration of how it was made.

https://doi.org/10.5876/9781646427819.c004

LIFE SKETCH

The Buenos Aires Biface was made in Belize some 1,300 years ago and exported to coastal Chiapas, a trip that now takes more than two days by bus. The trek would have been a journey of a month or more in 700 CE. For nearly three decades I have thought of BAB as a sacrificial knife, but it is too big for that. It was a ceremonial lancehead or scepter. The scepter came from the heart of a flattish boulder that weighed more than 40 kg and was at least 50 cm long in one dimension. It is technically "chert," but "flint" is the term most people use for the stone. In the 1530s Spanish conquerors called the stone *pedernal* and Yucatec Mayan called it *took'*. The flint nodule womb of the scepter had a thin shell of light tan cortex. This is A-grade rock of fine texture and beautiful variegated dark brown color. Flaked to a centimeter thinness, this stone looks like translucent honey. BAB was made by Maya artisans and exported to the Pacific Coast, where it was obtained by a priest-king who most likely was not Mayan. The scepter is Mayan by manufacture rather than by final ownership.

During the birthing process the heart of the flint nodule was unveiled by using large stones to hammer away encasing stone. Smaller stones or wood hammers were used to rough out a lanceolate form that weighed about 5–7 kg, and this was progressively worked down to a flat piece about 43 cm long, 11 cm wide, and 2 cm thick. At this point, the Maya artisan changed his tools and used wood and antler chisels and mallets to sculpt away more carefully the remaining layers of stone to reach the desired final form. Both faces of BAB were thinned until it was about 1.3 cm thick, and then its edges were dressed to straighten and trim them to the final laurel-leaf form. Its edge outlines were likely judged against a template to achieve perfect mirror symmetry; both edges are complex curves. The fine dressing of the biface was done with small punches. This was followed by pressure retouch with a deer antler or wood tool to give the scepter a perfectly continuous edge. Only a small section of the base departed from the symmetry of the whole, and this was to preserve evidence of its origin and perhaps accord respect to Mother Earth for the gift of the original stone. A small section of the flint shell was preserved on one margin of the base. This cortex was hard enough that the final flake taken from the piece cut through it. This basal-thinning flake, whether intended or not, was the artist's signature of complete mastery over his medium and technique. Basal thinning is the most dangerous blow in biface making, especially for long bifaces. That it was performed on BAB at its most delicate state is amazing. With removal of the final flakes and with dressed edges, BAB left home for adventures abroad.

BAB was made with engineering and magical precision. It was a cubit long (41.7 cm), precisely one-fourth of an ancient Mesoamerican fathom, or *brazada* (1.667 m). The lower-fourth of this scepter near the base has an arc curvature of a circle 85 cm in diameter, and the curvature of the remaining three-fourths is an arc from a circle 137 cm in diameter. These two arcs are from circles related in the golden proportion of 1.618, or phi. BAB was made to sacred proportions, with no room for error, as appropriate for its use as a symbol of royal power and authority. Given its purpose, it may have been a commissioned piece from the start. The scepter was not created for profane use. It only became mundane when yanked into our world in the 1990s.

About 6 cm of the tip of the lancehead was broken off when this piece was found. For 13 centuries it had rested flat in a Classic period grave. Its tip and edges were broken and battered in exhumation. All the modern modifications detract from its original grandeur. Modern scars are perpendicular to one face and expose a grayer flint beneath its bronzed and shiny skin. BAB was not ceremonially decommissioned. Flint lanceheads of the same age from the neighboring kingdom of Izapa (all much smaller than BAB) were burned and smashed when removed from sacred service. BAB was neither. It was accorded a different death and a reverential burial.

When new, BAB was a lean and mean cutting instrument 41.7 cm long that weighed in at about 540 g. It was likely mounted on a straight wooden staff just over 1.7 m long, with the whole ceremonial spear or hafted scepter measuring 2.08 m. This was a fathom plus one cubit measure, being five-fourths the measure of a normal man, a height appropriate for a king, even if just metaphorically. The top of the lance just below the stone tip was wrapped in jaguar skin. When the king stood before his subjects with scepter-spear in hand, the tip of the lancehead was nearly half a meter above his head. This difference in height was not so apparent to onlookers, because the feather headdress he wore made up most of the difference. Object and owner made it to the same grave. Only the lancehead made it out, but not unscathed.

Since 1992, BAB has been under my care as one of scores of artifacts picked up on archaeological survey in southern Chiapas. I was told nothing about the circumstances of its recovery. I realized as soon as I saw it that it was extra special and deserving of more attention than I could give it at the time. Not until three years ago could I make any time for it. In the twenty-five intervening years, I never forgot it.

SOME FORENSICS

I cannot swear in court that even half the obituary is true, but I hope it is. BAB is much longer than most flint lanceheads depicted on Maya sculptures or polychrome pots and may have been used differently than they. A few kings on sculptures are shown holding ceremonial spears with tips this large, but this is unusual. BAB could have been a scepter, pure and simple, and held in the hand unhafted, as some chipped Maya silhouettes are thought to have been (Clark et al. 2012). Fragments from large 3D-effigy incense burners from Late Classic Izapa

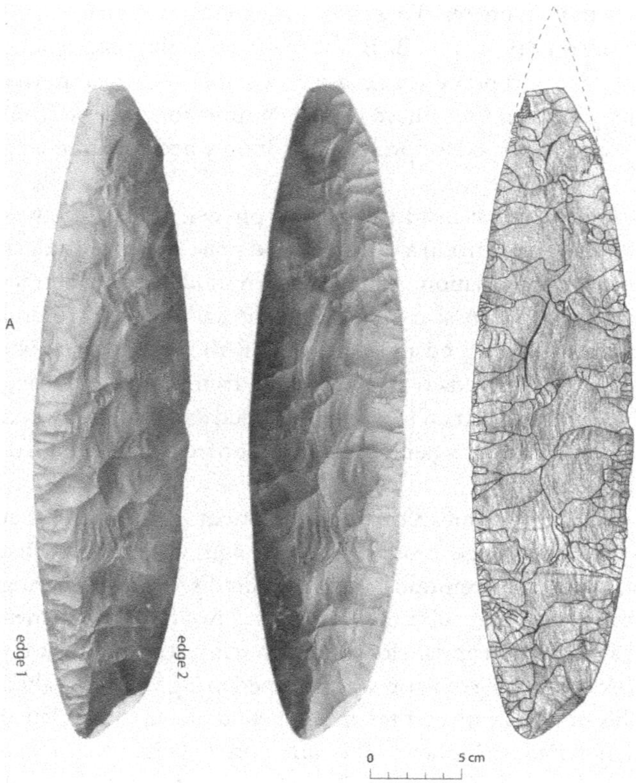

FIGURE 4.1. *Face A of BAB showing light on each edge and a pencil rubbing of the face.*

about 40 km east of Buenos Aires are of human right hands holding objects that look like lanceolate knives (Lowe et al. 1982, 240, fig. 12.13). These lanceolate objects are considerably shorter than BAB. Being a complete cubit in length, this flint object was literally the length of the lower arm from the elbow to the tip of the middle finger of the extended hand. It was also a half-palm wide and thus would have been hard to hold in the palm of one hand. It looks like a monstrous version of a sacrificial knife, or what is called a hypertrophic object. It is actually twice the size of sacrificial knives but nearly as thin.

Figures 4.1 and 4.2 show three views of each face of BAB in the best light. The two photographs in each figure show the biface lit up with light for each edge. These are to give every flake scar a decent chance of being seen. The accompanying drawings are pencil rubbings that show the character of each flake scar. I inked in the ridges to outline each scar in crisper detail. The rubbings portray the biface as slightly wider than it is because they represent rollouts of the slightly curved surfaces of each face. They show the topography of each flake scar.

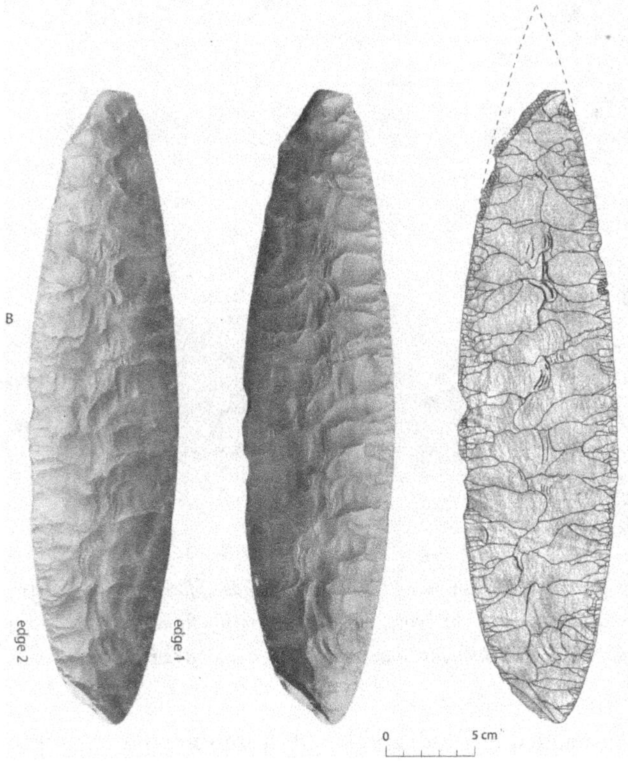

FIGURE 4.2. *Face B of BAB showing light on each edge and a pencil rubbing of the face.*

The biface is now 36.3 cm long and 8.5 cm wide at its widest point. It varies from 8.3 to 12.7 mm in thickness. It is thinner at the base and tip than in the middle. It averages about 12.0 mm in thickness. This gives a width-to-thickness ratio of 7.1. Both faces are gently curved from 20 to 30 cm in diameter, depending on width. It is 30 cm in diameter in the middle and 20 cm near both ends. When placed on a flat surface, the biface is stable on one face (Face B) and rocks slightly when placed on the opposite face (Face A), a characteristic that Antonio Prado (1999) taught me was characteristic of ancient Maya ceremonial flints (see Clark and Woods 2015).

Flake scars and sizes on the faces vary. The smallest percussion flake scars are near the base and the tip, and the largest ones are in the middle. Five types of flakes from three different techniques are evident on BAB. The largest flake scars are from indirect percussion, and the smallest resulted from different kinds of pressure. These flake scar types are illustrated in figure 4.3. Type 1 flakes were the largest removed during the final phase of production. Most Type 1 flakes had

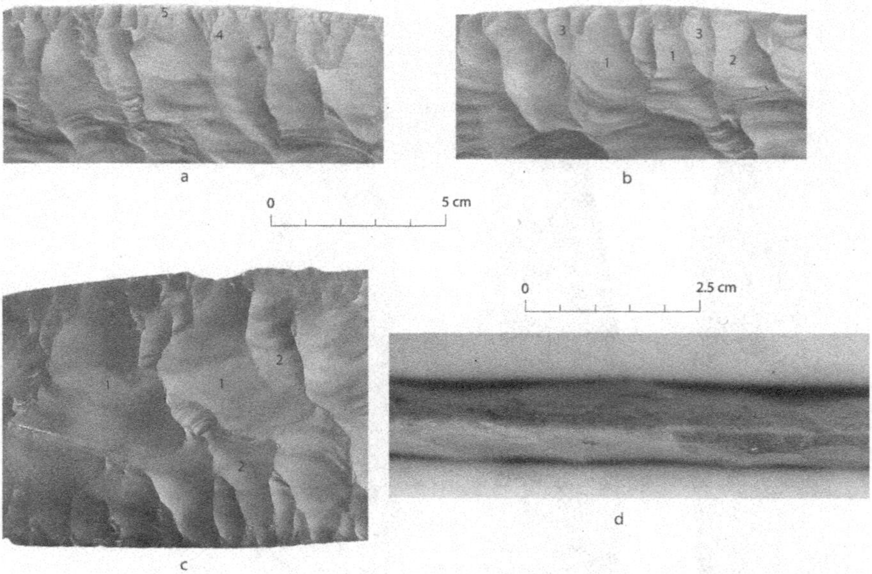

FIGURE 4.3. *Close-up photographs of BAB showing the different kinds of flake scars and attributes: (a) Type 4 and 5 pressure flake scars; (b) Type 3 indirect percussion flake scars; (c) Type 1 and 2 indirect percussion flake scars; (d) straight, sharp edge with microflake retouch.*

a conch shape as they expanded in width and length from the point of impact and trespassed the centerline of the biface. Type 1 flakes were spaced about 3 to 5 cm apart to allow them to expand at their distal ends. This spacing left raised delta-shaped islands on the face between the hollows left by the removal of large flakes (figure 4.4). These deltas were removed by indirect percussion. These intermediate indirect percussion flakes are Type 2 flakes. Smaller indirect percussion flakes were removed to retouch margins and remove smaller deltas. These are Type 3 flakes (figure 4.3).

The smallest visible flakes on BAB were removed by two types of pressure flaking (see figure 4.3a). The small, ribbon flakes removed small deltas and made the edge of the biface more regular. They cluster in triangular arrangements of the original deltas they removed. Most of these have pointed, feather terminations similar to tiny microblades. They can be made with a pointed antler tine by flaking up from the underside of the edge toward the upper face. One face has its edge retouched with tiny microflakes about a millimeter long and wide. Most of them have feather terminations. They were flaked with the edge of a flat tool in a scraping motion to the edge. Some were removed as groups rather than individually. These flake scars are all perpendicular to the edge. This scraping "pressure flaking" removed minor irregularities in the edge profile. Following

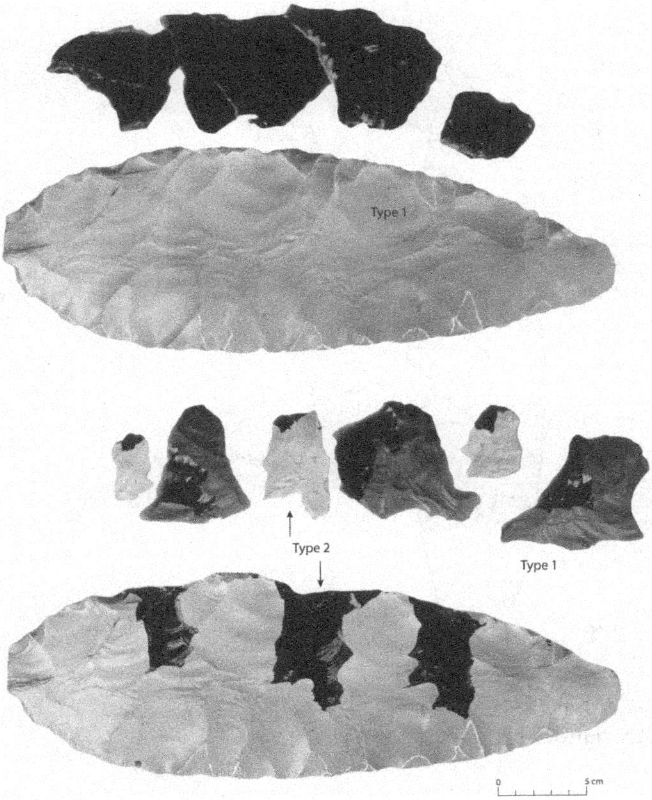

FIGURE 4.4. *Obsidian biface made by indirect percussion showing Type 1 and Type 2 flake scars.*

their removal, the edge was scraped with a stone to remove flakes not visible to the naked eye. These were microscopic hinge flakes that slightly blunted the edge. There is no evidence of edge modification from use.

Figure 4.5 illustrates the basic outlines of the flakes from both faces of BAB without the details of the topography of individual scars shown in figures 4.1 and 4.2. It also introduces a large biface from Copan that will enter the conversation shortly. Figure 4.6 shows the reference system for identifying individual flake scars on these bifaces by face, edge, and scar. By definition, bifaces have two faces and one continuous edge. For descriptive convenience, I divide the edge into two entities split along the principal axis or centerline, from base to tip. Every flake scar of each face and edge can be numbered; I only number Type 1 and 2 flake scars here. These drawings project the curved edges of BAB to estimate its original length. The very tip could have been more curved. If so, I have overestimated its length and numerological precision. These flake scars

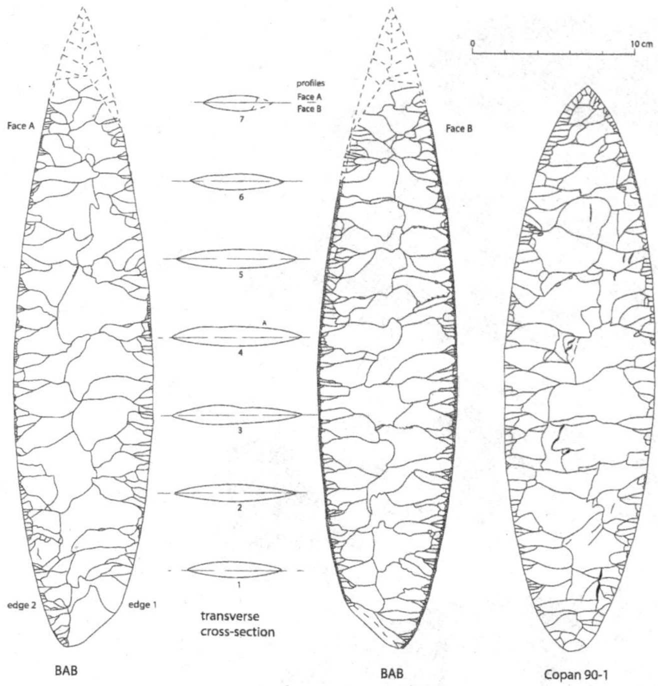

FIGURE 4.5. *Drawings of BAB and Copan biface 90-1 showing the types of flake scars and the transverse cross-section of BAB at 5 cm intervals.*

are slightly transverse to the face rather than perpendicular to it. This pattern indicates handedness and consistency in knapping habits.

Edge 1 of Face A has 21 Type 1 and 2 flake scars, and there were probably another 4 small ones on the broken portion, making a total of 25. Edge 2 has remnants of 19 of these scars, with another 4 estimated, for a total of 23. These are about the same size as those flaked from Edge 1. Most of Edge 2 was flaked after Edge 1. Flakes from Edge 2 undercut those from Edge 1. Flake scar A2-11, for example, is the largest on BAB and from one of the last flakes removed. It went under flake scars A1-12, A1-14, and A1-16. As apparent in figure 4.1, this flake ended in a slight hinge, and as it rolled out of its socket it tore some pinhead-size angular flakes from the distal end of the scar of Flake A1-15. Flakes A1-8 and A1-17 were also final removals, and they undercut flake scars from Edge 2. Also apparent on this edge is the randomness of fine retouch. Edge 2 in the area of Flakes A1-5 through A1-9 shows extensive indirect and pressure retouch that completely removed the negative bulbs of percussion of these flake scars. There are very few Type 5 flakes on this face. The fine marginal retouch was mostly unifacial and from Face B.

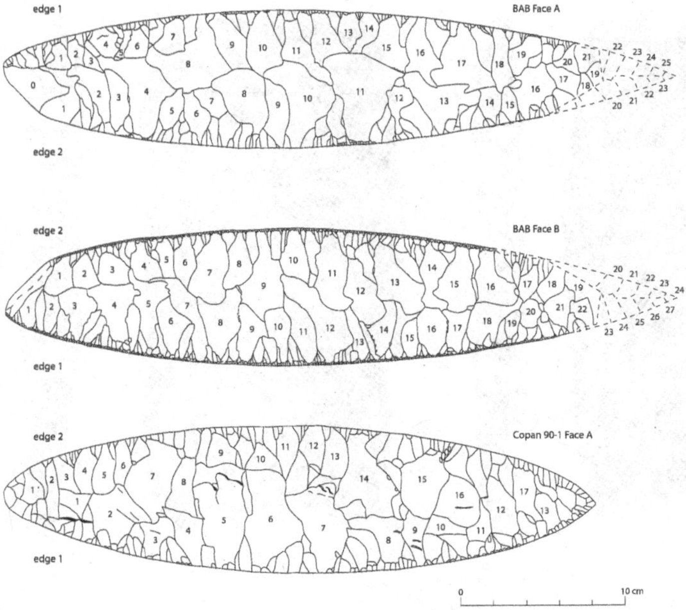

FIGURE 4.6. *Reference system of the flake scars on BAB and Copan 90-1.*

Face B is the comelier of the two faces and has more regular flake scars, orientations, and fine edge retouch. For the preserved portions of the biface, I count 19 Type 1 and 2 flakes originating from Edge 2 and 22 from Edge 1, with another 4–5 estimated for the missing tip. This is about the same number of flakes as estimated for Face A. Even with the different widths of flake scars on the different faces, over the course of 42 cm of edge, the flake scar widths average about 2 cm wide. Many of the large flakes are bell-bottomed in profile and expand in one direction like the toe of a boot. The flakes that left these marks almost certainly expanded in two directions, but the removal of flakes adjacent to them sliced away half of each scar. This is convenient because the bisection of scars reveals the flake removal sequence and the direction of knapping.

For Face B, Edge 1 was flaked from the tip to the base, and Edge 2 was flaked from the base back to the tip. The flaking of Face A was the opposite of Face B. Edge 1 of Face A was flaked from the base to the tip, and Edge 2 was flaked from the tip to the base. This pattern pertains to Type 1 flakes, not Type 2 flakes. These latter flakes removed the intersections of adjacent Type 1 flakes and thus reveal nothing about the overall knapping sequence (see figure 4.4). The Type 1 flakes on each face comprise complete circuits and bespeak of regularity of knapping habits, as does the regular tilt of all the flakes from both edges. The knapping was one edge at a time rather than one complete face at a time.

FIGURE 4.7. *Flaking sequence of an experimental obsidian biface showing the alternation of platforms on one edge to opposite faces.*

Much of Face A was flaked after Face B. This refers to the final sequence of large flake removals and not to the overall manufacturing process. Both faces of the BAB preform must have been flaked two or three times before this to create a flat and thin final preform of lanceolate shape. The flake scars evident on these faces are from the final finishing work. The edges of BAB are completely straight, centered, and sharp and have an angle of 28 degrees (figure 4.3d). These are characteristics shared by the best of Maya ceremonial flints. In a previous study of these flints, I suggested that this feature allowed artisans to flake perfect profiles (Clark et al. 2012). This is incorrect. The straight, thin, sharp edge was the principal feature of the silhouette. It was the last thing produced.

The major flake scars on each face are offset from each other, and those taken from the same edge to different faces are also (figure 4.7). This is a natural consequence of the need for platforms, deliberately created bevels on an edge. A sharp, unreinforced edge crushes under impact and leads to knapping mayhem. The sharp edge of a negative bulb of percussion is even more fragile. Removal of a flake takes away the material needed for making a platform for the face directly opposite it. On BAB, large flakes were spaced 3–5 cm apart. Their removal stranded elevated deltas between their negative bulbs of percussion. These deltas were thick sections of the original edge and were used as platforms for

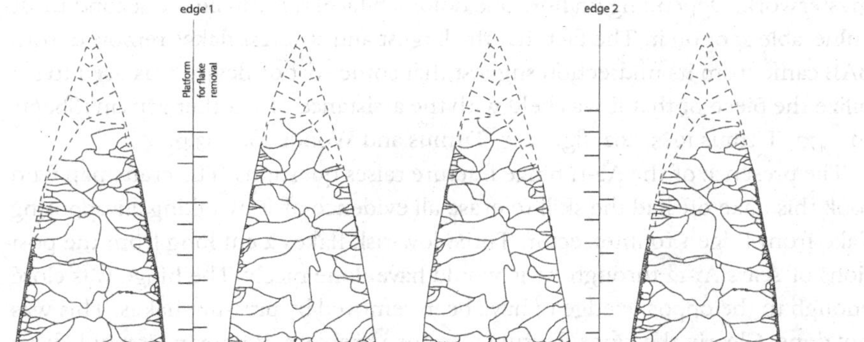

FIGURE 4.8. *Platform placements on the two edges of BAB, showing the alternation of flake removals from Face A to Face B.*

removing flakes from the opposite face. The alternate flaking on each face left a sinuous edge when viewed edge-on, as illustrated in figure 4.7. The large flakes removed from each edge of BAB were offset (figure 4.8). For the final series of major flake removals, Edge 1 of both faces was flaked before Edge 2, the last to be flaked by indirect percussion (see figures 4.5 and 4.6). For the final pressure finishing, Face B was flaked last.

As mentioned, Flake A2-11 ended in a slight hinge. Some analysts would count this as a knapping error. It was one of fifty large flakes removed from Face A and represents the equivalent of a golf score of one missed putt in fifty greens. I think this flake is evidence of consummate skill and ranks up there with Flake A2-0, the end-of-program basal thinning flake. Punching a flake over 5.5 cm long and 4.0 cm wide from the weak middle of a biface that was 1.3 cm thick is a masterwork. Depending on how one holds a biface, this can be the second-most-vulnerable spot on it. The fact that the largest and stoutest flakes removed from BAB came from its midsection suggest that some sort of device was used to stabilize the piece or that it was held with the assistance of another artisan (Sheets 2016, 72; Titmus 1985, 262, fig. 10.15; Titmus and Woods 2003, 134).

The presence of the A2-11 hinge fracture raises questions. The craftsman who took this flake off had the skill to erase all evidence of it by taking an opposing flake from Edge 1 to intersect it. Three low-risk flakes 2 cm long from the positions of scars A1-12 through A1-15 would have done nicely. The hinge was close enough to the opposite edge to have been removed by pressure flakes. This was not done. Clearly, the hinge fracture was not a problem. Removing it would have deflated Edge 1 by removing some of the edge as a platform and thus would have destroyed the perfect curve of the outline. I think the edge profile was the critical attribute being produced on BAB. There was no attempt to produce parallel or regular flake scars (either in size or inclination) on either face of BAB. Rather, the aesthetic of its countenance is of rusticity—scars of different sizes and shapes carefully fitted together to the same flat plane, much like the stones in a mountain lodge fireplace. The regularity and symmetry of this piece are in the overall shape in three dimensions, not its flake scars. Nothing on this biface disturbs the eye or touch as being out of place. Symmetry and harmony pervade the whole—with the deliberate exception of the slice of cortex at the base on Edge 2. The overall elegance of the piece directs attention to this testament of the mother rock by being the only thing different in shape, color, and texture. This is the only segment of the edge that is not sharp.

The only hypertrophic biface well published for the Maya area is Specimen 90-1 from Copan (Sheets 2016, 72–74). I made a drawing from the laser scan of this artifact to compare to BAB (see figures 4.5 and 4.6). I have not had the privilege of examining this object to verify my interpretation of flake terminations. Basic features of BAB and Copan 90-1 are listed in table 4.1. The Copan biface is shorter and wider than BAB and has more pronounced edge curves. It is also less carefully flaked and finished. The white chert it is made from may be more grainy than the fine brown flint of BAB. There appear to be some vugs or crystals in the center of Copan 90-1, but I cannot verify this from the image, and none are described for it. The edge curvature of Copan 90-1 is complex. Both the tip and the base have a curve 20 cm in diameter, and the main edge has a diameter of 101 cm. Each end is about one-eighth of the total, with three-fourths of the profile being the major arc. The ratio of the major arc to the minor arc is 5:1. As with BAB, a fourth of Copan 90-1 was devoted to the minor arc (table 4.1).

Payson Sheets (2016, 72) avers that Copan 90-1 was produced by indirect percussion, and this is a solid bet. He mentions two slight hinge fractures. In drawing

TABLE 4.1. Comparison of the Buenos Aires Biface with Copan biface 90-1

Characteristic	*Buenos Aires*	*Copan 90-1*
Color	dark brown, variegated	whitish
Texture	fine	fine
Cortex on base	yes	none reported*
Length	41.7 cm (est.)	36.0 cm
Width	8.5 cm	9.6 cm
Thickness	1.2 cm avg.	1.2 cm
Grams	540 (est.)	517
Major arc	137 cm	101 cm
Minor arc	85 cm	20 cm
Major/minor arc	1.61	5.05
Length/width	4.9	3.75
Width/thickness	7.1	8.0
# Type 1 flakes/edge	13 (for 36 cm preserved)	15
# Type 1/cm	0.36	0.42

* Zachary Hruby suspects there may be a touch of cortex under the cloth and pigment on Face B (personal communication, 2019).

this piece I counted 10, with two being obvious. The largest (A1-7) (see figure 4.6) is in the middle of the biface, similar to BAB A2-11. This is a curious match and a good clue of ancient holding techniques (discussed later). The middle of each biface was secure and could withstand a lot of force.

A striking difference between Copan 90-1 and BAB is that the Copan biface lacks Type 2 and 5 flakes. It has three kinds of flakes instead of five (figure 4.9). Most center scars are from Type 1 flakes, and the edge retouch is from Type 3 and 4 flakes—small indirect percussion flakes and even smaller ribbon pressure flakes. There are no clear Type 5 flakes. The two bifaces came from different workshops, and there were differences in how the different artists went about their work. The Copan biface has four small patches of flakes from an earlier stage of reduction. BAB has none of these. For Copan 90-1, Edge 1 was flaked from the tip to the base, and Edge 2 was flaked from the base to the tip. This edge rotation is similar to that on both faces of BAB. I suspect that Face A of the Copan biface was the last flaked. This flaking sequence would account for its greater flatness and the large flakes removed from it. As evident in the illustration published by Sheets (2016, 72, fig. A4), the edge of Copan 90-1 is closer to Face A than to Face B. The Copan biface is a splendid piece and clearly different from BAB in many details. It evinces much more indirect percussion retouch of its edges than apparent on BAB.

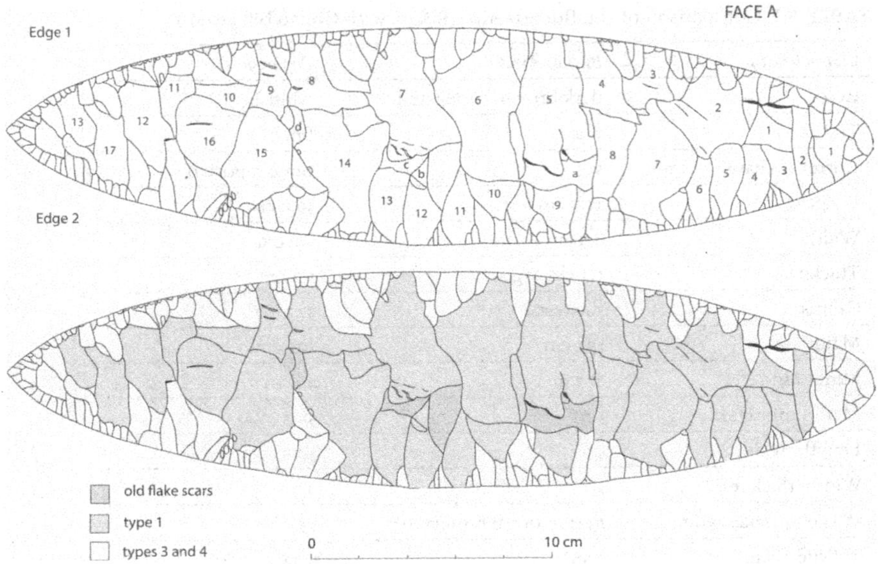

FIGURE 4.9. *Copan 90-1 showing flake types and flaking sequence. Drawing by Clark after Sheets (2016).*

MAKING IS SEEING, SEEING IS UNDERSTANDING

The paired couplets introducing this section summarize my approach to chipped-stone objects. I accept Don Crabtree's view that doing is knowing and that making something is the way to understand it (discussed in Clark and Woods 2015). I have added the missing middle term of the implied syllogism that making is understanding. Nothing focuses "seeing" as well as trying to make a perfect copy of something. One has to respect all details. All three predicates of the gerunds of the subheading have layers of meaning. Making, seeing, and understanding can be skin deep or deeper than the Grand Canyon. I have been looking at BAB since 1992, but I never really started to see it until gathering details for writing this chapter and for fine-tuning my experiments. Each attempt to duplicate a feature on BAB has changed my understanding and appreciation for it.

Replication experiments involve core philosophy and not just knapping fun. My question about BAB is actually trite: How was it made? Any first-year archaeology student can answer the question correctly: by percussion and pressure. So why all the fuss and years of investigation? A. V. Kidder, in his foundational study of stone artifacts for Mesoamerica, described an obsidian "Laurel-leaf blade" from Uaxactun, Guatemala, as having "symmetry of form, keenness of edge, and excellence of chipping" (1947, 25). This description has not been rivaled for poetic concision and accuracy. These features well describe Copan 90-1 and BAB.

But how did they come by these features? How does one make such an object? What is behind the excellence in chipping?

In my discussion here I attempt to move beyond description to understanding, but it is necessary to start with description. The best illustration and description of a large Maya chipped-stone biface are that by Sheets (2016) for Copan 90-1. I follow his lead for BAB and add information from replication experiments as a way to probe the piece and understand it from a creator's vantage. This gets at the question of ancient Maya flint craftsmanship and mastery, a topic reserved for closing comments.

An adequate description of BAB in an archaeological report would include a mug shot and basic statistics required by the online dating service—length, width, thickness, weight, color, edge sharpness, proclivities, and relative condition. For Copan 90-1, an excellent laser scan is provided, but this fails to show the chipping of the tip and base clearly. The accompanying description is mostly of the textiles and other organics that veil one of its faces. An analytical drawing would have helped. As a Mesoamericanist, I am delighted with the preservation of textiles, but as a rock guy I would like these removed ASAP to reveal the flake scars. At the moment, this biface must be appreciated largely on the basis of one face.

After providing the standard information for bifaces online, Sheets makes eight observations and adds three conjectures. He mentions that Copan 90-1 is thin for its length, is flat, has fine controlled flaking, largely lacks surface flaws such as step and hinge fractures (two shallow ones noted), has expanding broad flakes on its face and fine retouch flake scars on its edges, has a precision outline, and was covered with multiple layers of organic materials. The evidence that the Copan "eccentrics" were bundled and dressed for final burial is unprecedented information related to social functions and perceptions of these objects but is beyond my objective here of getting at the production sequence. Sheets considers the workmanship of this piece equal to that of the splendid flint silhouettes from this site. He speculates that Copan 90-1 was made from tabular chert rather than from a very large macroblade because there is no curvature in its longitudinal section. He further conjectures that Copan 90-1 was made by indirect percussion, likely by two artisans performing a classic mugging—one to hold the arms and the other to punch. Support for these conjectures is the regularity of the flaking and the thinness of the biface relative to its length and width. I will consider these attributes shortly. Sheets reconstructs the manufacturing sequence as starting with direct percussion, moving to indirect percussion with antler punches, and then finishing with pressure flaking with antler tools. I think he is correct and that the same sequence is part of BAB's biography.

Other than the color of the chert and its clothes, the description of Copan 90-1 fits BAB. The Buenos Aires Biface is flat and has a straight, centered edge; large, expanding flakes on its faces; fine retouched edges; and a precision outline.

Actual measurements vary for the two pieces. They are not identical twins but are close siblings. Not one of their attributes is remarkable by itself. It is the combination of them together on bifaces over 36 cm long and only 1.2 cm thick that makes these pieces stunning. I have no doubt that both bifaces were made by highly skilled artisans and master artists. But how skilled were they? What was the nature of their skill? What is the objective evidence of their level of skill? These are questions that can only be answered by channeling ancient spirits or through replication experiments. I have chosen Option 2.

Off and on the past three years, I have attempted to duplicate features of BAB. In the process, I have gained appreciation of the craftsmanship petrified in its clean lines and smooth surfaces. I started with the obvious features of length and thinness, and only in the course of writing this chapter have I paid much attention to the fine retouch. I considered it the easiest feature to duplicate. I was wrong. When I took a critical look, what I saw changed how I view BAB and how I should try to replicate it. For two years I was convinced BAB had been made by direct percussion. Experiments of the past two months convinced me otherwise. BAB was made by indirect percussion. Consideration of its major features provides evidence and the reasons I changed my mind.

Initial Stages of Manufacture

There is no evidence for either biface for the first stages of manufacture, so claims for them are hypothetical. For some Maya silhouettes, the nature of the banding of the flint is the best evidence as to whether the stone came from a nodule or was tabular (Clark et al. 2012). The banding on BAB shows that it was from a flattish stone. We can retire the notion that the longitudinal flatness of silhouettes is evidence they were made of tabular stone.

The flake scars on BAB represent the final series of the various flake types. These can be retrodicted to the immediately anterior moment. I have done this for Face A to get an idea of what it looked like prior to retouching (figure 4.10). The final-phase preform was about 2 cm thick and 10 cm wide. All it lacked was removal of the last layer. Depending on the raw material and its physical constraints, one or two stages from breaking into raw material to final preform should have been sufficient.

In my preliminary experiments the purpose was to get long, thin preforms without breaking them. I found tropical hardwood tools work best for obsidian. I have not experimented much with chert or flint because I do not have them available. Because fine flint is stronger than obsidian, thin flint bifaces should be easier to make. I have not verified this expectation. In my experiments I have spent much time trying to distinguish direct and indirect percussion flakes. This is rather trivial matter in the large scheme of things, but it is the basement requirement for making a legitimate replica. After I learned how to make big bifaces, I looked in on BAB

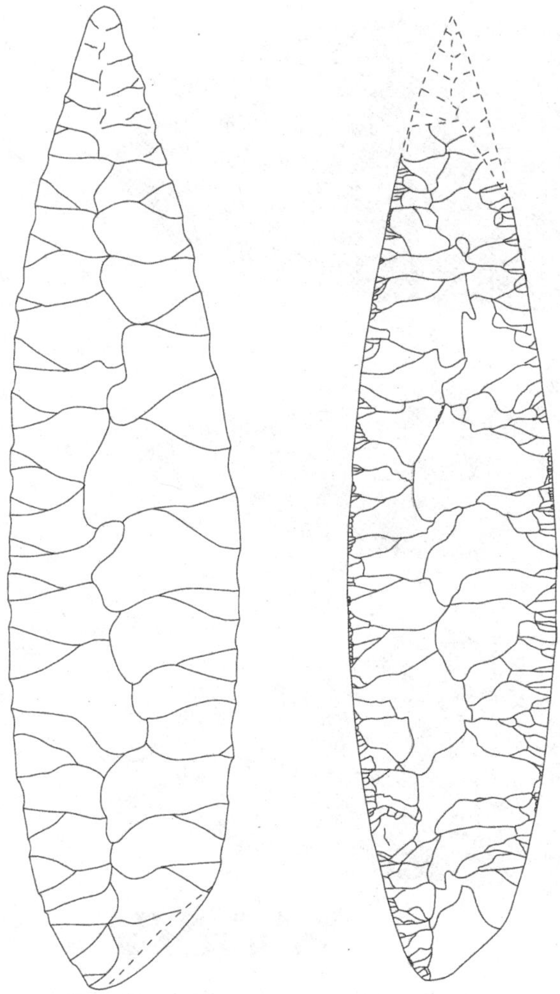

a b

FIGURE 4.10. *Reconstruction of BAB before the final indirect and pressure retouch.*

and noticed that my flake scars were a poor match. Mine were more regular. This made no sense. My bifaces were not in BAB's class, but my flake scars were.

Sheets's argument that Copan 90-1 was made by indirect percussion rests on the regularity of the work, with the implication being that such regularity cannot be achieved with direct percussion. This claim has to be specified. Differences in the two techniques can be seen by comparing the drawings of BAB and Copan 90-1 (see figure 4.5) with two direct percussion experimental

FIGURE 4.11. *Modern obsidian bifaces made by direct percussion and indirect percussion showing differences in flake scars: (a) direct percussion biface made with a hardwood billet (June 6, 2017); (b) direct percussion biface made with a hardwood billet (August 13, 2019); (c) indirect percussion biface made with hardwood punches and mallets (August 7, 2019).*

bifaces shown in figure 4.11. The flake scars on the ancient pieces are controlled chaos. The flake scars on BAB and Copan 90-1 show regularity in spacing but absolutely no regularity in scar shape or size. The coupling of regular spacing and small negative bulbs of percussion the same size (as appreciated by some traces) with flake scars of irregular size and shape is a marker of indirect percussion. The flake scars on my best percussion bifaces are regular in width and tend to be rectangular in outline. They also have wider negative bulbs of percussion from wider platform removals. Indirect percussion flakes, in contrast, tend to have smaller points of initiation and be conch-shape and flare at their distal ends, as evident in the two large flake scars on the indirect percussion experimental biface in figures 4.4 and 4.11c. The large flake scars on BAB and Copan 90-1 look like puzzle pieces in which each piece has a different outline.

It is hard not to be impressed by the regularity of these two Maya bifaces, but this is an impression based on attributes other than the scars on their faces. Compared to Teotihuacan giant bifaces or Danish daggers or Egyptian Gerza knives, the flake scars on these Maya bifaces look like the hunt-and-peck flaking of amateurs. These other bifaces are finished by pressure and that is a critical difference, but large percussion bifaces from many parts of the world have much more regular flake scars and patterns than the Maya pieces do. A characteristic that shows up clearly on BAB is that the size of percussion flake scars corresponds directly with width. The smaller scars are at the base and tip, and the larger ones are in the middle, where the biface is the widest. This is a remarkable pattern and an important clue of the production process. There are no flake scars indicative of an earlier reduction stage or shape. It is not remarkable that a Maya artisan would vary the force of his blows to remove flakes of different length, depending on the width to be spanned. This was done on BAB and Copan 90-1. I have looked at this pattern on BAB since 1992 and did not realize what I was looking at. The pattern of flake scars on BAB only makes sense for its size and shape. These are not decapitated flake scar remnants from giant, anterior forms such as seen on the stems of the largest Maya silhouettes. The flake scars on BAB and Copan 90-1 were tailor made to fit them. *BAB and Copan 90-1 were not shaped from thinned preforms; they were shaped preforms that were thinned.* Shape preceded thinness. Unlike the knapping strategy for large Maya flint silhouettes, the shape of the hypertrophic bifaces was not left until last.

In my experiments, I had the production sequence backward. For BAB, outline shape and form preceded thinness. I have to start over in trying to duplicate it. I did not recognize that the large flake scars on BAB were customized for space previously proscribed. The smaller flakes were to clean up after Type 1 and 2 flakes. One reason BAB looks so regular is that it has smooth and continuous surfaces. There are no disruptive negative bulbs of percussion queued along its edges. These were removed with Type 3 and 4 flakes. The negative bulbs of Type 3 flakes were removed by Type 4 flakes, and those of Type 4 were removed by Type 5 flakes.

What this means is that the immediately anterior stage was when most of the critical work was accomplished and the parameters established (see figure 4.10a). The final-stage preform was flat, thin, and had the outline in embryo. The edges were centered and straight, and the outline was within a centimeter from each edge, or less, of the final desired outline. The final flaking thinned and smoothed the piece. The anterior preform was already thin, just a bit rugged. This piece was finished from the inside-out. Type 1 flakes were spaced to intersect in their lower halves. These cleaned the middle of the biface on each face and established the final thickness and the lenticular cross-section. Remaining flakes were progressively shorter as the artisan worked toward the edge. Type 2, 3, and 4 flakes cleaned up the deltas and margins isolated by the removal of Type 1 flakes. Type 5 flakes finished the straightening and sharpening of the edge. Type 1 and 2 flakes appear to have been removed on BAB with a punch about 1.5 cm in shaft diameter. Type 3 flakes were removed by a punch of less than half this diameter. Type 4 flakes were removed with a pointed tool, most likely a deer tine. Type 5 flakes could have been removed by the flat of an antler punch or the flat of a hammerstone.

Retouch and Edging

My experiments have mostly concerned the big features of BAB, and I have not been concerned with the fine retouch (see James Woods, chapter 5). The pressure flake scars look rather unremarkable as things go. I only took a look at them in an effort to be comprehensive in my description of BAB. When I realized I had inverted the production sequence, this changed my appreciation of the marginal retouch of the piece and the significance of its straight, sharp edges. One feature of BAB that I had not replicated and did not even care about was its lenticular cross-section (see figure 4.5). I had read this as flat, but it is not really flat. The midline edge of BAB is virtually straight, centered, and sharp. I think the anterior, final-phase preform had edges that were straight in longitudinal cross-section and centered but not sharp. A final series of indirect percussion flakes was taken from these edges, and these flake removals created sinuous edges with missing lunate slices (see figures 4.4, 4.10, and 4.11). Type 1 and 2 flakes cleaned up the faces of BAB and produced a biface with mirror image curvatures on both faces.

The final-phase preform was probably more aesthetically pleasing than the intermediate form illustrated in figure 4.10. With removal of Type 1 and 2 flakes, the nearly finished biface had ragged edges and hollows alternating with deltas up and down its edges. The deepest parts of the negative bulbs had to be removed, the edge had to be straightened, and the deltas had to be debased and brought down flush with the flats of the flake scars flanking them on each side. The largest deltas were removed by Type 3 flakes, and the smaller ones by Type 4 flakes. All of these flake removals reduced the width of the biface. These flakes had to eat away enough edge to decapitate the negative bulbs of percussion of Type 1 and 2 flakes. One reason the Type 3 flakes appear to be rather randomly distributed on the faces might be that the cleanup operation was dictated by the size and depth of the negative bulbs or the failure of some shaping flakes to produce the right curve. Enough extra width had to be budgeted for on the final-phase preform to anticipate normal flaking problems. The spaced Type 1 flakes established depth markers all along the edge, which served like cuts in carpentry to establish the parts to be cut out.

I have just begun trying to replicate the retouch on BAB. It is more complicated than I thought (see Woods, chapter 5). The flake scars on BAB show that this was done with a bare minimum of observable alteration to the Type 1 and 2 flake ridges. I think the principal purpose of the retouch was to straighten the edge and remove lumps on the margins to achieve the smooth and regular transverse cross-section.

Indirect Versus Direct Percussion

One never knows where generative questions will come from. I compared BAB to Copan 90-1 as an afterthought. I became intrigued that both large bifaces had hinge scars in their midsections and mentioned that this was a dangerous spot to be ripping off large flakes—depending on the way a biface is held. Coupled with the observation that BAB had a final basal thinning flake, I concluded that the ancient artisans must have supported their bifaces during knapping differently than I had been supporting mine. In past experiments I tried various ways to reduce vibrations in knapping long bifaces. I got into a rut of thinking that jigs or other artisans' hands must have been necessary. I have no need for these with direct percussion because I hold bifaces against the outside of my left leg, and this absorbs the surplus energy of the blows. To investigate the question raised by BAB and Copan 90-1 about midbiface hinge scars, I tried a different holding method. I combined Gene Titmus's leather cradle for delicate indirect percussion work (see Titmus and Woods 2003) with lengthwise holding between my legs. Previously, I had held bifaces by their ends between my knees and nearly perpendicular to my legs rather than parallel to them.

My first experiment with the new holding position so exceeded expectations that I changed to it immediately. The new holding method required a change in how I use my tools. I have to prepare my platforms differently as well. Everything got easier and safer. Taking flakes from the base is not a huge problem, and taking large flakes from the middle of the biface turned out to be the safest place to do it. As a consequence of all this, I have changed my evaluation of some of the flake scars on BAB. They were not as risky as I thought and are not evidence of superman craftsmanship, just of different craft lore.

In terms of learning from past masterpieces, the main message is to heed things that do not fit one's expectations or experience. I still have not made a competent copy of BAB at its length, but I think I know the techniques for doing so. Questions raised and observations made in writing this chapter have moved me much closer. As a side benefit, the new way of holding bifaces duplicated the flaking sequence on BAB (see figure 4.7). Duplicating this attribute accidentally in the process of resolving other issues can be taken as evidence of getting closer to the ancient holding technique.

This brings me to the question of a two-person technique. It is nice to have an assistant to fetch drinks and hand one tools, but a second artisan probably was not needed for making these long bifaces. I leave the question open because I do not know enough to put it to rest. My newfound optimism has not yet confronted many catastrophic knapping failures. I have proven that endshock is still possible. I did make a 45 cm long final stage preform whose ends extended beyond the protective embrace of my legs squeezed together like a vise. The ends did not shake off because supporting the piece on each side of removed flakes was enough to suck out the extra energy of the indirect percussion blows.

At this point it is appropriate to mention that I am about twice the size of an ancient Maya craftsman and thus deploy a completely different leverage system in my work. It is not at all clear that what I do will translate to different somatic designs. My inseam is considerably longer than theirs and gives me substantially more padding for long bifaces. The other issue is that I changed my stroke, and it reduces risk of breakage, so the combination of a better system for sucking vibrations out of knapping blows and of reducing the vibrations of these blows in the first place is a win-win situation.

I mentioned that BAB shows evidence of handedness in the trend of its largest flakes. It also shows a regular sequence of flaking each edge and then rotation of the biface and repetition of the same procedures. I take these as evidence of a one-man technique (the masculine gender is intentional here and the only appropriate one for the ancient Maya flint masters). Neither of these would be expected if the biface had been held by someone else and the person with the punch took off flakes from a biface directly in front of him. My former holding position mimicked the relative positions of the biface in a two-man technique.

I only have a few days' experience with the new holding technique, but it changes everything. It allows more regularity in flaking and more regularity in outcomes. With this way of supporting bifaces, the most secure part is in the middle. It makes sense that the largest flakes on BAB and Copan 90-1 were from this section. To return to golf analogies, this is the place to hit the long drives; putting is for the tip and the base.

SKILL AND CRAFTSMANSHIP

Evaluations of skill are largely personal and thus a moving target. Most of us attribute higher levels of skill to objects we cannot ourselves make, and thus our evaluations require submission of our own CVs of knapping experience to be relevant to a wider audience. Another way is to look for mistakes and little flaws. As noted by Sheets, Copan 90-1 has few that are obvious, and BAB has even fewer.

These bifaces have clean lines and an overall regularity that exudes elegance. The regularity of repetitive motions speaks to skill born of practice. I give these two bifaces high marks on all the external and objective measures of skill. Unless and until I can match this work, I can say little more than express my awe of ancient craftsmanship. I suspect that there is much more to these pieces than I have picked up on, because I approached them as pieces of chipped stone and focused on the competence of the chipping. What about their overall designs?

I think a good knapper with the appropriate experience could make a copy of BAB in three hours if it were purely a technical project of getting the best biface from a given stone (see Woods and Titmus 1996). The final retouch probably took as long as the two preceding stages combined. One would want to factor in the learning time, of course, but these were not the products of great energy expenditure. Making a small shell bead of no consequence would take longer. These hypertrophic bifaces were not utilitarian tools but objects of art, symbols, and likely objects encoded with special lore. BAB was of a particularly significant length, and its edge curvatures have a proportion considered sacred in Medieval Europe and Mesoamerica. Making a long, thin biface with consistent face curvature, sharp edges, and continuous edge curves is difficult. Making one to a length and edge curves specified beforehand at least doubles the level of difficulty.

I cannot imagine any knapper in the world today who would not be impressed by the competence and artistry of Copan 90-1 and BAB. Beyond question they were made by artists in full control of their technique. A key part of their making most likely included ritual purification and other rites (Clark et al. 2012). This would fit with what we know about the Maya and royal arts all around the world. There were probably many more people among Mayan flintworkers who could make these things than were allowed to do so. Knapping knowledge and skill were not likely the critical production constraint. BAB and Copan 90-1 were the sorts of objects that required a license. Manufacture of these objects was not a side business for knappers churning out axes for the hoi polloi. The existence of these hypertrophic bifaces makes no sense outside a stratified society and production and consumption of royal symbols. These objects are few and far between, not because they were difficult to make but because few persons deserved to have them or use them and could authorize their creation.

REFERENCES

Clark, John E., Fred W. Nelson, and Gene L. Titmus. 2012. "Flint Effigy Eccentrics." In *Ancient Maya Art at Dumbarton Oaks*, edited by Joanna Pillsbury, Miriam Doutriaux, and Reiko Ishihara-Brito, 274–287. Washington, DC: Dumbarton Oaks.

Clark, John E., and James C. Woods. 2015. "Squeezing Life from Stones: The Human Side of Replication Experiments." In *Works in Stone: Contemporary Perspectives on*

Lithic Analysis, edited by Michael Shott, 197–212. Salt Lake City: University of Utah Press.

Kidder, A. V. 1947. *The Artifacts of Uaxactun, Guatemala*. Carnegie Institution of Washington, Publication 576. Washington, DC.

Lowe, Gareth W., Thomas A. Lee Jr., and Eduardo Martinez Espinosa. 1982. "Izapa: An Introduction to the Ruins and Monuments." *Papers of the New World Archaeological Foundation* 31. Provo, Utah.

Prado, Antonio. 1999. *El creador maya*. Guatemala City: Editorial Galaría Guatemala.

Sheets, Payson. 2016. "Appendix A: Individual Descriptions of Bifaces and Eccentrics." In *Protecting Sacred Space: Rosalila's Eccentric Chert Cache at Copan and Eccentrics Among the Classic Maya*, by Richard Agurcia Fasquelle, Payson Sheets, and Karl Andreas Taube, 70–133. San Francisco: Precolumbia Mesoweb Press.

Titmus, Gene L. 1985. "Some Aspects of Tool Notching." In *Stone Tool Analysis: Essays in Honor of Don E. Crabtree*, edited by Mark G. Plew, James C. Woods, and Mark G. Pavesic, 243–264. Albuquerque: University of New Mexico Press.

Titmus, Gene L., and James C. Woods. 2003. "The Maya Eccentric: Evidence for the Use of the Indirect Percussion Technique in Mesoamerica from Preliminary Experiments Concerning Their Manufacture." In *Mesoamerican Lithic Technology: Experimentation and Interpretation*, edited by Kenneth G. Hirth, 132–146. Salt Lake City: University of Utah Press.

Woods, James C., and Gene L. Titmus. 1996. "Stone on Stone: Perspectives on Maya Civilization from Lithic Studies." In *Eighth Palenque Round Table, 1993*, edited by Martha J. Macri and Jan McHargue. San Francisco: Pre-Columbian Art Research Institute.

5

Replicative Experimentation on the Effigy Flints of Copan, Honduras

JAMES C. WOODS

In an earlier paper, Gene Titmus and I summarized our experiments at replicating Maya silhouettes (2003). In that paper, a staged approach to making silhouettes was presented, progressing from large bifacial blanks to the finished forms. Although the process surely was more lineal with few clearly defined stages, we concluded that there was a need to progress from the larger details, which require robust percussion forces (see John Clark, chapter 4), to more judicious forces using indirect percussion, to delicate forces using pressure flaking. The focus of this chapter is on the final processes of adding delicate details to the silhouette forms.

The exact techniques and sequence of flake removal used in ancient times are not always possible to discern from flake scars alone. It can be demonstrated that some very large flakes can be detached using pressure flaking and, conversely, some very fine and regular flaking patterns made with a punch. There are many possible variables involved in producing fine details, including holding technique, manufacturing tool form, tool material, and angle of applied force. Some flake scars on ancient specimens simply cannot be used to conclusively indicate

https://doi.org/10.5876/9781646427819.c005

a specific tool or holding technique. However, what I do present in this chapter is the idea that regardless of the tools and techniques used, the production of delicate details was an exacting and precarious procedure wherein even slight error could result in breakage that could catastrophically alter the intended form of the silhouette.

My personal experience with silhouettes has been through working with Gene Titmus on replicas, examining specimens in museum collections, and referring to numerous illustrations in publications (see also Agurcia Fasquelle et al. 2016; Clark and Titmus 2012; Joyce 1932; Schele and Miller 1986, pl. 25, 26, 114; Stuart and Stuart 1993, 103; Woods and Titmus 1996). I have learned that even after knapping for forty-seven years, it is difficult to "look past the forest to see the trees."[1] In other words, it is a challenge to focus on the minute details of a silhouette when the overall spectacular nature of the ancient specimens is so commanding.

In this chapter, I do not attempt to differentiate a serpent tail from a lightning bolt, or argue the identity of one specific deity produced in profile. Instead, I look at small technological details, such as when an artist looks at the brush strokes of a famous painting, or a sculptor studies toolmarks on a famed bronze sculpture. I focus on the margin details of silhouettes, including decorative edges, facial details (nose, mouth, teeth, and lips), extremities (arms, fingers, legs, and tails), headdresses, and items attached to bodies or held in the hands (smoking mirrors, backpacks, and weapons).

Previous experiments have shown that most of the initial detailing on a silhouette could have been produced by indirect percussion to form what we called Element Isolation Notches (Titmus and Woods 2003, 135). Final features were then created using various techniques of detailed flaking to form what we called Element Designs (136). Our experiments demonstrated that even though it is true that the most catastrophic failures occurred early in the percussion stages, significant failures such as decapitations and amputations also occurred, even during delicate pressure-flaking.

NEW EXPERIMENTS AT PRODUCING SILHOUETTE DETAILS

To attempt replication of the ancient silhouette production process, large, chert bifaces would be used (for a discussion of the replication process, see Flenniken 1984). However, raw material in suitable sizes to produce blanks needed for silhouettes in the size range of the Copan specimens is nearly impossible for me to find, and I lack the experience and skill at working material in that size range. Thus, in order to get any practical experience at re-creating silhouette technical

1 Alejandro Pastrana (2016) has previously proposed the use of the term "silhouette" to replace the commonly used "eccentric." Grammatically and technically, "silhouette" is a more appropriate term. This chapter acknowledges that grammatical adjustment accordingly.

details, I worked with smaller bifaces. I looked closely at specific design elements on ancient silhouettes, primarily from the cache of chert silhouettes from Rosalila. I also occasionally referred to illustrations of silhouettes from other published sources, primarily to attempt replication of complete, though smaller silhouettes. For the experiments described in this chapter, I substituted the use of obsidian for the chert used exclusively in the Rosalila specimens. It was not possible for me to find chert of suitable quality and size to allow attempted replication of the Rosalila specimens. Though more fragile, obsidian has fracture properties similar enough to chert to allow an analogy. Working with smaller selected details of the massive Rosalila pieces allowed me to focus almost entirely on the final detailing processes. The experimental procedure used here is not unlike a painter doing studies in a sketchbook, or a sculptor making a scale maquette before producing a life-size sculpture.

For these experiments, I first photocopied illustrations from various published sources, then enlarged some details to actual size. I then cut out templates so I could transfer outlines to my replica bifaces. Replica obsidian bifaces were then coated with a flat spray lacquer to eliminate reflections and show surface topography. In this way, flake-removal details could be clearly distinguished and documented. In some multistep processes, experimental bifaces were repainted several times to photograph successive manufacturing steps.

CREATING SILHOUETTE DETAILS

I spent several months attempting to replicate a variety of silhouette design elements, not always successfully. I summarize some of those experiments here, starting from what I consider easier and then to more complex examples.

Oblique notches: Some silhouettes (figure 5.1) possess what are described as oblique notches (Agurcia Fasquelle et al. 2016, 66) or clapboard-style notches (Clark and Titmus 2012, 285). These are among the least-complicated design elements to replicate and rarely result in errors. This style of notching involves removal of multiple, small pressure-flakes to form each individual notch, requiring minimal force and allowing an opportunity to recover from any damaged notching-flake platforms. The main issues in making oblique notches is matching the size of the notches and spacing them properly along the length of the available margin. This planning is particularly significant, as it has been suggested by John Clark that the number of notches along a margin may have numerical significance (286). In my experiments, I found it useful to mark replica margins to establish and maintain notch spacing.

Direct regular notches: Many silhouettes have sections of margins with shallow, perpendicular notches formed in a regular pattern around headdresses or other design elements (figure 5.2). These are often defined as "direct regular notches" (Agurcia Fasquelle et al. 2016, 66) or simply "fine, perpendicular notches" (Clark

FIGURE 5.1. *Examples of oblique or clapboard-style notches on replicas.*

and Titmus 2012, 287). Direct regular notches present several challenges, making them more difficult than oblique notches. First, direct notches are generally made by removing single notching-flakes alternatively from each face. The unintentional rounding of a shallow notch platform can prevent further notching-flake removal and results in mismatched notches (figure 5.3). Second, to create direct regular serrations, the margin must be the proper thickness and quite regular. Third, care must be taken to avoid breaking off one serration, or the rhythmic

FIGURE 5.2. *Examples of direct regular notches on replicas.*

pattern will be lost. One common error in making serrated margins includes removal of an overly deep notching-flake by application of excessive outward force, which breaks off one or more teeth.

Facial features: Anthropomorphic facial features are an important aspect of most Maya silhouette designs (figure 5.4). As might be expected, their production was highly regularized as illustrated by Clark (Clark and Titmus 2012, 286). Examination of many faces on the ancient silhouettes shows the basic facial outline was formed along a gentle curving profile with a wide notch separating the nose from the upper lip, a narrow notch forming the mouth, and a receding line forming the chin. The most distinguishing features are the nose and lips. As simple as this appears, lips are a difficult feature to replicate. A common error in forming this type of face is inadvertent breakage of the oft-projecting lips. If you form the projecting lips first, then try to form the mouth last, it is easy to break off one or both lips. To prevent this, it is better to notch the narrow mouth opening first, then form the lips around the notch. Close examination of flake scars on ancient specimens confirms this sequencing.

Smoking celts: Smoking celts are common on Maya silhouettes (figure 5.5). They are challenging to create, because any error can result in the smoking

FIGURE 5.3. *Examples of direct notching errors: broken serrations, incorrect spacing, and rounded notch platforms.*

FIGURE 5.4. *Forming facial features on replica showing curved outline of facial features.*

FIGURE 5.5. *Replica of face with smoking celt.*

celt being rendered unidentifiable. This is such an important symbol that its absence can completely alter the identity of the character, and thus they are considered fairly difficult to produce (Titmus and Woods 2003, 139). To form a celt, small pressure-flakes must be detached using judicious application of force perpendicular to the design element mass to prevent breaking off delicate projections.

Deep decorative notches: Some silhouettes have deep, direct notches that expand inward from the margin (figure 5.6). These are difficult to replicate, requiring patience and lack of errors. The notch entry is narrow, and the insides of the notch expand, leaving curved projections between the notches that are very delicate. If one small detail is broken, the repeating pattern is disrupted. These notches can be formed in a pattern of diminishing size. One of the specific challenges I discovered in attempts to replicate decorative notches of this type was devising ways to hold the delicate silhouette while flaking (see below).

Headdresses: All silhouette faces have some form of headdress, ranging from simple to highly ornate (figure 5.7). Some of these headdresses are among the most complex details to replicate and result in frequent breakage, which is particularly problematic, since the small details of a headdress were, in my opinion, the last details to be applied to a silhouette. In forming a headdress, the order in which pressure flakes are removed, the direction of the applied pressure force, the form of pressure flaker, and the manner of holding must all be carefully

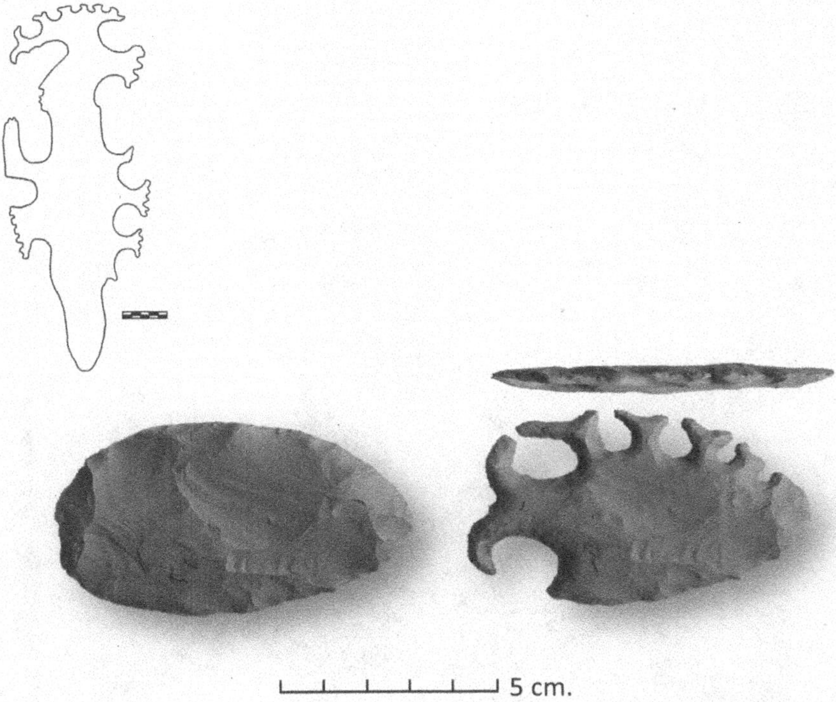

FIGURE 5.6. *Example of deep expanding notches.*

considered. I nearly abandoned my experiments with silhouettes early on as a result of so many failed headdress attempts.

Element isolation notches: Deep element isolation notches have proven for me to be the most difficult detail to produce without error (figure 5.8). In our earlier paper (Titmus and Woods 2003), we focused on indirect percussion. However, for this purpose, my recent work convinced me that indirect percussion sometimes imparted excessive forces on delicate silhouette features, especially those made late in the manufacturing process. So, I spent many hours experimenting with pressure flaking to produce these details. I found it very challenging to create deep notches into the thicker body of a silhouette but felt that the highly regularized flaking pattern observed inside deep element design notches on ancient specimens strongly suggested some use of pressure flaking. After using both techniques (figure 5.9), I now realize it is impossible to clearly distinguish small indirect flakes from large pressure flakes based on flake scar length or regularity. I learned that making deep element design notches into the thicker mass of a biface generally requires a combination of *both* a punch and pressure flaker, as one technique alone does not generally yield satisfactory results.

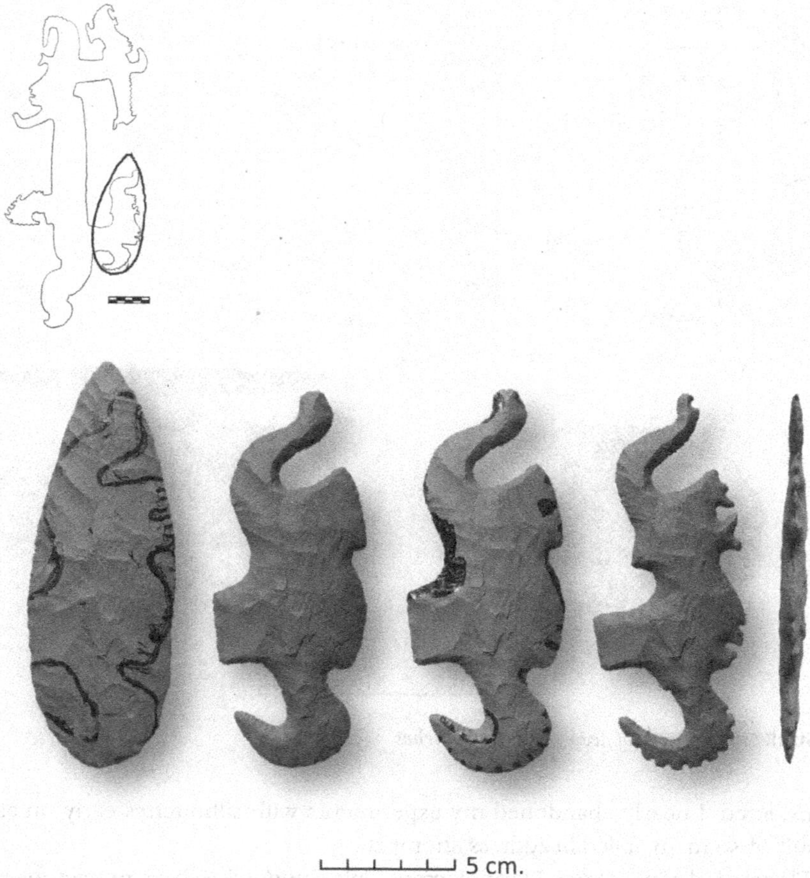

FIGURE 5.7. *Example of headdress on a replica.*

In thicker areas, the notching platform angle is less acute than notches near the thinner margins. A thicker cross-section can result in a near-obtuse notching platform angle, greatly increasing the amount of required force for flake removal. It presents more opportunity for inadvertently rounded flake platforms, excessively deep flake scars, opportunities for perverse fractures (figure 5.10), and increases in the chance of overshot flakes that might remove portions of the opposite margin. That being said, near-obtuse margins inside element design notches can provide excellent platforms for seating of a punch. Then (figure 5.11), removal of one or more flakes by indirect percussion can be followed up by more precise pressure-flakes. Last, working inside a deep element isolation notch presents a challenge for knapping tool placement. Slight leveraged pressure from a tool placed inside an element design notch can easily snap

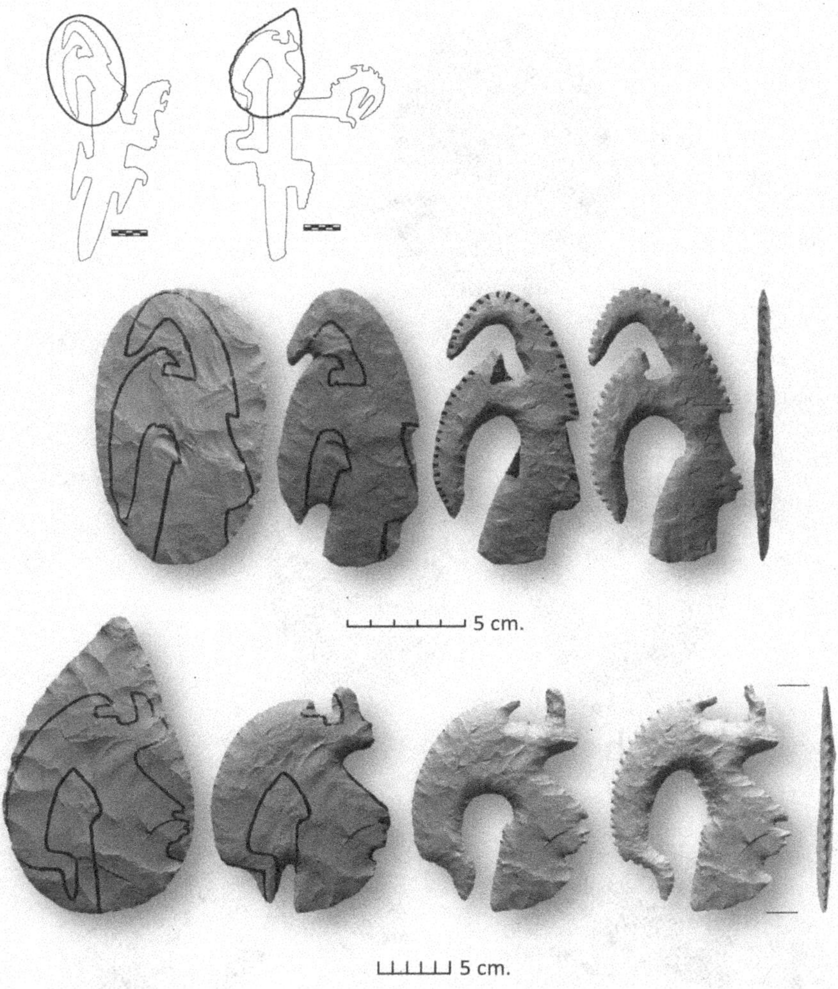

FIGURE 5.8. *Examples of headdresses on replicas formed by deep element isolation notches.*

off a detail (figure 5.12). Figures 5.11 and 5.12 were not based on pieces from the Rosalila cache. Both of these were small enough to allow an attempt to replicate completely, and both had deep element isolation notches that required use of both indirect percussion and pressure flaking.

Examination of flake scars on deep element isolation notches on many specimens shows that ancient knappers (figure 5.13) exerted strong inward force only when that force was toward the main body of the silhouette or larger design element. If too much force was applied away from the main body, say toward a headdress feature, breakage was almost guaranteed.

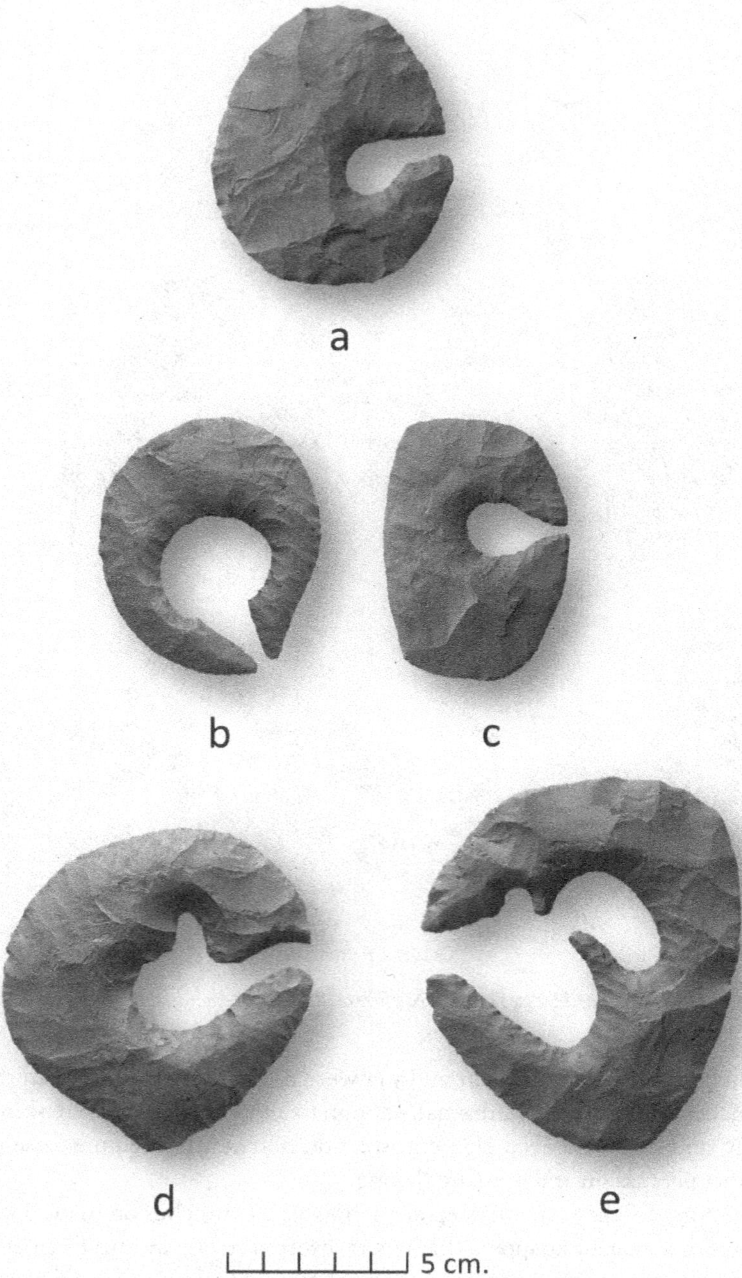

FIGURE 5.9. *Element isolations notches by: (a) indirect percussion; (b and c) pressure flaking; and (d and e) a combination of indirect percussion and pressure flaking.*

FIGURE 5.10. *Examples of perverse fractures during forming of deep element isolation notches.*

Applying strong inward forces (figure 5.14) either by pressure or indirect percussion required some rather unusual holding techniques for both hands. At times I felt like a guitar player reaching for a very difficult chord. I occasionally had to use multiple fingers, sometimes positioned through a design element notch, to adequately secure a preform. Finding suitable holding positions to prevent damage to the preform or injury to my hand or fingers was difficult on nearly all replicas shown here, regardless of size of the piece being worked. Awkward holding positions sometimes resulted in small lacerations. However, I learned to not wear gloves, as fabric or leather would snag delicate margin details. Instead, I would occasionally wrap parts of the preform in a soft, fine-weave cotton while leaving only the objective portions of the preform exposed.

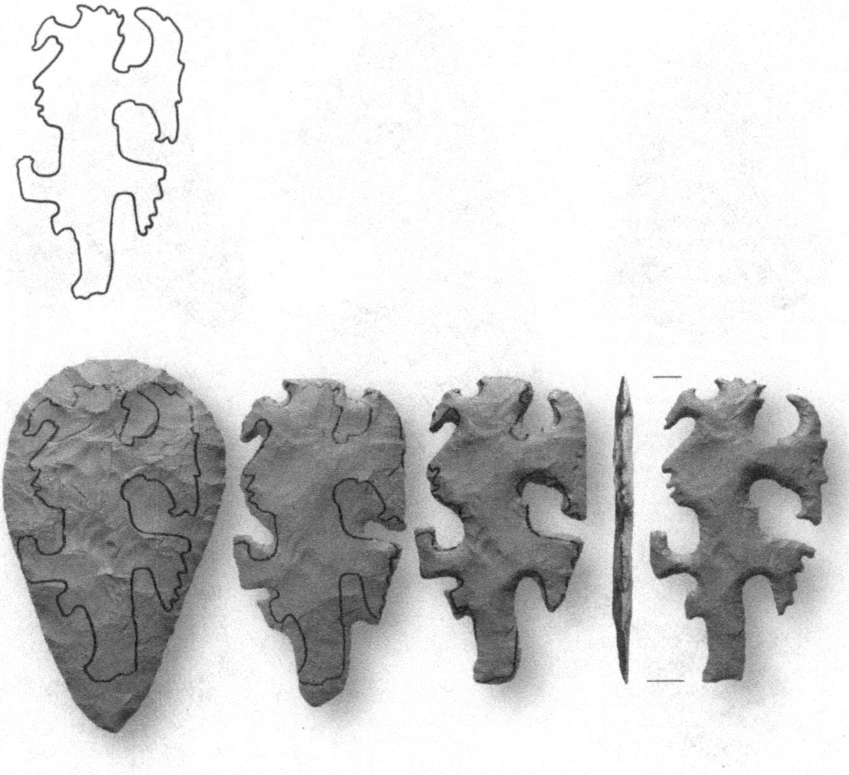

FIGURE 5.11. *Example of a replica formed primarily by deep element isolation notches made by a combinations of indirect percussion and pressure flaking.*

CONCLUSIONS

Preliminary experiments at reproducing small silhouette details have shown that errors on both ancient and modern specimens can reveal much about technology and design. Small technical errors can significantly alter the intended design turning a smoking celt into a horn, reducing the requisite number of notches on a headdress, or completely decapitating an important design element.

I also discovered errors could result from basic workshop issues. For example, it is very easy to snag and break a delicate margin detail with hands, fingers, tip of a knapping tool, gloves, clothing, or protective pad. Incorrectly setting a delicate silhouette replica on a cluttered workbench had catastrophic consequences. Thus, I had to learn to move slowly and keep my working area clean and fairly well organized. This conclusion offers interesting considerations for the likely accommodations and contexts of ancient workshops.

FIGURE 5.12. *Replica of Dumbarton Oaks specimen showing example of typical manufacturing error while attempting a deep element isolation notch.*

Ricardo Agurcia Fasquelle and colleagues (2016) have suggested that the delicate nature and lack of damage on the Rosalila specimens indicates that they were likely made nearby and not transported from afar. They cite the lack of damage on the cached specimens and suggest the few breaks that are evident were from manufacture, not transport (2016, 66). I would maintain that in many instances, it would not be possible to clearly distinguish between a break caused by handling during manufacture from a break caused during transport of a finished silhouette. Several replicas in my experiments were damaged from handling during and after manufacturing. To minimize such breaks, I learned to wrap some parts in fabric to provide sufficient protection. Fabric wrapping also helped prevent lacerations to my fingers, which was a common occurrence. Wrapped replicas seemed very durable and were moved in and about my workshop on numerous occasions. Thus, wrapping of silhouette replicas served at

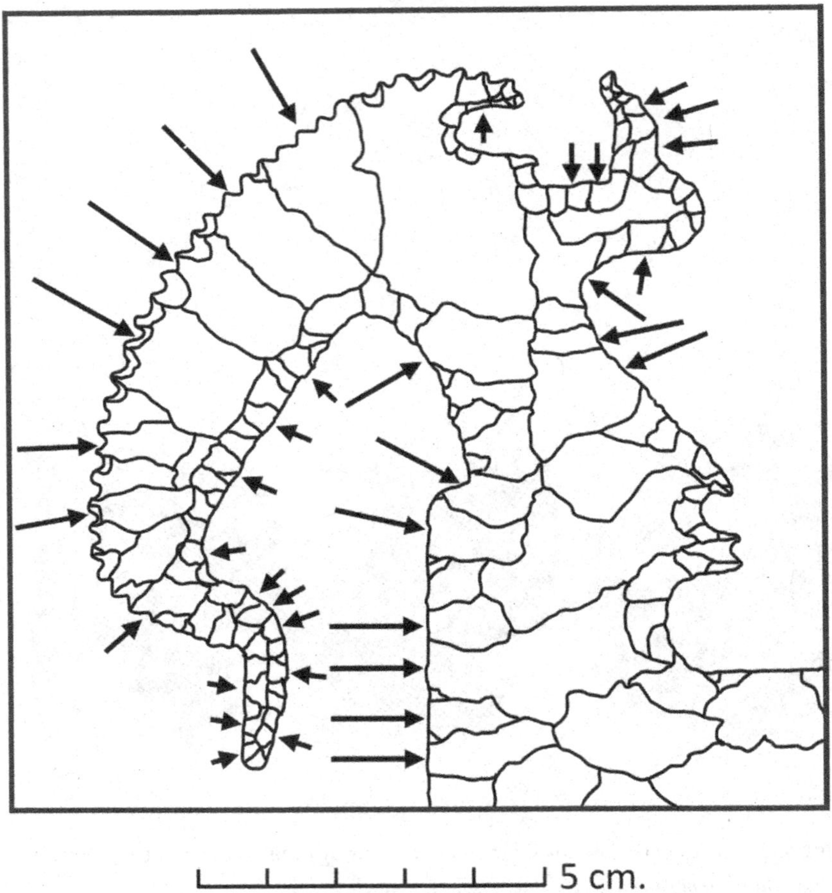

FIGURE 5.13. *Detail of Copan specimen showing flake scars of varying length; long arrows denote strong inward applied force, short arrows denote more-gentle upward applied force.*

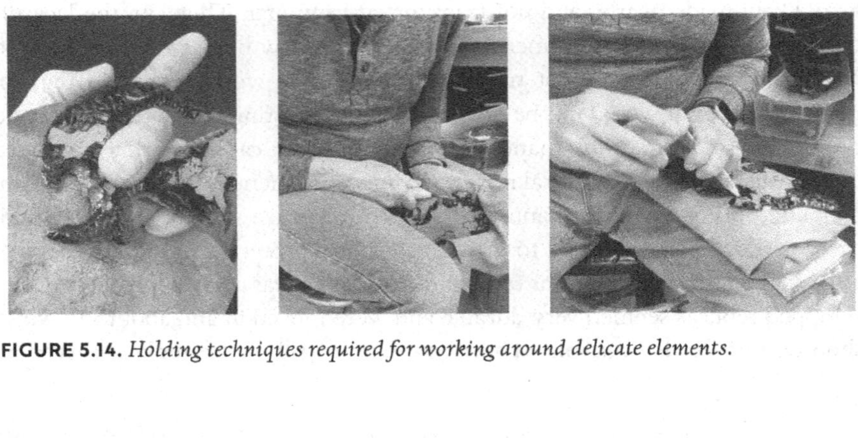

FIGURE 5.14. *Holding techniques required for working around delicate elements.*

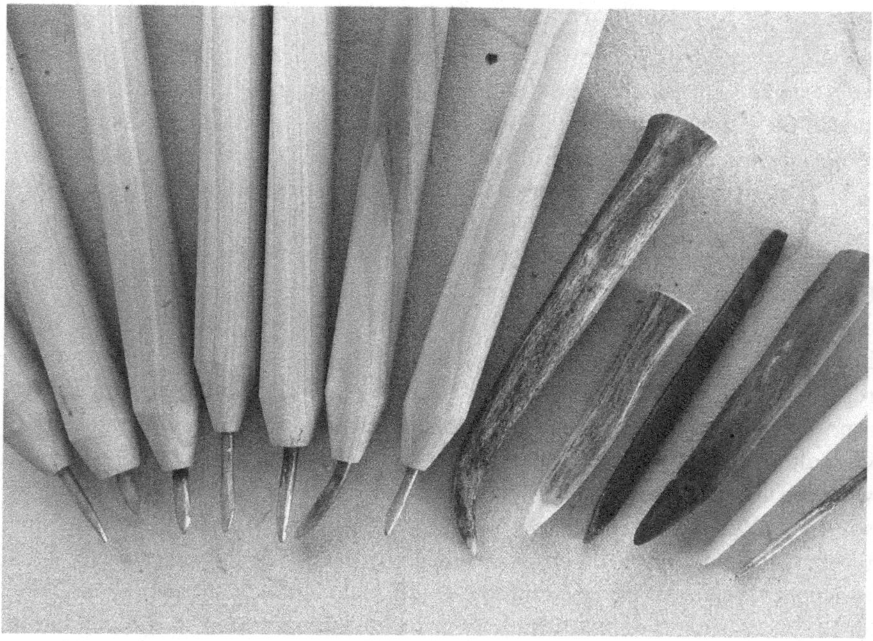

FIGURE 5.15. *Some tools used in this study.*

least three purposes: to minimize handling damage, prevent personal injury, and facilitate transport and storage. From my experiences, I would suggest that transporting wrapped silhouettes, even large ones, over long distances would not be a significant issue. Thus, it is important to note the association of fabric with silhouettes from Copan may suggest both practical and symbolic purposes.

I quickly learned that making a silhouette requires good vision. I was forced to use strong light, reading glasses, and sometimes an illuminated magnifier. Vision issues were likely faced by ancient knappers, as it is a common human condition to lose close vision starting in the early forties (www.aoa.org). This raises some interesting questions when proposing that many of the Copan eccentrics were likely made by a "master." If knapping is an accumulated skill (Clark and Woods 2015), then it is likely older craftspersons would be producing the most complex work.

Knapping tools must be of varied form and well maintained. My experiments required numerous tools of different configuration and size, and frequent tip maintenance (figure 5.15; see also Titmus 1985). For the work reported here, I used numerous tools, frequently shifting between them as needed.

My experiments, though they only represent a few months of trial and error, resulted in some observations regarding the value of attempting to replicate small details on ancient specimens. For example, flake scars on ancient specimens can be highly instructive. They indicate not only the amount of applied force but also the direction of that force, often revealing when the knapper was avoiding forces in a direction that could break off delicate margin details. Flake scars can sometimes also reveal the order of flake removal. This fact is important in the study of silhouettes, because flake scars suggest longer flakes were removed earlier in the process and short "retouch" flakes usually intersect those flakes, showing they were removed last. And, it is in this distinction that one can "read" the sequence of silhouette manufacture.

Manufacturing errors on silhouettes are also very instructive. As shown by Payson Sheets (2016), many errors are of little consequence to the finished design. In fact, some "errors," such as step fractures along margins, may have been intentional. Some errors that detach portions of the silhouette can be repaired with only minimal alteration to the intended design. Other breaks can render a silhouette character unidentifiable. Identifying repaired errors can be difficult. If a smoking celt is not present, was it not part of the original design or was a broken celt covered over by subsequent flaking? It is tempting to look at some silhouettes and speculate about missing or damaged design elements. This makes one wonder how much leeway ancient knappers had in adjusting designs to accommodate errors. One example would include specimen 90-3 (Sheets 2016, 80–81). Comparison of the headdress crest on this specimen to the headdress crests on the two matching pieces (Agurcia et al. 2016, fig. 20) suggests the possibility that the downward-projecting portion of the headdress may have been broken and subsequently altered to accommodate the missing piece.

In closing, after many months of experimenting in my comfortable lab at home with modern tools and conveniences, I am now more than ever humbled by the precision and consummate skill shown on many ancient specimens. I wish to acknowledge and thank the authors of the recent publication on the Copan silhouettes for their documentation of the specimens. The detailed photos, digitized scans, and descriptions have not only revealed the forest; they have also revealed the trees within the forest.

Acknowledgments. I wish to acknowledge the assistance of several people without whom this study would not have happened. I thank Zac Hruby for organizing this symposium and inviting me to participate. And, thanks to Hattula Moholy-Nagy, discussant for this symposium. Thanks also to John Clark for his essential input and encouragement. John also provided support for our trip to Copan to see the eccentrics on exhibit. Thanks also to Richard Hansen, who provided the opportunity to visit the National Museum in Guatemala. Thanks also to Alejandro Pastrana for allowing me to participate in his work at the Mountain of Knives, as well as his providing me with many important references and ideas pertaining to "silhouettes." Last, I thank Gene Titmus, longtime friend and colleague, who provided the initial inspiration for this study.

REFERENCES

Agurcia Fasquelle, Ricardo, Payton Sheets, and Karl Andreas Taube. 2016. *Protecting Sacred Space: Rosalila's Eccentric Chert Cache at Copan and Eccentrics Among the Classic Maya*. Monograph 2. San Francisco: Precolumbia Mesoweb Press.

Clark, John E., and Gene L. Titmus. 2012. "Flint Effigy Eccentrics." In *Ancient Maya Art at Dumbarton Oaks*, edited by Joanne Pillsbury, Mariam Doutriaux, Reiko Ishihara-Brito, and Alexandre Tokovinine, 273–287. Pre-Columbian Art at Dumbarton Oaks 4. Washington, DC: Dumbarton Oaks.

Clark, John E., and James C. Woods. 2015. "Squeezing Life from Stones: The Human Side of Replication Experiments." In *Works in Stone: Contemporary Perspectives on Lithic Analysis*, edited by Michael J. Shott, 197–212. Salt Lake City: University of Utah Press.

Flenniken, J. Jeffrey. 1984. "The Past, Present, and Future of Flintknapping, An Anthropological Perspective." *Annual Review of Anthropology* 13:187–203.

Joyce, T. A. 1932. "The 'Eccentric Flints' of Central America." *Journal of the Royal Anthropological Institute of Great Britain and Ireland* 62:xvii–xxvi.

Pastrana, Alejandro, and Silvia Dominguez. 2017. *Proyecto Yacimientos de Obsidiana, Temporada de Campo 2016–2017. Informe técnico al Pastrana Alejandro y Silvia Domínguez 2017, Proyecto Yacimientos de Obsidiana, Temporada de Campo 2016–2017*. Informe técnico al Archivo del Consejo de Arqueología, *INAH*. Archival manuscript, Consejo de Arqueología, INAH, Mexico City.

Schele, Linda, and Mary Ellen Miller. 1986. *The Blood of Kings: Dynasty and Ritual in Maya Art*. New York: George Braziller and Kimbell Art Museum.

Sheets, Payson. 2016. "Appendix A: Individual Descriptions of Bifaces and Eccentrics." In *Protecting Sacred Space: Rosalila's Eccentric Chert Cache at Copan and Eccentrics Among the Classic Maya*, edited by Richard Agurcia Fasquelle, Payson Sheets, and Karl Andreas Taube, 70–133. San Francisco: Precolumbia Mesoweb Press.

Stuart, George E., and Gene S Stuart. 1993. *Lost Kingdoms of the Maya*. Atlanta: National Geographic.

Titmus, Gene L. 1985. "Some Aspects of Tool Notching." In *Stone Tool Analysis: Essays in Honor of Don E. Crabtree*, edited by Mark G. Plew, James C. Woods, and Mark G. Pavesic, 243–264. Albuquerque: University of New Mexico Press.

Titmus, Gene L., and James C. Woods. 2003. "The Maya Eccentric: Evidence for the Use of the Indirect Percussion Technique in Mesoamerica from Preliminary Experiments Concerning Their Manufacture." In *Mesoamerican Lithic Technology: Experimentation and Interpretation*, edited by Kenneth G. Hirth, 132–146. Salt Lake City: University of Utah Press.

Woods, James C., and Gene L. Titmus. 1996. "Stone on Stone: Perspectives on Maya Civilization from Lithic Studies." In *Eighth Palenque Round Table, 1993*, edited by Martha J. Macri and Jan McHargue, 479–489. San Francisco: Pre-Columbian Art Research Institute.

6

Was the Elaborate Chert Eccentric from San Andres, El Salvador, Made by the Rosalila, Copan, *El Maestro*?

PAYSON SHEETS

Maya eccentrics have puzzled scholars for a century. Fortunately, the careful excavation and analysis of the extraordinary Rosalila cache from Copan, Honduras (figure 6.1), sheds some light on their nature, function, and importance to the Maya (Agurcia Fasquelle et al. 2016). In this chapter I refer to them as "eccentrics" and realize that we do not have a clue as to what the Maya called them. They could be called by various other etic terms; however, in keeping with my previous publications on these objects, I retain the term "eccentric" here. There are many eccentrics in private and museum collections, some even as elaborate as these from Copan, but the overwhelming majority were looted and thus without chronological or cultural context. Even the few that have been excavated from other sites rarely had any organic materials preserved, and often any painting was removed prior to exhibition. Almost all were vigorously scrubbed for public viewing, removing what the Maya undoubtedly believed were essential aspects of the painting, bundling, use, caching, and deposition of these objects.

The striking similarities between an elaborate chert eccentric from San Andres, El Salvador (figure 6.2), and the finest six eccentrics from the Rosalila

https://doi.org/10.5876/9781646427819.c006

FIGURE 6.1. *Map showing locations of San Andres in El Salvador, and Copan in Honduras. Map by Payson Sheets.*

cache, from Copan, Honduras (Sheets 2016a), raises the possibility that both were made by the same person. I referred to this supreme knapper as "the master," or "El Maestro," because of the extraordinary skill evident in manufacture of the six finest eccentrics from the Rosalila cache. I referred to the skilled knapper who manufactured the slightly less magnificent, three other eccentrics as *the apprentice* (Sheets 2016a, 2016b). I am confident that El Maestro manufactured the six-largest elegant eccentrics and that an assistant, "the apprentice," made the remaining three. El Maestro at Copan may well have made the San Andres eccentric and perhaps done it a few years earlier than those from Copan. All these eccentrics date to the Late Classic period, but more precise dating has been elusive. Circumstantial evidence is presented here to better date these eccentrics.

The twelfth ruler of Copan, 628–695 CE, initiated major expansions of influence in the Maya area in the final decades of the seventh century, especially at two major sites in El Salvador: Chalchuapa and San Andres. Architectural, sculptural, and artifactual evidence witnesses that expansion, and the San Andres eccentric likely was made during the last years of his reign. That is an objective of this chapter, to try to develop some kind of context and possible relationships to this otherwise academically "floating" artifact. I argue that Ruler 12 may have arranged for the expert knapper to visit San Andres and fabricate the eccentric there. All these eccentrics have so many thin, delicate projections, and they are so long and thin that I believe they are too fragile to have been transported any significant distance. Rather, I think they were made adjacent to where they were cached. I believe they were neither curated nor used in public ceremonies prior to their being cached. Admittedly, there is no direct evidence of this in the form of debitage or production failures.

FIGURE 6.2. *Eccentric chert from San Andres, El Salvador. Photo by Payson Sheets.*

The Rosalila cache of nine eccentrics has been radiocarbon dated to the eighth century, during the Late Classic (Agurcia Fasquelle et al. 2016, 49) but perhaps after the San Andres cache, since the technical, stylistic, and morphological similarities are so great. Ruler 12's reign ended a few years before that span of radiocarbon dates, so if he made the arrangements for the San Andres eccentric, then it would predate the Rosalila cache. A few decades after the early part of the radiocarbon range, four eccentrics were manufactured for the Hieroglyphic Stairway, and they are securely dated to 755 CE. They differ significantly from the San Andres and Rosalila eccentrics, and might have been made by the next generation of immensely skilled knappers.

THE SAN ANDRES CHERT ECCENTRIC

San Andres, formally known as Campana San Andres, is the largest archaeological site in the Zapotitan valley of central El Salvador (Black 1983). It had a population of about 10,000 people during the Late Classic period. In the center is an elevated large plaza, upon which various pyramids and other structures were built. In 1977 Jorge Mejía was excavating in Structure 7, under the general guidance of Stanley Boggs. Mejía discovered the chert eccentric (see figure 6.2) in a cache during his excavations. The cache also included *Spondylus* shell, stingray spines, and Copador ceramics (Bello-Suazo 2009, 84; Cobos 1994, 35; Moisa 2013, 62). Unfortunately, neither he nor Boggs published anything about the excavations, the cache, or the eccentric. Various searches of museum collections and archives have found no field notes or written descriptions or interpretations. The result is that the eccentric is barely more contextually grounded than the looted eccentrics in various museum and private collections. However, its extraordinary nature, and the knowledge that it came from San Andres, justifies consideration and comparison here.

The San Andres chert eccentric is 43.4 cm long and only 1.4 cm thick (see figure 6.2), and thus is longer and thinner than the average of the finest eccentrics from Rosalila. The chert is of high quality for flaking, with no imperfections. No major flaking errors were detected, but two flakes got close to hinge fracture terminations, and there may have been a problem with the animal head portion of the eccentric (see later in this section). The base terminates with a large patch of white cortex, which is a hallmark of all of the large effigy flints from this workshop. The principal figure may be depicted as seated on a throne, by two downward projections on either side of what would be his lower torso. A reference to his feet might be the small forward-facing projection.

Behind the figure is a long, upward projection with twenty-six notches, leaving thirty small projections. Because the notches were made directly into the edge, the projections may be referencing feathers. Immediately above that is a long downward projection with twenty-nine notches, also likely referencing feathers. A very small error evidently occurred midway in the notching series, having removed two or three additional notches. Above those two long projections behind the principal figure is an anthropomorphic head in profile, with pursed lips. To the left of this head is a long projection upward, along the line of the principal figure's head that curves forward. Along the top of it are twenty-nine pressure-flaked notches, leaving thirty tiny projections that probably represent feathers.

Between the top element and the principal figure's forehead is an animal head, probably depicting a deer. It appears that the nose of the animal was intended to be longer and to narrow with distance, but a flaking error occurred. The evidence is that the nose does not end in a feather edge, clearly desired for all edges of these elaborate eccentrics. Rather, it ends in a flat, abrupt facet. The principal figure's face is carefully flaked, with a long straight line of the forehead ending in pursed lips, much like the ones in the headdress but more explicitly rendered.

The arm extends below the face, ending in an upturned hand. Four notches define four fingers and a thumb.

For decades since its discovery and exhibition, the San Andres eccentric has fascinated visitors to the national museum in San Salvador. For archaeologists it has been frustrating to relate it to anything else, or get an idea about to what time it dates. The discovery and publication of the Rosalila cache in Copan provide some intriguing clues, which are pursued in the following section. Artifacts demonstrating Copan-Salvadoran contacts provide some other clues. Jeb Card and Marc Zender (2016) summarized the architectural, stylistic, and artifactual evidence of Copan's influence and contacts with Salvadoran sites. They include an artifact with Ruler 12's name on it, given as a gift to an elite at Chalchuapa, and could well include arranging for the visit of the supreme knapper, whom I call "El Maestro," to manufacture the San Andres eccentric.

COPAN AND THE ROSALILA CACHE

The most elaborate complete building ever excavated at Copan, and perhaps in the entire Maya world, was named Rosalila by archaeologists. Its construction began in the late sixth century CE, and it functioned, probably as an ancestral shrine and burial place of the city's dynastic founder, for over a century until it was carefully entombed by the early mid-eighth century (Agurcia Fasquelle et al. 2016, 8). It was of such great importance to the Maya that they did not destroy the uppermost portions of a building to be decommissioned. Rather, they covered the lavishly painted structure with thick layers of white plaster, carefully covered the entire building with fill, and then finally emplaced the cut stone masonry of the final structure. The cache containing the nine eccentrics was emplaced on the floor of the building during the long and elaborate process of its entombment.

The cache was contained within round river stones, forming a circle about 40 cm in diameter (Agurcia Fasquelle et al. 2016, 51). It contained elements of water, land, and air, in the form of marine shells and stingray spines, jaguar bones, and feathers. All of the bifaces and the eccentrics received some painting, evidently cinnabar, and all were wrapped individually with fine cotton cloth, usually in a gauze style. The cloth was usually colored green or blue. Many items had a rough brown final wrapping, derived from bark or agave. The wrapping evidently helped create sacred bundles (Sheets 2016a). Leaning up against the circle of stones were three chert bifaces, of which one was magnificently crafted, 36 cm long and only 1.2 cm thick (Sheets 2016b, 72; also see chapter 4 here). The interpretation of these eccentrics is that they were supernaturally powerful artifacts manufactured to guard sacred space (Agurcia Fasquelle et al. 2016). The very material from which they were made has enormous sacred power, as the Maya believed that obsidian and chert were formed by lightning sent by K'awiil or Chahk (see Taube

2016, 9–18). This belief is still predominant among the Maya today. I have heard many Ladinos and Maya in Guatemala refer to obsidian as *piedra de rayo*, which translates as "stone of lightning." The profile depiction of K'awiil on many of the other chert eccentrics adds even more supernatural power for guarding sacred space. The nine wrapped eccentrics were placed standing, leaning against the stones, and facing west. That direction symbolizes ending, as that is the direction in which the sun sets.

FIGURE 6.3. *Eccentric chert from the Rosalila cache, Copan, Honduras. Artifact 90-5. Photo by Payson Sheets.*

The nine eccentrics refer in their number to the underworld. I divided the nine into two groups: with six being of truly exceptional manufacturing skill and three others of almost that extraordinary level (Sheets 2016, 64). I suggested that all nine were made from tabular chert but now am not so sure of that. I asserted that the six were made by "El Maestro" and the three were made by the "apprentice." The three by the apprentice were quite impressive, averaging 31.2 cm in length, perfectly flat, with a well-formed principal figure face. All three have some manufacturing difficulties, with flakes ending in step or hinge fractures. Each has one well-formed principal figure face but no more secondary faces, in contrast to the multicephalic depictions in the elegant eccentrics by El Maestro. All the notching was done with a pressure flaker of a large diameter tip (~6 mm), with the exception of the lips and fingers, which were done by a small-diameter pressure flaker (cf. Woods, chapter 5). Both the technological skill and the style of depiction are so different from the other six, and from the San Andres eccentric, that these three receive no further consideration here.

The six eccentrics by El Maestro are truly extraordinary. They are very large indeed, averaging 41.3 cm in length and they are completely flat, and evince very few flaking difficulties. Because Artifact 90-5 (figure 6.3) is one of the two most similar to the Salvadoran eccentric, I will focus on it as a test case, and not include the other five in detailed comparisons. It is 35.5 cm long, impressively thin at 1.65 cm, and almost error-free in flaking. Thus, it is shorter than the Salvadoran eccentric, and the other five most elaborate ones from Copan.

THE COMPARISON OF ARTIFACT 90-5 FROM COPAN AND THE SAN ANDRES ECCENTRIC

Generally speaking, the two eccentrics are quite similar in early- and late-stage flaking, crisp outlines, and their impressive length and width-to-thickness ratio. Hereafter I refer to the San Andres eccentric as "SA" and the Rosalila Artifact 90-11 from Copan as "C." Both principal figures are rendered in similar fashion, with almost straight lines connecting the hairline to the tip of the nose. Fine pressure flaking created both specimens' lips and chin, with SA's notching being deeper, which creates the impression of pursed lips. The central figures of both eccentrics have an arm extending outward and upward, probably in an important and symbolic gesture. Each features the principal figure apparently seated on a throne, and each has a small projection forward, above the throne, that may indicate the principal figure's legs. The heads and torsos of both principal figures are rendered in very similar fashions, with long straight foreheads all the way to the nose, a round chin, and a straight back and angular shoulder. Each one has cortex at its very base, and both bases end with a flat facet, with only a slight amount of cortex on C and an extensive amount on SA. Cortex is largely calcium carbonate and softer than the chert, therefore rendering it challenging to maintain at the base. The C principal figure has an elaborate K'awiil smoking torch above the forehead of an anthropomorphic head on the extended arm, unlike anything on SA. Behind each principal figure are two long headdress extensions that are rather similar. The top one on both artifacts extends downward to a point, with fine pressure notching along the full length of its outside. The SA notching is direct, presumably depicting feathers, while C's notching is oblique, likely referencing lightning. Each artifact has a downward projection of the headdress, decorated with fine pressure flaking. The SA eccentric has an ascending projection that terminates very close to the descending extension, in contrast to C.

These eccentrics are multicephalic, with an anthropomorphic face in the headdress of SA, and in the hand of C. The SA specimen has pursed lips, and C does not have defined individual lips on either anthropomorphic face. In this regard this eccentric is unusual in the Rosalila cache. All other eccentrics, elaborate or not quite so elaborate, have carefully flaked pursed lips. It is possible that in a rush to finish the lithics, the final flaking of the lips was overlooked. Compared to all other eccentrics, the flat tab of the lower face appears unusual. C has a K'awiil smoking torch above the forehead, and oblique notching probably referencing lightning. SA has an animal head above the forehead of the principal figure, probably a deer. Above that animal head is a projection and a forward flare decorated with direct notching.

The San Andres and Copan 90-11 eccentrics are so similar in percussion and pressure flaking, direct notching, and morphological details that they well could have been manufactured by the same lithic specialist. It is also possible that they were made within the same lithic familial lineage, a generation from each other. Other archaeological evidence, presented later in the chapter, indicates that the SA eccentric may have been created before Rosalila.

CHRONOLOGY AND COPAN EXPANSIONISM

The Rosalila cache was dated using pertinent radiocarbon assays. Ricardo Agurcia Fasquelle et al. (2016, 49) dates the cache to sometime between 710 and 775 CE. Clues as to when the cache was deposited within this broad range of time can be gathered from a more general knowledge of when Copan interaction with central and western El Salvador occurred. Card and Zender (2016) note the intensity of Copan and Salvadoran sites sharing Copador, a name combining Copan and Salvador, ceramics in the early part of the Late Classic period. Two large Salvadoran sites share styles of monumental architecture and sculpture with Copan; they are San Andres and Tazumal (2016, 279).

Card and Zender (2016) provide extensive documentation of the height of Copan's contact with other sites in Guatemala, Belize, and those two sites in El Salvador. That occurred during the reign of K'ahk' Uti' Witz' K'awiil (Ruler 12), which lasted from 628 to 695 CE. Card and Zender (2016, 284) note that his regional influence occurred during the latter part of his reign, during the second half of the seventh century. A miniature flask was made in Copan, inscribed with his name, and discovered at Tazumal, Chalchuapa, El Salvador (Card and Zender 2016). Assuming that Ruler 12 was alive when the gifting to Tazumal royalty occurred, and his expansionistic activities may have included assisting in El Maestro's visit to San Andres, that could point toward the final decade of the seventh century. If the San Andres eccentric was manufactured in the beginning of the 710–775 CE range presented by Agurcia et al. (2016, 49), then it could have been El Maestro who made that eccentric and the others at Rosalila, since only fifteen years separate them. Of course, San Andres's eccentric could be earlier in Ruler 12's reign, and Rosalila later in Agurcia's time span, with a generation separating the two. In either case, the rather short time elapsed between the end of Ruler 12's reign and the beginning of the 710–775 CE range could explain the strong similarities in extraordinary skill, along with some relatively minor differences in depictions of the eccentrics. The lack of this extraordinary knapping of lithics after the founding of Classic Maya Copan up to the Rosalila cache, and then between Rosalila and the Hieroglyphic Stairway, and after that, indicates that the Maestro family/lineage was not resident in Copan. It is likely the home of the Maestro lineage is in the Petén or Belize, or even farther north in the Yucatan Peninsula. I believe it is likely that the same Maestro family lineage is responsible

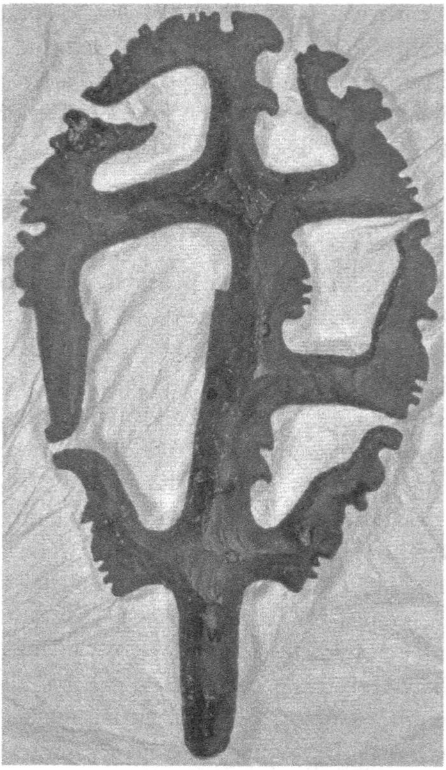

FIGURE 6.4. *Eccentric chert from the dedication of the Hieroglyphic Stairway, Copan, Honduras, AD 755. Photo by Payson Sheets.*

for the San Andres eccentric and all the late Classic eccentrics at Copan. The source of the fine chert, how it was percussion flaked, transported, and finer flaked at the loci of caching all need further research. How the knappers were invited, hosted, and compensated would be wonderful to understand.

Hruby (personal communication, 2019) notes the lack of fine brown chert debitage that would support local manufacture of the Rosalila eccentrics and suggests the artifacts could have been manufactured elsewhere and carried, carefully packed, to Copan. Having worked with these eccentrics, and analyzed them in detail, I doubt that they were manufactured finished and then transported any significant distance, because of their fragility. However, there might be an intermediate interpretation: that roughouts and preforms were manufactured at a distance from Copan, were carefully packed and carried in more durable forms than finished artifacts, and then had the final indirect percussion and pressure flaking done at Copan. That would have resulted in dramatically lower amounts of debitage. And debitage still carried religious significance because of its creation, and so it may have been given special discard.

OTHER LATER ECCENTRICS AT COPAN

Four other elaborate chert eccentrics have been discovered at Copan (see figure 6.4 for one of them). Gustav Stromsvik found one in rubble of the Hieroglyphic Stairway, and David Stuart found three below the altar at that same feature (Fash 2001, 147). They date to 755 CE (9.16.4.1.0 in the Mayan Long Count calendar) at the dedication of that feature, during the reign of Ruler 15. They are significantly smaller than the Rosalila and Salvadoran eccentrics; the maximum length of Stromsvik's eccentric is 18.5 cm. Perhaps the source of the excellent chert was partially exhausted. I interpret the burst of eccentric creation for

Rosalila, and then for the stairway, with no evidence of comparable lithic production before, between, or after these two, as evidence that both bursts were created by itinerant specialists. Styles had changed considerably in the six or more decades separating the two bursts, but flaking control, principal figure handling, and expertise seem to point to the same tradition or family lineage. One wonders what inducements may have been offered for these extraordinary craftspeople to walk hundreds of kilometers, with their heavy tools, chert, and materials, to perform such spectacular production to be immediately buried and begin protecting sacred space. The evidence for quick burial is that they were individually painted, wrapped in layers of colored cotton, wrapped in a coarse material, and emplaced within the cache with the other materials, and then capped with fill inside Rosalila as it was being buried. They were not accessible for removal and reburial.

SUMMARY AND CONCLUSIONS

During the final decades of Ruler 12's rule at Copan, he expanded his influence to the north, west, and south into El Salvador. His rule ended in 695 CE. His influence is particularly clear at the two biggest sites in the country: Chalchuapa and San Andres. A flask with his name on it was gifted to elites at Chalchuapa. The eccentric at San Andres probably was a part of his expansion southward and was likely created during the later years of his reign. The cache of nine eccentrics was emplaced in Rosalila sometime between 710 and 775 CE. Given the striking similarity of the San Andres eccentric and the finest six of the nine from Rosalila, I believe they date to early in that time span. They could have been created by the same person, or perhaps a generation apart. The four eccentrics from the Hieroglyphic Stairway are securely dated to 755 CE but represent a slightly different knapping style from the earlier examples. The San Andres eccentric and the finest six from Rosalila share more similarities than differences, and appear closer in time to each other, than do the four from the Hieroglyphic Stairway. It seems quite possible that the San Andres eccentric dates to the early 690s, Rosalila eccentrics date to the early 710s, and the Hieroglyphic Stairway eccentrics surely date to 755 CE. If that is correct, it is possible the San Andres and Rosalila eccentrics were created by the same person, and the Hieroglyphic Stairway specimens by the next generation, a few decades later.

Acknowledgments. I heartily thank Ricardo Agurcia for inviting me to analyze the most extraordinary cache I have ever seen. I consider myself immensely fortunate to have had that opportunity. And I thank him for correcting some errors that I made in an earlier draft. I appreciate Zachary Hruby inviting me to his fascinating seminar at the Society for American Archaeology meetings and to contribute a chapter to this book. His sharp editorial eye has markedly improved this chapter.

REFERENCES

Agurcia Fasquelle, Ricardo, Payson Sheets, and Karl Taube, eds. 2016. *Protecting Sacred Space: Rosalila's Eccentric Chert Cache at Copan, and Eccentrics Among the Classic Maya*. San Francisco: Precolombia Mesoweb Press.

Bello-Suazo, Gregorio. 2009. *Museo Nacional de Antropología Dr. David J. Guzmán, El Salvador*. San Salvador: Fundación Empresarial para la Acción Social.

Black, Kevin. 1983. "The Zapotitan Valley Archeological Survey." In *Archeology and Volcanism in Central America: The Zapotitan Valley of El Salvador*, edited by Payson Sheets, 62–97. Austin: University of Texas Press.

Card, Jeb, and Marc Zender. 2016. "A Seventh-Century Inscribed Miniature Flask from Copan found at Tazumal, El Salvador." *Ancient Mesoamerica* 27:279–292.

Cobos, Rafael. 1994. *Síntesis de la Arqueología de El Salvador (1850–1991)*. Colección Antropología e Historia 21. San Salvador: Consejo Nacional para la Cultura y el Arte, Dirección General del Patrimonio Cultural, y Ministerio de Educación.

Fash, William. 2001. *Scribes, Warriors and Kings: The City of Copan and the Ancient Maya*. London: Thames and Hudson.

Moisa, Ana Claudia. 2013. "Análisis de la Cerámica Copador procedente de cuatro Sitios Arqueológicos de la fase Payu del Occidente y Centro de El Salvador: Tazumal, Joya de Cerén, San Andres y Madreselva." Licenciatura tesis, Departamento de Antropología, Universidad Tecnológica de El Salvador, San Salvador.

Sheets, Payson. 2016a. "Lithic Technology." In *Protecting Sacred Space: Rosalila's Eccentric Chert Cache at Copan and Eccentrics Among the Classic Maya*, edited by Ricardo Agurcia, Payson Sheets, and Karl Taube, 61–67. San Francisco: Precolombia Mesoweb Press.

Sheets, Payson. 2016b. "Appendix A: Individual Descriptions of Bifaces and Eccentrics." In *Protecting Sacred Space: Rosalila's Eccentric Chert Cache at Copan, and Eccentrics Among the Classic Maya*, edited by Ricardo Agurcia, Payson Sheets, and Karl Taube, 71–133. San Francisco: Precolombia Mesoweb Press.

Taube, Karl. 2016. "The Symbolism of Eccentrics in Classic Maya Religion." In *Protecting Sacred Space: Rosalila's Eccentric Chert Cache at Copan and Eccentrics Among the Classic Maya*, edited by Ricardo Agurcia, Payson Sheets, and Karl Taube, 9–36. San Francisco: Precolombia Mesoweb Press.

7

Eccentric Production Techniques and Caching Practices from Structure A9 at Xunantunich, Belize

KELSEY J. SULLIVAN AND JAIME J. AWE

Ancient Maya eccentrics are ritualized lithic implements defined in various ways by archaeologists. Eccentric typologies focus on implements chipped into geometric, zoomorphic, and other more esoteric forms. Some typologists, however, exclude other cached lithic objects, such as whole or fragmented obsidian prismatic blades and exhausted blade cores, as well as chert cobbles and debitage, which are typically associated with utilitarian tool production sequences. We now know that utilitarian objects were often combined with, and likely served similar functions as, the more seemingly arcane symbolic forms found in Maya caches. Throughout the history of Maya archaeology, data collection techniques and research foci have become more detailed and context oriented, which allows current researchers to ask different kinds of questions of recent archaeological finds. Thankfully, persistent interest in eccentric lithics and caches over the years has left behind a relatively thorough record of these items and as a result, much of the archived eccentrics can be analyzed and compared to those discovered today.

Eccentrics are often studied in relation to their possible sociopolitical, ideological, and cosmological significances in ancient Maya society. These lithic

https://doi.org/10.5876/9781646427819.c007

implements reflect striking and divergent patterns in style, technique, and symbolism from other more utilitarian forms and from different eccentric traditions over time and space. Given their context and expertise in manufacture, it is evident that they played a role in dynamic social processes related to the expression of power and cosmologies. Discussion of the production techniques employed to make eccentrics, however, is often cursory within the literature. A more thorough exploration of these techniques enhances our understanding of the various use-lives of eccentrics and how these implements articulate with other lithic production processes. A deeper sense of these production-related elements may allow for discerning the roles and identities of their producers, as well as their consumption and use, primarily by elites.

Technological studies of eccentric production have focused closely on the techniques used to create the famous forms, such as the anthropomorphic flints found at Copán (Agurcia Fasquelle et al. 2016; Titmus and Woods 2003). Gene Titmus and James Woods (2003) provided a significant contribution to the study of eccentric production with their experimental approach to understanding the production techniques applied to make the elaborate silhouette eccentrics recovered from Copán and Quiriguá. However, much more detail needs to be regularly afforded to these features of all eccentrics to broaden our understanding of differential production sequences and how these differences may relate to sociopolitical and economic structures, such as access to raw materials, and collectives of artisans or other regional variation, as well as differences in expressions of cosmologies, as discussed by Zachary Hruby (2007).

Other experimental approaches to eccentric production have explored standardized production templates (Clark and Woods 2015) and drilling as a method of preforming perforated discs (Sullivan 2017). The results of these studies suggest that multiple techniques were required to make these instruments and that even "some simple chipped forms match or exceed the artisanship, balance, and symmetry of the most complex eccentrics" (Clark and Woods 2015, 205). Replicative experimentation, which is inherently sensuous, experiential, and phenomenological, provides unique insights into ancient technological practice and ontology. A good place to start is with the basic forms of eccentrics, the fundamental production sequences of which provide various pathways to finishing these ideologically charged implements.

In this chapter, we discuss the technological aspects of a variety of eccentrics recovered from Xunantunich, a primarily Late and Terminal Classic site located in the Upper Belize River Valley of western Belize (figure 7.1). The large assemblage of eccentrics at this site demonstrates an incredibly wide array of production techniques, as well as levels of craftsmanship, used to make these implements. Ritualized lithics were cached in most of the buildings in the Late Classic civic-ceremonial core. In one particular structure, Structure A9, two

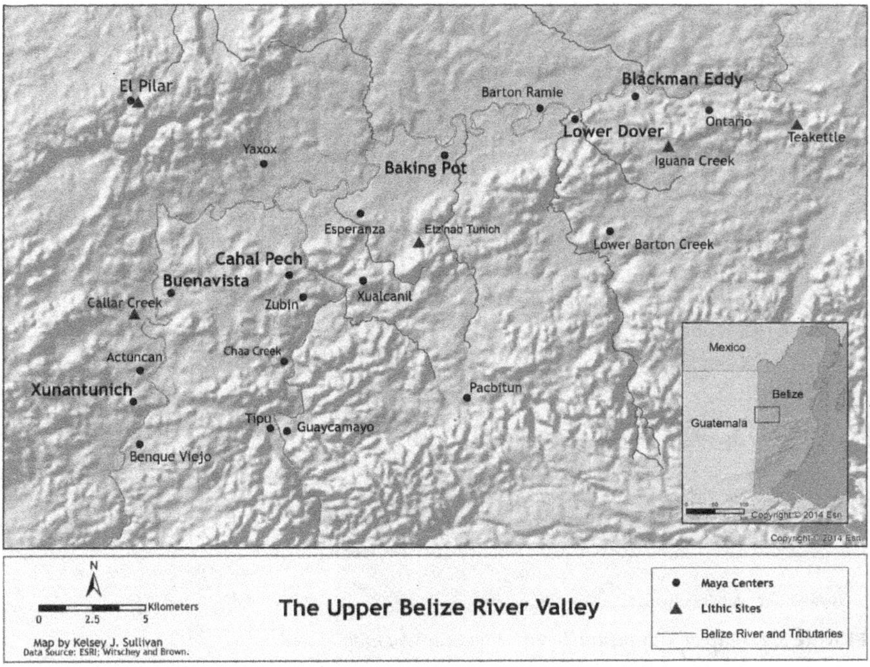

FIGURE 7.1. *Map of ancient Maya sites in the Upper Belize River Valley.*

caches were discovered. The caches contained a range of eccentrics, produced through simple production techniques, including unifacial denticulate flakes, fine and coarsely chipped bifacial chert eccentrics, and obsidian eccentrics related to the prismatic blade reduction sequence (Sullivan 2017; Tilden et al. 2017). We focus on the artifacts from these two caches to examine unique production techniques not observed in other eccentrics at the site. The close examination of these ritualized lithic implements demonstrates the intertwining of utilitarian and ritual lithic production sequences, illuminating the varied nature of the eccentrics recovered in Str. A9 at Xunantunich.

XUNANTUNICH AND THE BELIZE RIVER VALLEY

The Belize River Valley is located in Central Belize. The valley supported nearly continuous settlements, punctuated by medium-sized centers, which developed in the Preclassic period and persisted through the Late Classic and into the Terminal Classic in some cases. The architectural fluorescence of several sites occurred in the Classic period, especially in the Late Classic, including the rapid expansion of the large hilltop civic-ceremonial plazas of Xunantunich's site core.

Xunantunich is located approximately 3 km north from the modern city of Benque Viejo, adjacent to the western border with Guatemala (figure 7.1). The

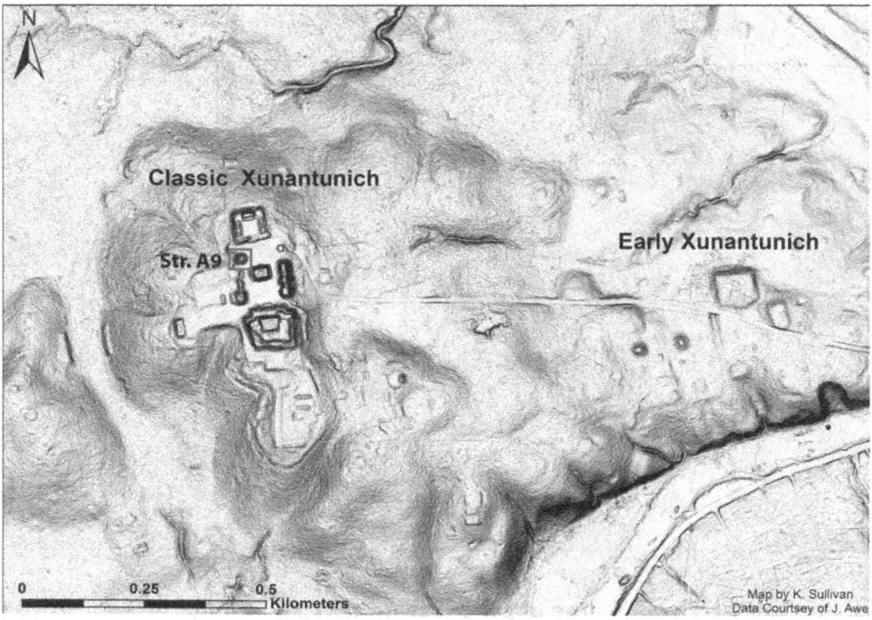

FIGURE 7.2. *Map of Xunantunich with Structure A9 highlighted.*

site is divided into two sectors, Early Xunantunich and Classic Xunantunich (figure 7.2). While Early Xunantunich developed during the Middle Preclassic (Brown et al. 2017), the expansion of Classic Xunantunich into a primary valley center would not occur for another thousand years, during the Late Samal and Early Hats' Chaak Phases between 600 and 730 CE (Ashmore 2010, 46; Audet 2006, 90; LeCount et al. 2002, 41; Slocum 2018). The rapid expansion of Classic Xunantunich during the Late Classic differs significantly from the long-term development and occupation of many sites in the Belize River Valley, which seem to have accumulated prestige and legitimacy over a comparatively long period of time (Awe 1992, 2008). It has been argued that the rulers of Xunantunich sought to legitimize the status of the site and its elites during this late development through an intensive caching campaign with a focus on ritualized lithic caches (Jamison 2010; Sullivan 2017).

The Belize River Valley, and particularly Xunantunich, offers a superlative study area for examining ritual caching behaviors thanks to the large corpus of lithic caches collected over 100 years. These types of caches are usually found at larger valley sites, including Xunantunich, Actuncan (McGovern 2004), Buenavista del Cayo (Ramos Ponciano 2017), Cahal Pech (Conlon 2013), Baking Pot (Bullard and Bullard 1965; Ricketson 1931), and Barton Ramie (Willey et al. 1965). Single caches, however, have also been recovered from a few minor centers in the valley, including Benque Viejo and Chaa Creek, as well as from peripheral groups such as Tzinic and the Martinez group near Cahal Pech (Connell 2003; Gann 1918; Iannone 1993; Powis 1993; Sullivan 2017). Caches from Xunantunich are of particular interest, as thirty-four caches, containing over 500 eccentrics in total, were placed throughout the site core (table 7.1). While this quantity of eccentrics is not uncommon at major sites throughout the Maya Lowlands, such as Tikal (Moholy-Nagy 2008), it is particularly high for sites in the Belize River Valley.

Research at Xunantunich began at the end of the nineteenth century, and eccentric caches have been routinely recovered up to the present day (Audet 2006; Braswell 1995; Brown et al. 2017; Feely 2019; Gann 1918, 1925; MacKie 1985; Ramirez 2023; Santasilia and Tilden 2016; Stewart 1953; Sullivan 2017; Thompson 1942; Tilden et al. 2017; Watkins et al. 2019). The total number of eccentrics recovered at Xunantunich, well over 500, surpasses all other Belize River Valley sites by a wide margin (approximately 200 eccentrics in total have been found in the rest of the valley). In an effort to make caching data available for future researchers, we have adapted Jamison's cache table (2010, table 6.2) to create an inventory of context and contents of eccentric finds at the site, including caches, isolated finds, and eccentrics within burials (table 7.1). Because of the challenging nature of working with published data, especially from excavations in the late nineteenth and early twentieth centuries, some caches lack specific details. In addition, many new caches have been recovered since the time of Jamison's 2010 work, which have been added to the table presented here (Brown et al. 2017; Feely 2019; Ramirez 2023; Santasilia and Tilden 2016; Sullivan 2017).

STRUCTURE A9, PLAZA A-II

For this study we focus on eccentrics recovered from excavations of Structure (Str.) A9 in 2016, conducted by members of the Xunantunich Archaeology and Conservation Project, a sub-branch of the Belize River Valley Archaeological Reconnaissance (BVAR) Project under the direction of Dr. Jaime J. Awe, but also draw comparisons with specimens collected in previous excavations. Str. A9 is a pyramidal temple that is 10 m tall and 25 m across the base. It is located on the northwestern side of the civic-ceremonial Group A (figure 7.2; Slocum 2018, 33). Group A is divided into three courtyards, from South to North, Plazas A-I, A-II, and A-III. Str. A9 comprises the west side of the central courtyard or Plaza A-II. This structure was investigated by T. W. F. Gann (1925) and Jamison (2010) prior to the 2016 XAC/BVAR excavations (Slocum 2018; Tilden et al. 2017). These most recent investigations revealed two hieroglyphic panels flanking the central stairway and a large tomb within, as well as two eccentric caches in front of the stairway (Awe et al. 2020; Slocum 2018; Sullivan 2017; Tilden et al. 2017). These caches are the focus of this chapter.

TABLE 7.1. Context and contents of eccentrics caches from Xunantunich

Location	*Cache Context*	*Cache Contents*	*References*
Plaza A-I, Stela 1	Substela cache	3 chert eccentrics	Gann 1925, 63
Plaza A-1, "on the south side of the largest pyramid," likely Structure A1	Substela cache	2 chert eccentrics	Gann 1918, 97
Plaza A-1, Gann's Structure D (Structures A2, A3, and A4)	Summit cache	64 chert and obsidian eccentrics (fine chipped to denticulate flakes)	Gann 1918, 96
Plaza A-1, in platform in front of Structures A2, A3, and A4	Structural fill cache	CACHE B: "Several" chert eccentric, "resting on the fragments of a shallow dish" CACHE C: 34 eccentrics and 2 obsidian blades	Stewart 1953, 3
Structure A1	Structural fill cache	Chert eccentrics	Jamison 2010, 141 (table 6.2)
Structure A3	Substela cache	12 chert eccentrics	Gann 1925, 57
Structure A3	Structural fill cache	XUN-15-A3-CA1: 9 chert and 9 obsidian eccentrics	Santasilia and Tilden 2016, 126–127
Structure A4	Summit caches	CACHE 1: 5 red chert crescents and 2 *Spondylus* shells CACHE 2: 5 red chert crescents CACHE 3: 11 chert eccentrics and 1 jade bead	Audet 2006, 140
Structure A4	Cache above burial within the floor of penultimate structure	10 chert eccentrics, 5 obsidian eccentrics, 4 quartz or chalcedony eccentrics, as well as 1 obsidian blade core, 2 chert blades, 1 limestone biface, and fragments of 2 Saturday Creek Polychrome dishes	Audet 2006, 146

continued on next page

TABLE 7.1—*continued*

Location	*Cache Context*	*Cache Contents*	*References*
Structure A4	Eccentrics in burial within the floor of penultimate structure	7 chert eccentrics	Audet 2006, 146
Structure A7	Summit cache	GANN CACHE: 35 chert and 5 obsidian eccentrics, human femur and fibula, 100+ chert cores, 2 small blocks of jade, 7 marine shells, a flattened stone with dots, a block of obsidian, chert hammerstones, and 2 celts WATKINS ET AL. ECCENTRICS: 9 chert eccentrics were recovered on the summit between 2017 and 2018.	Gann 1925, 53–54; Watkins et al. 2019
Structure A9	Substela cache	XUN-16-A9-CA1: 28 chert eccentrics: 14 bifacial and 14 denticulate flake eccentrics	Sullivan 2017, 89–92; Tilden et al. 2017
Structure A9	Cache in front of stairway axis	XUN-16-A9-CA2: 9 obsidian eccentrics (8 unifacial core preparation flakes, 1 exhausted blade-core eccentric), branch coral, 5 marine shells, 5 freshwater shells, polished and raw jade, pyrite, and hematite	Sullivan 2017, 93-101; Tilden et al. 2017
Structure A14	Caches in front of stairway axis	CACHE 1: 9 chert eccentrics CACHE 2: 8 chert eccentrics and 1 jade bead	Audet 2006, 151
Structure A15	N/A: Isolated find	1 small anthropomorphic chert eccentric	MacKie 1985

continued on next page

TABLE 7.1—*continued*

Location	*Cache Context*	*Cache Contents*	*References*
Northeast Walkway/ Chaaca Patio	Disturbed cache	1 obsidian eccentric	Keller 2010, 186, 205
Group B	N/A: Surface find	1 notched obsidian eccentric	Thompson 1942, 27
Structure D6, Stela 11	Substela cache	8 obsidian eccentrics	Braswell 1995, 217–219
Structure D6	Within burial in the structure platform	1 fragment of an "eccentrically chipped obsidian from the Ixtepeque source"	Braswell 1995, 219
Structure D6	Cache in front of stairway, on central axis	11 chert eccentrics	Lytle 2017
Structure D6	On top and within burial at the summit	"At least" 20 chert and obsidian eccentrics	Brown et al. 2017, 60; Lytle 2017
Structure D6	Within the Late Classic structural fill	17 chert eccentrics	Lytle 2017
Structure D16	Within the Terminal Classic structural fill	2 chert eccentrics	Braswell 1995, 231
Ballcourt 2	Ballcourt Alleyway, central axis	CACHE 1: 14 eccentrics surrounding large lip to lip vessels and 28 eccentrics and four stingray spines within the vessels CACHE 2: 9 chert and obsidian eccentrics CACHE 3: 24 chert and obsidian eccentrics, resting on a bed of jute shells CACHE 4: 14 chert and obsidian eccentrics	Feely 2019

Context of Str. A9 Caches

Cache 1 (Xun-16-A9-CA1) was discovered below Stela A4 (Helmke and Awe 2016a, 2016b; Slocum 2018; Tilden et al. 2017). While substelae caches are not common in the Belize River Valley, they have been recovered at Xunantunich (i.e., Strs. A1, A3, A9, and D6; Gann 1925; Sullivan 2017) and Buenavista del Cayo (Ramos Ponciano 2017). These caches are understood to have been placed during the erection of the monuments, since they were often deposited in small square or circular constructions, which would have supported the bases of the monuments.

Cache 1 contained fourteen bifacial chert eccentrics in association with fourteen denticulate flakes (figure 7.3) and a single eroded ceramic sherd (not pictured). It was discovered at the base of uncarved Stela A4 located directly in front of Str. A9 in Plaza A-II (Sullivan 2017; Tilden et al. 2017). Placed in a small stone enclosure set on bedrock, the cache was interred during the erection of the large monument. The cache materials may have been bound in cloth or placed directly into the construction materials, since they were tightly packed together. Due to the repetitive nature of this practice, both at Xunantunich and the greater Maya Lowlands, it is clear that a connection between eccentric lithics and stelae was important.

Another common caching context at Xunantunich is in front of structures, in line with the central axis of their stairways. Cache 2 (Xun-16-A9-CA2) was recovered directly below the first step of the stairway in front of Str. A9 and contained nine obsidian eccentrics (Slocum 2018). At Xunantunich, caches in this context usually have nine eccentrics or multiples of nine. Caches found in this context are interpreted to relate to the dedication of the structure (Audet 2006; Ramos Ponciano 2017; Sullivan 2017).

The contexts of both caches, located on the plaza side of the structure, suggest a possible connection between the rituals and public performance. Although the location of a cache in an area considered to be public is not conclusive evidence of the ritual being performed in front of crowds (Chase and Chase 1998), we interpret them to have a performative function, to both create and communicate the important status of this structure and likely the individual interred within Str. A9.

REPLICATIVE TECHNOLOGIES FOR THE STUDY OF ANCIENT MAYA ECCENTRICS

To enhance the study of eccentrics, specifically the technological aspects of production, collaborative efforts between the BVAR project and the Lithic Casting Lab at Northern Arizona University, run by Dr. Francis E. Smiley, resulted in the production of eccentric lithic casts from three caches (Xun-15-A3-CA1, Xun-16-A9-CA1, and Xun-16-A9-CA2) recovered at Xunantunich in 2015 and 2016 (Santasilia and Tilden 2016; Sullivan 2017; Tilden et al. 2017). Forty-one unique

FIGURE 7.3. *Cache 1 (Xun-16-A9-CA1) from Structure A9, exposed in situ below the large uncarved Stela A4 (photo by K. Sullivan).*

casts of these ancient Maya eccentrics from Xunantunich add important educational opportunities, as part of the 1,500 casts in the collection, which spans millions of years of lithic technology from around the world.

These cost-effective resin casts are made in silicone molds, which can be used to produce hundreds of copies. While the original color of artifacts may be challenging to replicate, the high precision of the molds allows for the rendering of detailed flake scars and surface damage, including scratching and other evidence of use-wear, providing incredibly useful tools for technological studies. Moreover, these casts, and other replicative technologies, such as photogrammetric models, provide a broad range of educational applications to a range of stakeholders, through interactive digital and hands-on media. It is our hope that the study of these casts, along with publicly accessible 3D models, will provide new insight into the sequences of production of eccentrics, while simultaneously expanding access and educational opportunities.

The present research relies on a combination of artifact analysis and the use of the casts from both Caches 1 and 2 from Str. A9. Limited artifactual analysis in the field, complemented by more intensive laboratory analyses of the replicas, allowed for a more thorough examination of the technological features of these implements. Based on the study of the original artifacts, as well as these replicas, we discuss the range of production techniques used to produce eccentrics found in the two Str. A9 caches.

ECCENTRIC PRODUCTION TECHNIQUES

Ancient Maya craftspeople produced eccentrics primarily of obsidian and local cherts, although occasionally they used quartz, chalcedonies, and fine imported cherts. They employed a suite of knapping techniques, including initial bifacing with direct hard- and soft-hammer percussion, elemental notch production with indirect percussion, and pressure flaking to produce finishing details. Most eccentrics could be produced by a single craftsperson, but in the case of large and highly complex pieces, the process likely required additional personnel for stabilization and support during the application of percussive forces (Titmus and Woods 2003).

Eccentrics from Xunantunich were made from medium- to fine-grained local cherts and obsidian, primarily from the El Chayal source in Guatemala. The production technologies and eccentric forms used at Xunantunich are quite common throughout the Maya Lowlands. Most eccentrics from Xunantunich are bifacially chipped eccentrics, made of both chert and obsidian, which vary substantially in size and complexity. Other more expedient forms of these implements have been recovered. Eccentrics from Caches 1 and 2 are discussed in detail in the following section, based on raw materials and the techniques used to make them.

Chert Eccentric Production-Techniques in Cache 1

Local cherts are the most common raw materials used to produced eccentrics at Xunantunich, which is a common tradition throughout the Belize River Valley and elsewhere in the Maya world. At Xunantunich, several lithic production techniques were employed in the creation of eccentrics, which vary greatly in their style and quality. Cache 1 featured twenty-eight chert eccentrics, which demonstrate multiple crafting strategies. In the following subsections, we discuss these techniques in depth.

Denticulate Chert Flakes

The simplest chert eccentrics recovered in Cache 1 were denticulate flakes (figure 7.4). These eccentrics are relatively simple to make, but complement, enhance, or even replace more complex and energetically expensive eccentrics. Medium to large flakes removed with hard- and soft-hammer direct percussion in typical biface reduction, particularly biface thinning flakes, provide great blanks to design simple silhouette eccentrics through pressure retouch. Flakes used in this manner may have been produced specifically for this purpose. However, it is also possible these flakes are directly related to bifacial eccentric reduction sequences or curated from other biface production or general core reduction activities. The skill required to produce the necessary flakes, as well as to apply pressure retouch, is similar to those needed for utilitarian tool production, and many

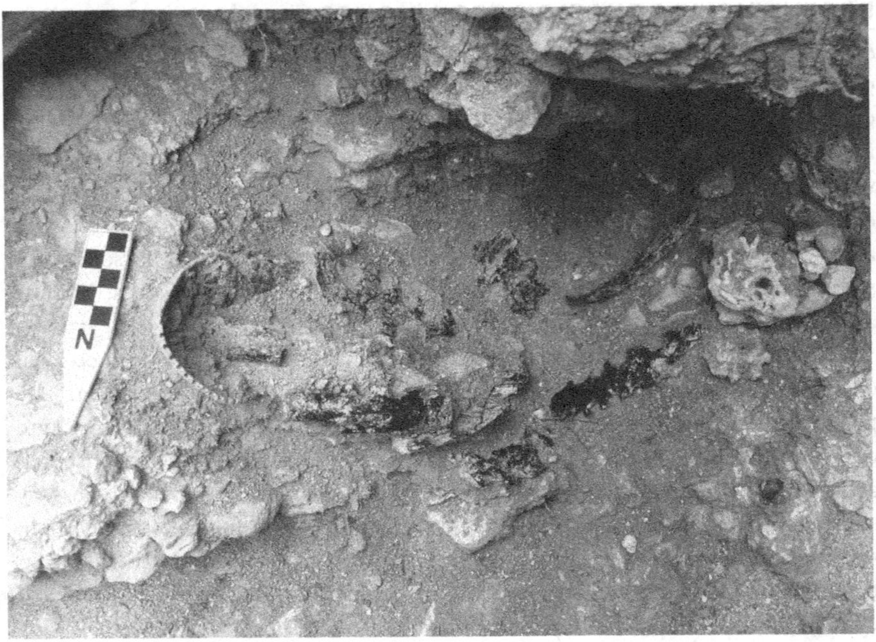

FIGURE 7.4. *Cache 2 (Xun-16-A9-CA2) from Structure A9, found in situ in front of the central stairway (photo by K. Sullivan).*

people could be taught to do this with relative ease. Although the production techniques used to design denticulate flakes are generally simple, the crafting of flakes for painting or incision, such as those eccentrics found at Tikal and Piedras Negras, may have required specific ritual knowledge (Hruby and Ware 2009).

Primarily, these denticulate flakes are unifacially pressure flaked on all or most margins, though at least one flake has evidence of limited bifacial pressure flaking. Remnant cortex is present on the dorsal faces of four of the pieces; however, all bare previous removal flake scars. The forms of these eccentrics are abstract, though two flakes exhibit a profile form (far right of figure 7.5), similar to the elaborate anthropomorphic eccentrics, which may refer to particular deities or ancestors. As has been demonstrated by many other researchers (Agurcia Fasquelle et al. 2016; Hruby 2006; Hruby and Ware 2009), eccentrics may often have been painted, incised, and/or bound in textiles. Thus, while simple ritualized lithics appear easy to produce, specific knowledge may have been required to create outlines and proportions appropriate for these treatments. Unfortunately, the denticulate flakes from Cache 1 were washed during laboratory processing, so an evaluation of the presence of traces of pigment was not possible.

The production of these eccentrics suggests that their significance does not always lie in the artisanship of the implement but sometimes in the ritual

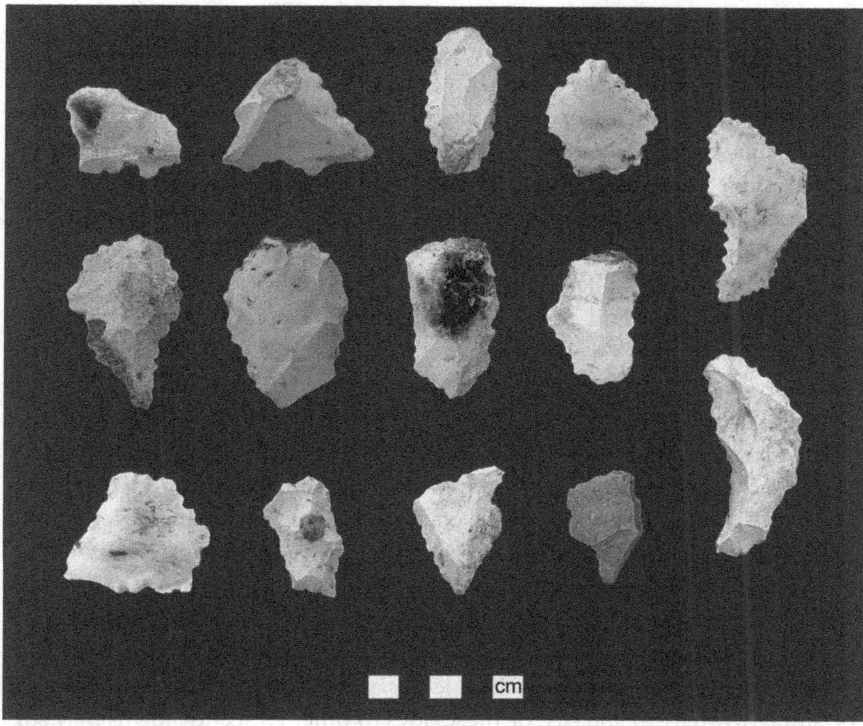

FIGURE 7.5. *Denticulate chert flake eccentrics from Cache 1 (Xun-16-A9-CA1) from Str. A9 (photo by K. Sullivan).*

production and use of the tools. This may also relate to the producer themselves, a person who holds *esoteric production knowledge* (Hruby 2007), or other forms of restricted ritual knowledge or experience, who may apply this knowledge to these materials, in an action or performance which is more significant than the particularities of the final product.

Bifacial Chert Eccentrics

Bifacially chipped chert eccentrics are the most common type of eccentric at Xunantunich and throughout the Belize River Valley. Chert eccentrics from Xunantunich are made of cherts, ranging in quality. Eccentrics at Xunantunich include common forms such as serrated and smooth crescents, serrated and undulating laurel-leaf bifaces, scorpions and other zoomorphic forms, anthropomorphic forms, bifurcates, and circular and oval disks, chipped into various quadripartite cross forms, similar to K'an and K'in crosses (Ramos Ponciano 2017; Sullivan 2017). In addition, perforated discs, featuring these quadripartite and circular holes in the center of relatively flat oval bifaces, have been recovered in at least six separate caches throughout Groups A and D (Audet 2006; Brown et al. 2017; Santasilia and Tilden 2016). These forms are common throughout caches from the Belize River Valley and the southern Maya Lowlands.

In Cache 1, fourteen bifacially chipped chert eccentrics were placed on top of the denticulate chert flake eccentrics just discussed. The production techniques used to make these implements include standard biface reduction through direct percussion with hard and soft percussors, as well as more complex techniques, such as deep notching and marginal serration through a combination of indirect percussion and pressure flaking. The eccentrics range in complexity, though most are standard eccentric forms, such as crescents, laurel-leaf biface forms, and a scorpion. Two are retouched flakes, though unlike the denticulate flakes, they feature indirect percussion notching and more significant bifacing than the flakes.

While most of these tools display standard methods of lithic reduction, one was crafted through the perforation of a relatively thin biface (figure 7.6, bottom row, center). Perforations of this style could be made by exploiting lower-quality materials in the center of cobbles through indirect percussion or preforming through drilling (Sullivan 2017). The single perforated eccentric in Cache 1 has evidence of a bending fracture across the flat end, suggesting that this specimen was repurposed from a thin oval biface that broke in manufacture or repair. This form of eccentric has been found in at least six caches throughout Xunantunich. The similarity in form and size of these discs to other crescentic forms may suggest an association in production sequences. The analysis of these perforated discs, especially those with broken ends, suggests that eccentric manufacture is related to other forms of biface production at the site.

As discussed by John Clark and James Woods (2015, 205), these eccentrics have a single flat face, with a more convex opposing face. Antonio Prado, a Guatemalan architect, proposed that this flat face provides stability for producing eccentrics based on "physical drawings of desired forms" (2015, 205). While the authors remained skeptical of this hypothesis, due to the fluid nature of tool faces during the reduction of bifaces, it is clear from the examination of Prado's large collection of eccentrics, as well as the artifacts discussed here, that this morphological feature of bifaces was a consistent feature of eccentrics. Furthermore, while stability may relate to production, it may also relate to the use of these implements. This feature hints that these ritual implements may have been displayed on flat surfaces prior to or during their interment. This needs to be studied in much more depth through artifact analysis and replicative experimentation.

The artisanship observed in bifacial chert eccentrics from Xunantunich is highly variable. While some caches feature finely chipped bifacial eccentrics that are incredibly small and thin, such as Cache 1 from Str. A3 (Santasilia and Tilden 2016), others from Str. A9 are thick and more crudely chipped (see figure

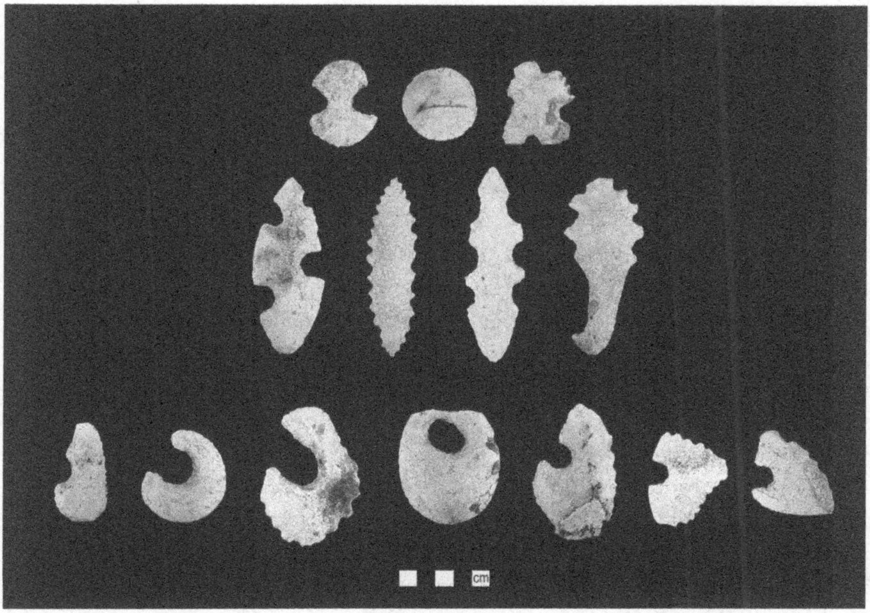

FIGURE 7.6. *Bifacially chipped chert eccentrics from Cache 1 (Xun-16-A9-CA1) from Str. A9 (photo by K. Sullivan).*

7.5). There are great thematic similarities in eccentric forms and production techniques at Xunantunich. However, the wide variation in the quality of the artisanship and expression of style in these objects suggests that many individuals were involved in the production of these implements.

Obsidian Eccentrics Production Techniques in Cache 2

Obsidian in Mesoamerica was an important resource traded over long distances throughout the Maya Lowlands primarily from Highland Guatemala but in more limited quantities from Central Mexico. Since lithic materials are quite heavy, their transport was a significant investment, which makes them economically, and perhaps politically, important.

Technologies associated with obsidian focus on prismatic blade production and, to a much lesser extent, biface production and expedient tool industries (Clark 1985; Clark and Bryant 1997). The ancient Maya used implements from all these obsidian industries within ritual contexts. Obsidian was reduced with great efficiency, and often pieces were curated or repurposed to maximize the use-life of the material. Eccentrics made of obsidian at Xunantunich reflect the integration of eccentric production and prismatic blade making. The incorporation of these varied production processes resulted in the more efficient use of obsidian, repurposing materials that would otherwise be considered waste in the blade production process.

The most frequent type of obsidian eccentric blank, especially at Xunantunich, is the exhausted blade core, which was reworked into more complex forms. In addition to exhausted obsidian blade cores, core preparation and other thinning flakes were reworked through pressure flaking into simple denticulate eccentrics. Other obsidian technology found in ritual contexts at the site include unmodified prismatic blades and thin lanceolates (Santasilia and Tilden 2016; Slocum 2018; Sullivan 2017). In general, however, obsidian eccentrics are found in slightly smaller quantities than chert eccentrics in caches at Xunantunich (table 7.1). The nine obsidian eccentrics found in Cache 2 display both typical and less-conventional production techniques (discussed in the following subsection).

Obsidian Macroblade Eccentrics

One obsidian production technique associated with eccentrics at Xunantunich is the bifacial notching of macroblades. Although examples of retouched obsidian flakes are not uncommon in Mesoamerica (Hruby 2006, 2007; Moholy-Nagy 2008), eccentrics from Cache 2 at Xunantunich are unique. In front of Str. A9, Cache 2 (XUN-16-A9-CA2) was recovered below plaza Floor 4, on the central axis of the stairway of the structure (Slocum 2018; Tilden et al. 2017). In addition to nine obsidian eccentrics (figure 7.7), the cache contained a range of marine and freshwater shells, as well as small fragments of jadeite, pyrite, and hematite (Slocum 2018; Sullivan 2017).

Eight of the nine eccentrics were made on large percussion blades ranging from approximately 70 to 120 mm in length and 25 to 45 mm in width. Zoomorphic forms were created through bifacial pressure flaking on lateral flake margins. Four of these eight macroblade eccentrics retain small amounts of cortex on their dorsal surfaces. Of these, three have eraillure flake scars on their ventral surface, just below a broad-lipped platform, suggesting the use of a fairly hard percussor. While only select tools have eraillure scars, all the blades have pronounced lipping of the platforms. In addition, all eight eccentrics have evidence of scratching on the platforms. Platform scratching is a simple and common, less labor-intensive way to prevent slippage for direct or indirect percussion blade removal than other techniques such as grinding and pecking (Clark 1985). Bulbs of percussion are variable, ranging from diffuse to slightly pronounced. The flake characteristics, specifically related to platform and bulb morphology, suggest that the macroblades were removed from the core with precursors made of hardwood or medium hard stone.

Artifact analysis in the field revealed two pairs of these eccentrics whose margins refit (figure 7.8). Casts of the eccentrics made refitting much easier, as it could be done anywhere, at any time, with no risk to the original artifacts. After

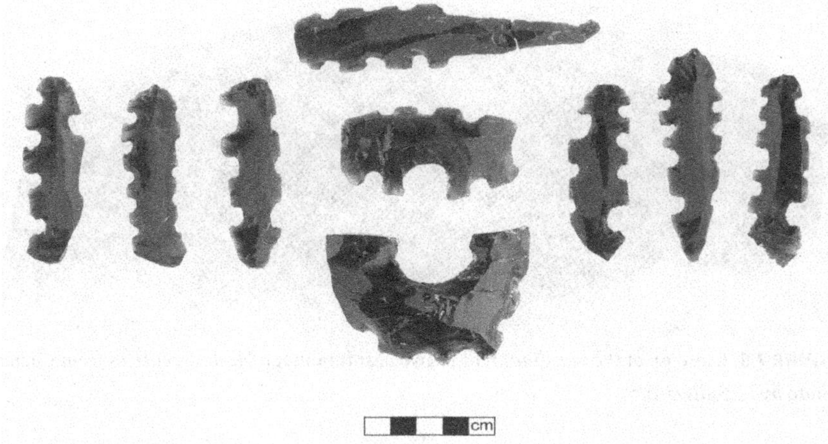

FIGURE 7.7. *Obsidian eccentrics from Cache 2 (Xun-16-A9-CA2), recovered in front of the central stairway of Str. A9 (photo by K. Sullivan).*

working with these casts quite a bit, we were able to refit the eccentrics. The refits revealed that the eight macroblades were removed from a single polyhedral core, likely in the preparation of a prismatic blade core (figure 7.9). Based on the angle of the refit eccentrics, the core was approximately 120 mm in diameter and 120 mm in length. However, because the artifacts exhibit only small bits of remnant cortex, it is more likely that the core was slightly bigger, prior to its decortication. Due to the lack of cortical obsidian debitage in any excavation at Xunantunich, it is likely that the core was partially reduced prior to its acquisition.

The nature and degree of the integration of utilitarian and ritual tool production processes remain unclear. It is evident, however, that technological overlap in the production processes existed in many cases. The macroblades used in Cache 2 may have been a by-product of utilitarian blade production, although it is also possible that a more complex relationship existed between utilitarian and ritual tools and their production. These macroblades may represent a form of sacrifice, where otherwise useful pieces of obsidian were used specifically for the production of eccentrics.

Cache 2 is the sole example of retouched macroblade eccentrics at Xunantunich, and the production of these reflect techniques not observed in other assemblages from other sites in the Belize River Valley (Conlon 2013; Iannone 1993; Powis 1993; Santasilia and Tilden 2016; Sullivan 2017). While retouched obsidian flakes have been recovered from ritual contexts at other sites, their production has not been as obviously linked to the prismatic blade production sequence, as

FIGURE 7.8. *Close-up of the marginal refit of two obsidian macroblade eccentrics from Cache 2 (photo by K. Sullivan).*

in this case. This is yet another example where eccentric production is clearly related to other utilitarian lithic production at Xunantunich, as observed with the chert eccentrics from Str. A9.

Exhausted Blade-Core Eccentric

In addition to the eight macroblade eccentrics in Cache 2, one thick, denticulate crescent (Cat. No. Xun-16-A9-CA2-i) was recovered in the cache as well (see figure 7.7). It is 85 mm wide by 61 mm tall and 15 mm at its thickest point. This piece shows evidence of being produced from an exhausted blade core or at minimum a core that had two to three prior blade removals from one side. The top of each side appears to be remnant platforms, which display minimal evidence of scratching. The platform may have been rejuvenated fairly close to the end of its life as a blade core. In addition, many step fractures near the platforms' edges suggest that the piece was no longer fit for successful prismatic blade removal, due to their thinness, prompting the repurposing of the piece. The eccentric was designed through the application of direct percussion bifacing on both faces and notching through indirect percussion on the distal edge. Minimal pressure flaking was applied for finishing details. Similar to the bifacial chert eccentrics from Cache 1, this piece has a stable face (Clark and Woods 2015). One face is concave but rests on the distal edge and platform of the right side, allowing the eccentric to lay on a flat surface without wobbling. It is unclear whether this eccentric is made from the same objective piece as the other macroblade eccentrics, though the materials have comparable appearances and geochemical compositions.

DISCUSSION AND CONCLUSIONS

The production techniques used to make eccentrics at Xunantunich reflect those from other sites in the Belize River Valley and elsewhere in the Maya Lowlands.

FIGURE 7.9. *Reconstruction of obsidian blade core from which the obsidian macroblade eccentrics from Cache 2 were removed (photos by K. Sullivan). This was done by refitting casts of the eccentrics.*

The remarkable number of eccentrics in comparison to other valley sites, however, demonstrates that these implements are symbolically related to Maya ideology and that they served a critical role in performances of ritual and political power at Xunantunich.

While the investment in the production of outstandingly fine tools, such as the large eccentrics cached at Copán, Altun Ha, and Lamanai, demonstrates a significant focus on the style and quality of the implements, eccentrics from Xunantunich are considerably simpler, especially those from Str. A9. This difference likely relates to raw material size and quality available in different regions, but we believe that it also reflects differential needs of those using these ritual implements. At Xunantunich, the vast number of caches were placed during rapid construction of the Classic site core, demonstrating the importance of consecrating this newly built space, not necessarily on the appearance of the eccentrics. The need for, or use of, a large number of these objects may have reflected the integration of utilitarian and ritual lithic production.

In this chapter, we discussed production techniques used to craft eccentrics recovered from Str. A9, which display variable design choices from other eccentrics at the site. Chert eccentrics from Cache 1 include standard bifacially chipped implements, as well as denticulate chert flakes. From Cache 2, we examined nine obsidian eccentrics, eight of which were removed from a single polyhedral core. Eccentrics from both caches relied on more utilitarian and expedient techniques than other ritualized lithics found at the site. Moreover, the eccentrics from these two caches demonstrate overlapping technological processes and point to the integration of the production of utilitarian and ritual lithic artifacts. This research is a starting point for more extensive explorations of the intertwined nature of lithic production processes.

Eccentric lithics cached by the ancient Maya served ritual purposes that are still poorly understood. The skills needed to produce eccentrics were both simple and complex in terms of flintknapping techniques. The power to wield these productive forces and to evoke the divine in the deposition of them, however, appears to rely on esoteric production knowledge, which enabled ritual specialists to enhance both ritual and political prestige of civic-ceremonial centers, as well as individual structures (Hruby 2006, 2007; Ramos Ponciano 2017; Sullivan 2017). At Xunantunich, this was especially important, as the city was rapidly constructed and rose to the apogee of its political influence in the Belize River Valley in the Late Classic period. The significant number of eccentric caches (table 7.1) suggests that the tools' power and the performance of these acts, likely in front of large crowds by elite ritual specialists, were more significant than the quality and craftsmanship of the objects themselves.

Acknowledgments. Thank you to the National Institute of Culture and History and the Institute of Archaeology, Belize, for permission to conduct research at Xunantunich. We would like to thank the members of the Xunantunich Archaeology and Conservation Project, particularly Douglas Tilden, Tia B. Watkins, Hannah Zanotto, Diane L. Slocum, Gregory Allen, Shane Dvorak, Dr. Catharina Santasilia, Jorge Kan, Merle Alfaro, and all the local collaborators from San Jose Soccutz. To Dr. Lucas Martindale Johnson, thank you for always providing important insight and identifying that the macroblades from Cache 2 refit, as well as for conducting the portable X-ray fluorescence (pXRF) analysis of obsidian artifacts. We would like to thank Dr. Zachary Hruby for providing thoughtful and thorough feedback on this chapter. Special gratitude is extended to Dr. Francis E. Smiley for casting the Xunantunich eccentrics at the Northern Arizona University's Lithic casting lab. This research was made possible by the generous funding from the Tilden Family Foundation, the Belize River Valley Archaeological Reconnaissance Project, and Northern Arizona University's Ray

Madden Research Award in Anthropology and Kenny Acord Award for Graduate Students in Archaeology.

REFERENCES

Agurcia Fasquelle, Ricardo, Payson Sheets, and Karl Andreas Taube. 2016. *Protecting Sacred Space: Rosalila's Eccentric Chert Cache at Copan and Eccentrics Among the Classic Maya*. San Francisco: Precolumbia Mesoweb Press.

Ashmore, Wendy. 2010. "Antecedents, Allies, Antagonists: Xunantunich and Its Neighbors." In *Classic Maya Provincial Politics*, edited by Lisa J. LeCount and Jason Yaeger, 46–64. Tucson: University of Arizona Press.

Audet, Carolyn. 2006. "Political Organization in the Belize Valley: Excavations at Baking Pot, Cahal Pech, and Xunantunich." PhD diss., Department of Anthropology, Vanderbilt University, Nashville.

Awe, Jaime J. 1992. "Dawn in the Land Between the Rivers: Formative Occupation at Cahal Pech, Belize and Its Implications for Preclassic Developments in the Maya Lowlands." Unpublished PhD diss., Institute of Archaeology, University of London, London.

Awe, Jaime J. 2008. "Architectural Manifestations of Power and Prestige: Examples from Classic Period Monumental Architecture at Cahal Pech, Xunantunich and Caracol, Belize." *Research Reports in Belizean Archaeology* 5:159–174.

Awe, Jaime J., Christophe Helmke, Diane Slocum, and Douglas Tilden. 2020. "Ally, Client, or Outpost? Evaluating the Relationship Between Xunantunich and Naranjo in the Late Classic Period." *Ancient Mesoamerica* 31(3):494–506.

Braswell, Jennifer B. 1995. "Investigations at Group D, Xunantunich, Belize: A Nonroyal Elite Corporate Group." In *The Xunantunich Archaeological Project: The Xunantunich Site Core: 1994 Research and Work*, edited by Richard M. Leventhal, 214–247. Ms. on file, Institute of Archaeology, University of California, Los Angeles, and Belmopan, Belize.

Brown, M. Kathryn, Whitney Lytle, Zoe Rawski, Victoria Ingalls, and Alessandra Villarreal. 2017. "Understanding the Preclassic Ritual Landscape in the Mopan Valley: A View from Early Xunantunich." *Research Reports in Belizean Archaeology* 14:53–64.

Bullard, William R., and Mary Ricketson Bullard. 1965. "Late Classic Finds at Baking Pot, British Honduras." *Art and Archaeology Occasional Papers* 8. Toronto: University of Toronto Press.

Chase, Diane Z., and Arlen F. Chase. 1998. "The Architectural Context of Caches, Burials, and Other Ritual Activities for the Classic Period Maya (as Reflected at Caracol, Belize)." In *Function and Meaning in Classic Maya Architecture*, edited by Stephen D. Houston, 239–332. Washington, DC: Dumbarton Oaks.

Clark, John E. 1985. "Platforms, Bits, Punches and Vises: A Potpourri of Mesoamerican Blade Technology." *Lithic Technology* 14(1):1–15.

Clark, John E., and Douglas Donne Bryant. 1997. "A Technological Typology of Prismatic Blades and Debitage from Ojo de Agua, Chiapas, Mexico." *Ancient Mesoamerica* 8(1):111–136.

Clark, John E., and James C. Woods. 2015. "Squeezing Life from Stones: The Human Side of Replication Experiments." In *Works in Stone: Contemporary Perspectives on Lithic Analysis*, edited by Michael J. Shott, 197–212. Salt Lake City: University of Utah Press.

Conlon, James M. 2013. "Archaeological Investigations of Structure B-3, Cahal Pech: A Preliminary Report." In *The Belize Valley Archaeological Reconnaissance Project: A Report of the 2013 Field Season*, edited by Julie A. Hoggarth and Jaime J Awe, 60–69. Belmopan, Belize: Institute of Archaeology.

Connell, Samuel V. 2003. "Making Sense of Variability Among Minor Centers: The Ancient Maya of Chaa Creek, Belize." In *Perspectives on Ancient Maya Rural Complexity*, edited by Gyles Iannone and Samuel V. Connell, 27–41. Los Angeles: Cotsen Institute of Archaeology Press at UCLA.

Feely, Cassandra. 2019. "Game On: Investigations of Ballcourts 1 and 2 at Xunantunich, Belize." Unpublished MA thesis, Department of Anthropology, Northern Arizona University.

Gann, Thomas W. F. 1918. *The Maya Indians of Southern Yucatan and Northern British Honduras*. Washington, DC: Smithsonian Institution, Bureau of American Ethnology Washington.

Gann, Thomas W. F. 1925. *Mystery Cities of the Maya*. Mystic Travelers Series. London: Gerald Duckworth.

Helmke, Christophe, and Jaime J. Awe. 2016a. "Death Becomes Her: An Analysis of Panel 3, Xunantunich, Belize." *PARI Journal* 16(4):1–14.

Helmke, Christophe, and Jaime J. Awe. 2016b. "Sharper than a Serpent's Tooth: A Tale of the Snake-Head Dynasty as Recounted on Xunantunich Panel 4." *PARI Journal* 17(2):1–22.

Hruby, Zachary X. 2006. "The Organization of Chipped-Stone Economies at Piedras Negras, Guatemala." PhD diss., University of California, Riverside.

Hruby, Zachary X. 2007. "Ritualized Lithic Production at Piedras Negras, Guatemala." *Archaeological Papers of the American Anthropological Association* 17(1):68–87.

Hruby, Zachary X, and Gene Ware. 2009. "Painted Lithic Artifacts from Piedras Negras, Guatemala." In *Maya Archaeology*, vol. 1, edited by Charles Golden, Stephen Houston, and Joel Skidmore, 76–85. San Francisco: Mesoweb Press.

Iannone, Gyles J. 1993. "Ancient Maya Eccentric Lithics: A Contextual Analysis." MA thesis, Department of Anthropology, Trent University, Peterborough.

Jamison, Thomas R. 2010. "Monumental Building Programs and Changing Political Strategies at Xunantunich." In *Classic Maya Provincial Politics*, edited by Lisa LeCount, 122–144. Tucson: University of Arizona Press.

Keller, Angela H. 2010. "The Social Construction of Roads at Xunantunich, from Design to Abandonment." In *Classic Maya Provincial Politics*, edited by Lisa LeCount, 184–208. Tucson: University of Arizona Press.

LeCount, Lisa J., Jason Yaeger, Richard Leventhal, and Wendy Ashmore. 2002. "Dating the Rise and Fall of Xunantunich, Belize." *Ancient Mesoamerica* 13(1):41–63.

Lytle, Whitney. 2017. *Transformations Within an Ancestor Shrine: New Discoveries from Group D, Xunantunich, Belize*. Paper presented at the 82nd Annual Society for American Archaeology Meeting, Vancouver.

MacKie, Euan Wallace. 1985. "Excavations at Xunantunich and Pomona, Belize, in 1959–1960." *BAR International Series 251*. Oxford: British Archaeological Reports.

McGovern, James O. 2004. "Monumental Ceremonial Architecture and Political Autonomy at the Ancient Maya City of Actuncan, Belize." PhD diss., University of California, Los Angeles.

Moholy-Nagy, Hattula. 2008. *The Artifacts of Tikal: Ornamental and Ceremonial Artifacts and Unworked Material, Tikal Report 27A*. Philadelphia: University of Pennsylvania Museum of Archaeology and Anthropology.

Powis, Terry G. 1993. "Special Function Structures Within Peripheral Groups in the Belize Valley: An Example from the Bedran Group at Baking Pot." In *The Belize Valley Archaeological Reconnaissance Project: Progress Report of the 1992 Field Season*, edited by Jaime J. Awe, 212–224. Peterborough, ON: Trent University.

Ramirez, Estevan. 2023. "Investigating Preclassic Occupation in the Site Core of Xunantunich." MA thesis, Northern Arizona University, Flagstaff, AZ.

Ramos Ponciano, Marcos Eduardo. 2017. "Eccentric Caches of Buenavista Del Cayo: Contextual Analysis and Cosmological Significance." MA thesis, Department of Anthropology, San Diego State University.

Ricketson, O. G. 1931. "Excavations at Baking Pot, British Honduras." In *Contributions to American Archaeology* 1(1):1–28. Washington, DC: Carnegie Institute of Washington.

Santasilia, Catharina E., and Douglas Tilden. 2016. "2015 Excavations of the Eastern Triadic Shrine at Xunantunich, Belize." In *The Belize Valley Archaeological Reconnaissance Project: A Report of the 2015 Field Season*, edited by Julie A. Hoggarth and Jaime J. Awe, 118–138. Waco, TX: Institute of Archaeology, Baylor University.

Slocum, Diane L. 2018. "Contextualizing Xunantunich in the Late Classic Upper Belize Valley Through Investigations of Structure A9." MA thesis, Department of Anthropology, Northern Arizona University, Flagstaff.

Stewart, Michael C. 1953. *Preliminary Reports on the Seasons 1952–1953 Work at Xunantunich*. Unpublished Field Notes, courtesy of the Steward Family, on file at the Belize Institute of Archaeology, Belmopan.

Sullivan, Kelsey J. 2017. "Caching It In: Local Patterns in Ancient Maya Ritual Caches of Eccentric Lithics Within the Belize Valley." MA thesis, Department of Anthropology, Northern Arizona University, Flagstaff.

Thompson, J. Eric S. 1942. "Late Ceramic Horizons at Benque Viejo, British Honduras." *Contributions to American Anthropology and History*, No. 35. Publication 528. Carnegie Washington, DC: Institution of Washington.

Tilden, Douglas, Diane L. Slocum, Kelsey J. Sullivan, and Jaime J. Awe. 2017. "Xunantunich Archaeology and Conservation Project 2016: Investigation of Structure A-9." In *The Belize Valley Archaeological Reconnaissance Project: A Report of the 2016 Field Season*, edited by Julie A. Hoggarth and Jaime J. Awe, 316–345. Waco, TX: Institute of Archaeology, Baylor University.

Titmus, Gene L., and James C. Woods. 2003. "The Maya Eccentric: Evidence for the Use of the Indirect Percussion Technique in Mesoamerica from Preliminary Experiments Concerning Their Manufacture." In *Mesoamerican Lithic Technology: Experimentation and Interpretation*, edited by Kenneth G. Hirth, 132–146. Salt Lake City: University of Utah Press.

Watkins, Tia B., Douglas Tilden, and Jaime J. Awe. 2019. "Architecture and Activity at The Ceremonial Center of Xunantunich: Results from the 2018 Field Season." In *The Belize Valley Archaeological Reconnaissance Project: A Report of the 2018 Field Season*, edited by Claire E. Ebert, John P. Walden, Julie A. Hoggarth, and Jaime J. Awe, 249–270. Flagstaff: Department of Anthropology, Northern Arizona University / Waco: Institute of Archaeology, Baylor University.

Willey, Gordon R., William R. Bullard Jr., John B. Glass, and James C. Gifford. 1965. *Prehistoric Maya Settlements in the Belize Valley, Peabody Museum of Archaeology and Ethnology, Papers* 54. Cambridge, MA: Harvard University.

8

Comparison of Chert Silhouettes from the Late and Terminal Classic Eastern Lowlands

ZACHARY HRUBY, CHRISTINA HALPERIN, CATHARINA SANTASILIA, AND JAIME J. AWE

In recent years, several lithic caches have been discovered at Ucanal, Guatemala, and Xunantunich, Belize, located in the eastern portion of the Maya Lowlands (figures 8.1 and 1.7; Halperin and Garrido 2016, 2020; Slocum 2018). In 2016 we decided to compare caches from each site to determine whether there were any similarities between them, and thus determine whether there may have been a shared cultural *tradition* or perhaps a *community* or *constellation of practice* among elite knappers separated by political boundaries and geographic space. We were interested to know whether different kinds of esoteric knowledge associated with the production of ceremonial lithic artifacts was shared between Classic Maya polities and, if so, what kinds of things were shared. The production of utilitarian bifaces, flakes, and blades often varies little over great distances in the Maya world, but the more elaborate silhouettes (i.e., eccentrics) do. With all of their stylistic and technical particularities, they provide an excellent opportunity to examine the possibility of shared lithic production knowledge (i.e., *hexis* [i.e., learned, embodied knowledge]; Bourdieu 1977) and practice across different sites and polities. As Jean Lave (2008) notes, however, communities of practice

https://doi.org/10.5876/9781646427819.c008

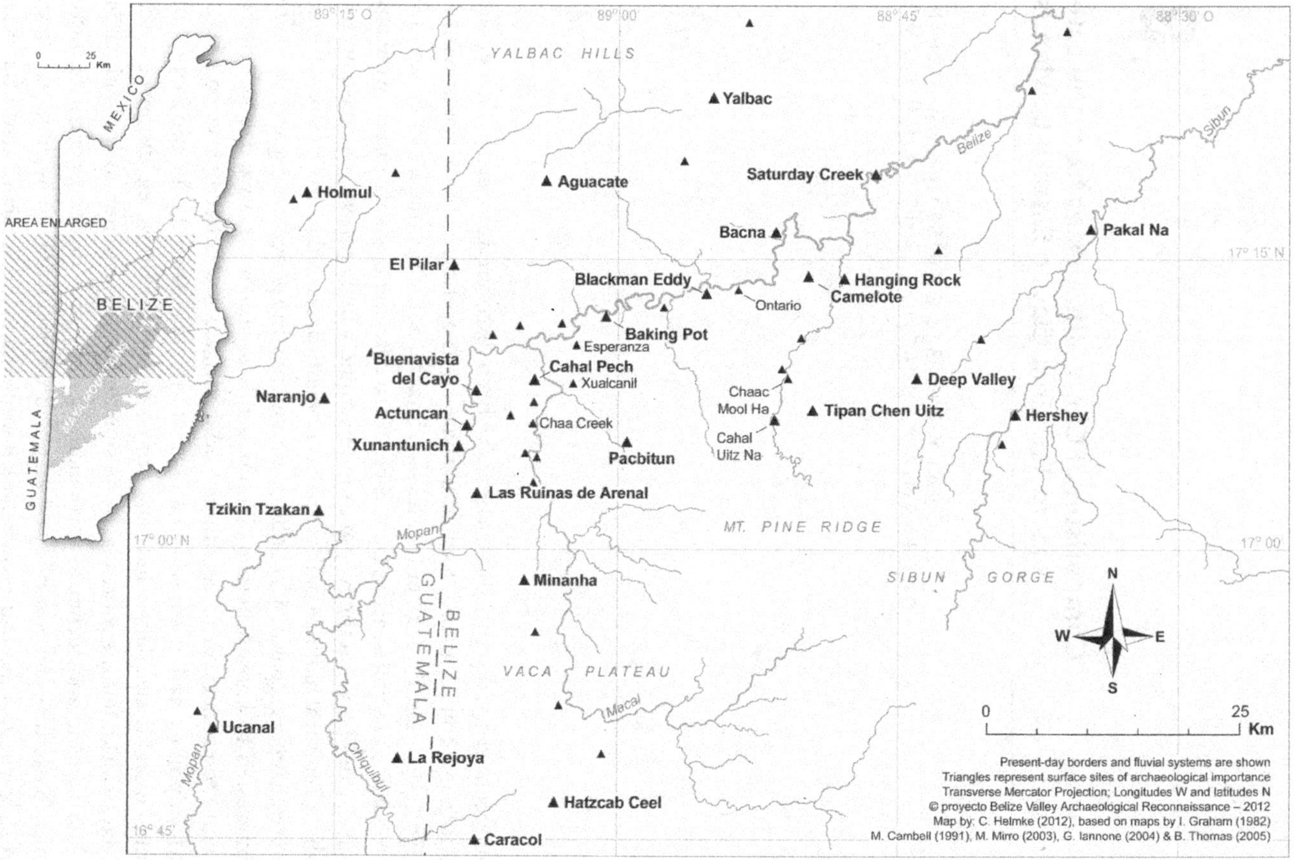

FIGURE 8.1. *Map of Eastern Petén and Central Belize.*

are not something to discover but rather a way of looking at social relationships (see also Roddick and Stahl 2016; Sassaman and Rudolphi 2001). We use this theoretical perspective to flesh out possible social relationships understood through these enigmatic objects and the people who made them.

Many communities of knapping practice appear to have been tethered to their local geology, to a certain extent, as part of their social identity. Cache vessels and cache offerings were often made locally and affirmed local rather than foreign affiliation. The likely goal was to tie local lineages to local creation myths meant to legitimize ideologies that placed the king, his lineage, and the city as the center of the universe and creation itself. Chert silhouettes were perhaps even more tied to home, since chert materials were often found right below the people's feet (see Roddick and Janusek 2018 for an Andean example). Nevertheless, some chert silhouettes were virtually identical between cities. Given that the vast majority of these ritual tools were not produced for exchange, it is possible that knowledge of what and how to produce them was shared between crafters. Like Maya writing, esoteric knowledge production was likely spread through the fissioning of lineages and the creation of new political centers by people who carried the practices with them, as well as close alliances and intermarriage between royal and noble houses from distant polities. Given the elite and royal contexts of chipped-stone silhouettes, we believe their production was tied to power (Hruby 2006, 2007) and as such require some consideration of power and access to production knowledge (Crown 2016; Inomata 2001).

Although the sites of Ucanal and Xunantunich were politically independent from one another, they are located close to one other (~50km) and situated along the same river course (the Mopan River). In addition, they were also tied together through similar histories and interaction spheres. For example, the majority of Pine Ridge Carbonate incurved rim bowls at Ucanal are Mount Maloney black slipped types (Halperin et al. 2020), which was the ceramic of choice for not only quotidian cooking and serving of food but caches and offerings at Xunantunich during the Late and Terminal Classic periods (LeCount and Yeager 2010). Additionally, the two sites both possess hieroglyphic monument blocks that originally decorated one of the building facades at Caracol (Helmke and Awe 2016a, 2016b; Martin 2017). Both centers were client states of the larger polity of Naranjo during the end of the seventh century and beginning of the eighth century CE, although they later asserted their respective political independence during the Terminal Classic period (ca. 810–950 CE; see Awe et al. 2020). Thus, the examination of three Late and Terminal Classic caches from these sites provides an opportunity to not only document these newly discovered caches but identify possible communities of practice surrounding these ceremonial objects and their changes over the course of the Late to Terminal Classic period. Although we lack important data points regarding chert source locations and specific production locales, and consequently do not have direct data on all the links of the *chaîne opératoire*, we address their implications through the lens of communities of practice, brokers, and boundary objects (Lave and Wenger 1991; Wenger 1998). Thus, this chapter explores the utility of this approach for ceremonial lithic studies at this early stage of analysis.

Our data consist of chert silhouettes from two Terminal Classic caches from Ucanal and a Late Classic cache from Xunantunich. We use partial data from other sites as well, in order to situate these Eastern Lowland sites in the broader Classic period silhouette horizon. The information from other sites is incomplete, because we rely on reconstructed metrics, technology, and outline from photographs and drawings. Most significant, we draw from the site of Tikal, which features a large corpus of chert silhouettes from the entirety of the Classic period, and to a lesser extent on a small published corpus of silhouettes from Naranjo. The work of William Coe, who created a detailed typology and classification of eccentrics and caches from Tikal, has been indispensable, and we frequently refer to the study, which was published with Hattula Moholy-Nagy in 2008. The Tikal study, a project that really began with his (Coe 1959) work at Piedras Negras, Guatemala, is essential reading to comprehend our analysis below.

COMMUNITIES AND CONSTELLATIONS OF PRACTICE

The eccentrics or silhouettes contained in these caches were difficult to make, if not for the technical aspect of the more elaborate bifaces, then for the often precise proportions that must be understood to finish the most simple of them (see chapter 1 for a general discussion). In most cases the modified flake silhouettes were backdrops for paintings (Hruby and Ware 2009), which is part of what makes the interpretation of their iconography difficult (Hruby 2007). This situation begs the question: What did the knappers actually know, and did all the knappers control the same esoteric religious knowledge to create these rather diminutive god effigies? Did a single knapper or artisan create all of the chert silhouettes for elite caches? There is good evidence, for example, to suggest that the artisans who made the flint silhouettes were not the same people who made the obsidian ones (Hruby 2007). In other words, were the production of items for caches made by single artisans or were they group activities, directed by one or a few coordinators?

In terms of knowledge production and sharing, Maya royals, and elites in general, were likely extremely finicky and guarded about who got to own, know, and carry out these practices. One of the reasons for this restrictive behavior and the likely creation of sumptuary laws was that many sensitive forms of ritual required the production of sacred items and their proper use, for example in the divination and calculation associated with important events, as well as in the impersonation of powerful supernatural beings. The second reason is directly related to the maintenance of the status quo vis-à-vis upwardly mobile people eager to increase their social status. One of the primary ways that Maya elites maintained their position was by controlling esoteric religious and crafting knowledge (Hruby 2007; Inomata 2001). Certainly, the most potent of these practices was writing, which tied together creation mythology, local genealogies, and the recording of history—the very foundation of ruling ideologies of the Classic Maya. The guarded nature of writing practices is underscored by the fact that the majority of people in Maya society were illiterate, and holy books were reserved for the ruling elite. Other crafts were part of these restrictive practices of knowledge production. As Takeshi Inomata and his colleagues (2001) have demonstrated, the Maya palace was not simply a place of luxury but also a locus of craft production for making the most prestigious items, such as pyrite mirrors and carved stones, directed and owned by the elite. Hruby has also argued that ancient Maya silhouette knappers may have used their knowledge as a form of symbolic capital (*sensu* Bourdieu 1977) to cement their position in society and gain higher status (Hruby 2006, 2007; Hruby and Ware 2009). However, it is how the guilds, *oficios* (see Hruby 2007; McAnany 2010), or communities of knapping practice organized themselves within political centers and between them that is the focus of this chapter.

Communities of practice are a useful way to explore local and regional variability in technology and iconography but through the lens of individual practitioners who share production knowledge and histories through practice as situated learning (i.e., *legitimate peripheral participation*; see Lave and Wenger 1991; Wenger 1998). Emergent communities of practice were first studied in depth by Jean Lave and Etienne Wenger (1991) and understood as "mutual engagement, a joint enterprise, and a shared repertoire" of doing and making things (Wenger 1998, 73). These theories can aid archaeologists in problematizing (1) different kinds of skills from more simple to more complex as they are shared in more or less restricted ways; (2) the possibility of membership in more than one community of practice, and (3) how technologies and design elements were transmitted between individuals and between communities of practice through brokers and boundary objects (see also chapter 1). Archaeological models that focus on different artifact traditions and horizons, as well as interaction spheres (Blanton 1976; Freidel 1979), have the effect of removing individual artisans from the processes of social change or stasis (see Joyce 2017), or essentializing all this complexity into a master-apprentice relationship.

Communities of practice have been used to examine pottery making in the Southwest (Mills et al. 2013), the Belize Valley region of the Maya Lowlands (Jordan et al. 2020), and Western Africa (Gosselain 2008, 2012), among other regions of the world (Roddick and Stahl 2016). Such ideas have not been employed much in the realm of lithic studies, perhaps because projectile point styles and production techniques often did not change much over great distances and over long periods of time. Nevertheless, for ritualized lithics, they provide an opportunity to think about distinctions and overlaps between symbolic forms and production techniques similar to the way communities of practice approaches have considered the highly visible decorative styles on pottery in conjunction with the knowledge of paste recipes and practices of forming vessels.

Concepts of communities and constellations of practice have been expanded to address the relationships, or relationality, between power, political turmoil, and crafters (Roddick and Stahl 2016). Such a focus is relevant to Mesoamerica, especially the Classic Maya, since the entirety of the Classic period can be characterized as expanding and contracting alliances between largely antagonistic city-states. Indeed, one of the greatest conundrums of Maya archaeology is understanding how material and immaterial culture traits can be so homogenous over such a wide area vis-à-vis political disunity. For example, while we can explain how some art styles, motifs, and writing conventions were shared between polities through the capture or sequestering of foreign scribes and carvers, we cannot explain how all of these practices are shared. We know from epigraphic studies that some artisans and scribes traveled between the capital city and its subsidiaries as a way to entangle these smaller sites through the gifting of symbolic and material capital. There is no better example than the Usumacinta region, wherein scribes and carvers were sent out to create monuments that touted political relationships between the center and the periphery (e.g., Houston 2021). There is little doubt that these master artisans shared at least some of these practices with local stone workers and acted as "brokers" in terms of practice theory (Wenger 1998). We could easily interpret this state of affairs as constituted by a community of practice among master and lower-status, or novice, artisans within political centers with constellations of practice existing between cities in the same political hierarchy.

These processes, however, do not explain how writing and art could be so similar between antagonistic polities or across the entirety of the Maya world. Were there writers' conferences where people got together and shared techniques and mythologies? Or was freedom of movement more acceptable for some kinds of crafters? It may be that greater freedom was given to those on religious pilgrimages, and thus people were allowed to communicate with foreigners and share information on these journeys. In any case, other than the client-patron relationships, hierarchical subsidiary relationships, and the capturing of artisans, other modes of sharing knowledge and learning were likely in play. It is clear that these were highly conservative traditions, except for, interestingly, the small areas in which creativity was encouraged. In hieroglyphic writing, for example, the structure of the sacred language and texts did not vary much, nor did the content of the writing. Scribes were nevertheless encouraged to express creativity in the execution of these texts, as one can easily witness in the full figure glyphs of Copan, or the twelve different ways to write the third-person pronoun **U** glyph at Palenque (figure 8.2). It was this restricted creativity that led to some

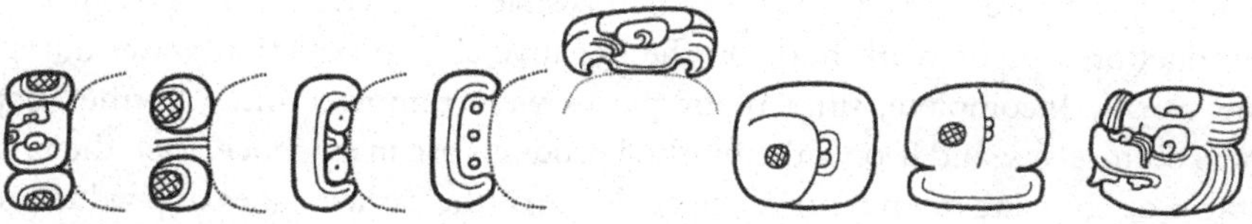

FIGURE 8.2. *Eight different versions of the third-person pronoun U glyph in Maya writing (Drawings by Alexandre Tokovinine).*

of the most beautiful and elaborate artworks in the Maya world. We argue that the production of eccentrics likely followed similar guidelines.

In past work, Hruby (2006, 2007) focused on how ancient Maya knappers may have been organized into something akin to a European guild system but situated in Mesoamerican ontologies, in the context of what was termed in early Colonial times as *oficios* (Clark and Houston 1998). The importance of this concept has also been noted by Patricia McAnany (2010), who used it in her pathbreaking study of ancient Maya economics. Although the term is Spanish, it characterized people in the same community of practice, and yet so little is actually known of this form of social organization that it still lacks specificity. The *oficio* rubric clearly demarcates individuals who belong to and share in mutual engagement and whose identities were both tied to one another through technical knowledge and through religious belief and social identification. The analogy falls short, partially since it remains a vague concept due to a lack of ethnohistorical data but also because it does not provide many avenues for understanding different levels or scales of complexity of practice and how that production knowledge was shared, without relying on analogies from other regions or even continents. The communities of practice concept may be better equipped to understand Pre-Contact communities of practice, without forcing crafters into an imperfect, albeit useful, archaeological analogy. In turn, the notion of *constellations of practice* provides a framework for thinking about wider, more diffuse sharing practices in which multiple communities of practice are linked by shared histories and brokers or boundary objects, such as circulating finished products (Roddick and Stahl 2016, 9–10; Wenger 1998).

To examine communities of practice among ancient knappers, one must start from the bottom up—a task to which subtractive lithic technologies are well suited. Lithic analysts must outline raw material locations and the *chaîne opératoire* of different modes of production practiced by one or more crafters. Like other crafts, there are some pros and cons to reconstructing *chaînes opératoires* of chipped-stone manufacture. On the plus side, if production debitage can be located, then one can determine with some accuracy what someone was making and where the material likely came from (e.g., via chemical and visual analyses). In rare cases, it may also be possible to determine the output of a production area or workshop. On the negative side, production waste dumps are usually incomplete, since many pieces were removed, often by the knappers themselves, and if only the finished products are in evidence, then the final stages of chipping often remove the evidence of the earlier stages of production. Consequently, the entire production sequence may not be known. Indeed, there has only been one possible silhouette workshop found in the Maya area: Colha, Belize (Shafer and Hester 1991).

In Mesoamerica, as elsewhere in much of the Americas, the entirety of the *chaîne opératoire* for most lithic industries is not known, or is likely based on some ill-informed, or at least Eurocentric, assumptions of efficiency, supply and demand; individual-centered production; and secular practice disconnected from religiously informed ontologies. Most archaeologists interested in chipped-stone production, as well as replicative experimenters, work alone. They are interested in figuring it out by themselves; finding sources; publishing their findings alone or coauthored with distinct knowledge sets per article; and sitting alone, knapping, and working on techniques that can be carried out by one person. Traditional American archaeologies are male centered, with visions of the lone hunter making their own equipment, completely self-reliant. As is often the case, these lithic specialists may have focused on Paleo-Indian peoples—highly mobile hunting and gathering bands with extremely low populations—who may very well have been self-reliant in the extreme. In many ways this "lone-knapper" approach has permeated much of our reconstructions of past lithic production models (see chapter 3). Thus, while we may get close to replicating how a flake was detached from a core, less focus has been placed on the relations of production, distinct *hexis* and practice sets, and the implications of these more complex forms of situated learning and social organization.

TECHNOLOGICAL TYPES AND *CHAÎNE OPÉRATOIRE*

The *chaîne opératoire* begins with resource acquisition. For most Maya cities, chert, chalcedony, and flint availability was a local matter. Chert availability may have been a deciding factor in settlement strategy, since it is more difficult to build a city of limestone without chert tools to cut, hammer, peck, and carve its building blocks. There are many types of landforms from which the Maya extracted their raw materials, but the most common were from outcrops found above the soil, *bajo* mud or soil matrix as float, and river or creek contexts where the material collected from recent or ancient alluvial processes. Each of these have different implications for the organization of production. In the first example, the source may represent a "boundary object," which according to Wenger (1998) are things or places that act as a liminal spaces between different communities of practice. Andrew Roddick (2016), for example, has suggested that based on ethnographic data clay sources in the Titicaca basin were places in which different communities of potters intersected with each other. Through the application of stylistic and technological analysis, Jilian Jordan et al. (2020) recently demonstrated a similar practice among Late Classic period potters in the Belize River Valley.

In any case, local knappers were undoubtedly involved in resource extraction, or at least in assaying the nodules brought to their workshops. In many ways choosing a viable nodule involves the greatest breadth of knowledge, since the color, quality, and consistency of the stone must be surmised to achieve the desired end product. In addition, the knapper must think three-dimensionally and envision the finished product "within" the stone. If the stone is too small, or has too many undulations, then the desired end may simply not be achievable. This process is important and speaks to resource scarcity, since, of a hundred nodules available at a given source, only one or two may fit the correct size and color requirements of the end product. Of course, this process is more important for the very large silhouettes, but it involves considerations that all knappers must have had to make.

The selected nodules then had to be decorticated through direct percussion. Far from a mindless task, removing the decortication flakes and macroflakes was all done with specific goals and products in mind. The majority of lithic caches Hruby has studied over the last twenty-five years suggest that the larger nodule reduction flakes were reserved to make smaller secondary silhouettes that often flanked larger ones in the cache context, the latter of which usually took a central position. The most common centrally placed silhouettes are the sun or discoidal forms, or sometimes the largest of the serpent or lightning forms. Usually a medium-sized nodule (20–30 cm in length), reduced to make medium-sized silhouettes (10–15 cm), also yields two to three secondary eccentrics that fall in the small- (<10 cm) to medium-size range. These decisions were being made actively as the nodule reduction was happening. Once these blanks were produced, either a rough bifacial core or a macroflake, then the knappers would continue with obvious direct percussion strategies to further remove overhangs and thin edges, and well as remove primary biface-thinning flakes.

The next stage is less known, a transitional or middle-stage reduction process in which indirect percussion is used. This process is discussed in chapters 3, 4, and 5 and appears to have been a much more widespread practice in Mesoamerica, and the Americas in general. The process involved using an antler or wooden punch to thin the biface, create even-flake patterning, and maintain the desired width of the biface. It is easier to create a greater thickness-to-width ratio, which is common in ritualized lithic artifacts from the Maya area. The process is also often easier, with fewer errors if it is done by more than one person—someone to hold and position the core, another to place the punch, and the other wielding the percussor. Indirect percussion can be done by a single person, but it isn't easy. The end result of this stage is the silhouette preform, which usually consists of a celtiform, discoidal, or laurel-leaf format. These products were also placed in caches at most Maya sites, and the basic forms carried their own symbolic load (Hruby 2006, 2007).

The preform was then ready for shaping and notching into the final format. In general, the basic technological types of chert silhouettes in the Maya area include *fully bifaced*, *mostly bifaced*, *partially bifaced*, and *modified flake* (figure 8.3a–d). While some of these type names may seem quibbling, each one indicates very different approaches, resource constraints, and skill level to produce. *Fully bifaced silhouettes* are the most time consuming to produce, and the surface features little to no evidence of the blank or nodule from which it was produced (figure 8.3a). They may also include indirect and pressure notching and were finished by pressure. The type implies the production of a fully bifaced preform *before* the notching and contouring is carried out. *Mostly bifaced silhouettes* are those that still retain a noticeable portion of the original blank or nodule on their surfaces (figure 8.3b). Although finely crafted, they may also feature some of the original arc, or curvature, of the flake from which they were made. They were probably manufactured with the same level of skill as the fully bifaced examples, and it was the constraints of the smaller blank that kept the knappers from making a full biface. This type may indicate a lack of resources, with subsequent reliance on smaller-than-desired blanks for production.

Partially bifaced silhouettes are easier to make but still require knowledge of direct and indirect percussion to notch and finish properly, not to mention proper proportions and measurements (figure 8.3c). Only the edges were reduced bifacially, and in some cases not on all edges. This type may indicate resource scarcity or a lack of knowledge or skill by the knapper making the piece. The *modified flake silhouettes* are the simplest type and only require basic pressure work to complete (figure 8.3d). Again, the measurements and proportions must still be known to make these, and there is probably also a symbolic component to the kind of flake chosen to make them. All four are commonly found together in Maya caches, and those studied in this chapter are no exception. These four types could all have been made by the same person, and the more elaborate ones may have been made by more than one person. Importantly, it is on the fully bifaced examples that individual hands and styles may be more easily identified, because there are more details and flake patterning to compare and the final product is less dictated by the flake or blank used to make the final product. Finally, all of these stages of production could have been accompanied by chanting and bloodletting, and preceded by ritual preparations of fasting and abstention from sexual activity (Clark 1991; Clark and Houston 1998; Hruby 2006, 2007).

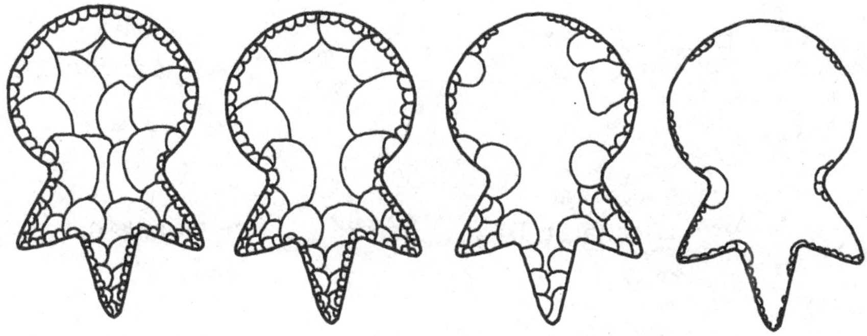

FIGURE 8.3. *The basic technological types of chert silhouettes in the Maya: (a) fully bifaced silhouette; (b) mostly bifaced silhouette; (c) partially bifaced silhouette; (d) modified flake silhouette*

THE UCANAL AND XUNANTUNICH CACHES

The Ucanal and Xunantunich caches analyzed here were all closely associated with stelae, having been found in front or just to the side of stelae and thus likely deposited at the moment of monument erection. At both sites, a few of the eccentrics were not found associated with the main group of eccentrics, which may indicate some regional pattern of deposition that is not yet understood. The Ucanal cache Cache #1-1 contained 9 chert silhouettes, Ucanal Cache #20-1 contained 9 cherts and 11 obsidians, and Xunantunich cache A-3-1 featured 9 chert and 9 obsidian silhouettes.

Context: Ucanal Groups J and K

Ucanal Cache #1-1 (UCA1C-11-3-78; Artifact numbers UCML-22 to -28) was excavated by the Proyecto Arqueológico in 2016. The cache was found in Group J, an elite residential and administrative complex (figure 8.4). It was interred at the base of Ucanal Stela 19, which was poorly preserved, with only its base remaining in situ and its upper portion missing during the time of excavation (Cruz Gómez and Garrido 2016).

No evidence of inscriptions or carvings were noticeable on the base of the monument. The stela is located directly in front of (to the east of) a low shrine or altar consisting of a small, square platform approximately 30 cm high at the center of Group J. Seven total chert silhouettes were found in situ in three rows within the final phase of platform fill. While ceramics in the fill immediately surrounding the cache were from mixed time periods (Late Preclassic to Late Classic), more recent excavations of the Group J platform indicate that the final phase of the entire platform underwent a major remodeling during the Terminal Classic period (Halperin and Garrido López 2019; Perea and Dubois-Francoeur

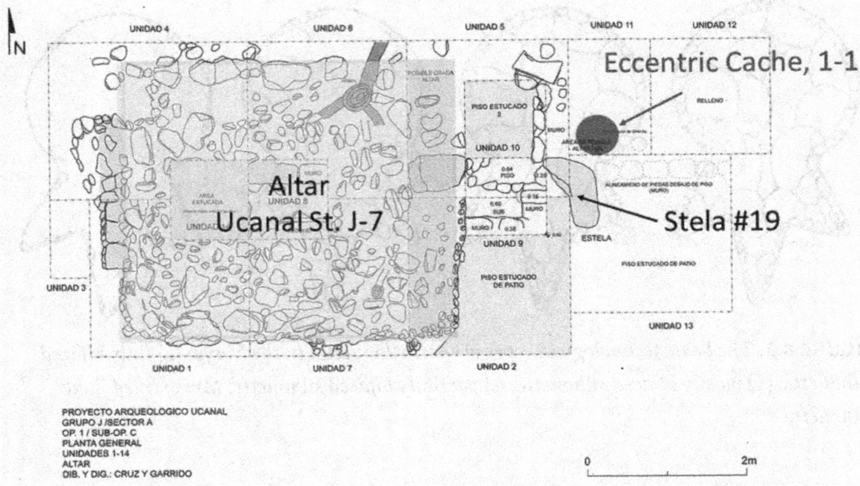

FIGURE 8.4. *Archaeological context of Ucanal Cache #1-1 and the remains of Stela 19 in Group J.*

2020), dating the erection of Stela 19 and the burial of Offering #1-1 to this later time period.

Interestingly, two complete silhouettes (UCML-29 and UCML-30) from Ucanal Cache #1-1 were found near the ground surface above the level of the eroded plaza floor sealing the lithic cache (figure 8.5; UCA1C-12-1-80, UCA1C-11-2-77). Because the two "misplaced" silhouettes are of the same material, a rose-ecru translucent chert, as two from the in situ cache, they were likely produced alongside the cached eccentrics. Although the reasons two of the eccentric pieces were not placed with the others are difficult to discern, such finds point to the performativity of the caching process, whereby the act of placing the eccentrics helped realize particular ritual goals and whereby the eccentrics themselves may have a particular animacy or power that dictated particular actions.

An additional eccentric cache, Offering 20-1, was recently excavated in 2019 and found directly associated with newly discovered Stela 29, located in front of a large temple-pyramid in a public ceremonial group, Group K (López López 2020). The stela, presently the oldest-known stela with inscriptions at the site to date (Halperin and Martin 2020), dates to the Terminal Classic period, with 879 CE as a probable corresponding period-ending date. The stela was erected on a small masonry platform (1.5 × 1.2 × 0.3 m) containing late-phase Terminal Classic ceramic sherds. Offering 20-1 was placed within the small stela platform in front of the stela but slightly offset to the east from its center axis.

Although the chert and obsidian silhouettes were intermixed, it is clear that some were purposefully placed close to each other at the moment of deposition.

FIGURE 8.5. *Position of the chert silhouettes of Ucanal Cache #1-1 in relation to Stela 19. (photo by C. Cruz-Gómez)*

For example, chert silhouettes B-4 and B-5 not only were found directly on top of each other but were made of the same source material. Chert silhouettes B-8 and B-9 were also found directly next to each other in the cache and were made of the same source material (see below).

The Context Xunantunich Structure A-3

Cache A3-1 was discovered on the centerline of Structure A3, a large west-facing pyramid in the main plaza at Xunantunich. Structure A-3 is approximately 16 m tall and 24 m wide, and represents the central pyramid of the site's eastern triadic shrine. The structure appears to have been built in one construction phase during the end of the Late Classic period. Excavations behind the first original step, which was located on the plaza floor, revealed the A-3-1 cache (figure 8.6). The cache was aligned with Stela A1, which stood 50 cm in front of the cache and the first step. It contained 18 objects: 9 chert eccentrics and 9 obsidian eccentrics. Three were located on their own circa 10 cm away, while the rest were all clustered together in one big pile. They exhibited no sign of paint or other marks of decoration.

The location of Cache A3-1, along the central axis and base of a nondomestic structure (A-3), is a pattern that is consistent with many caches containing chert silhouettes at Xunantunich. Cache A14-1, for example, was axially located at the base of Structure A14 and contained two discrete sets of nine objects each (Audet 2006, 151). The first set of cached items included nine chert silhouettes. The second set, located 30 cm to the south of the first set, contained eight chert silhouettes and one jade bead (figure 8.7). Directly west and across the plaza

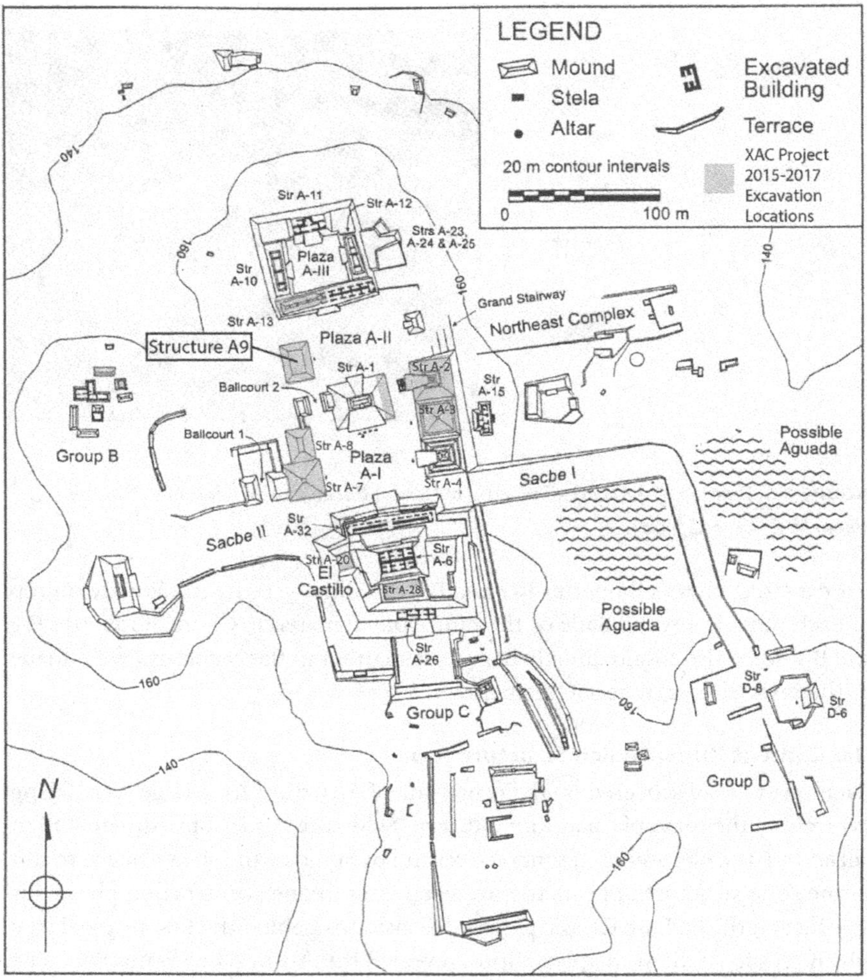

FIGURE 8.6. *Map of Xunantunich showing the location of Structure A-3.*

from Structure A14, excavations along the axial base of Structure A9, and its associated stela (Stela 4), recently uncovered two other caches (Awe et al. 2019; Slocum 2018; Sullivan 2017). The first cache (A9-1), which was discovered below Stela 4, contained twenty-eight chert silhouettes. The second cache (A9-2) was located at the base of the axial stairway and contained nine obsidian silhouettes, several marine shells, a number of freshwater snail (*jute*) shells, a fragment of branch coral, and several small fragments of jadeite and pyrite, plus five pieces of a red mineral that is either hematite or cinnabar.

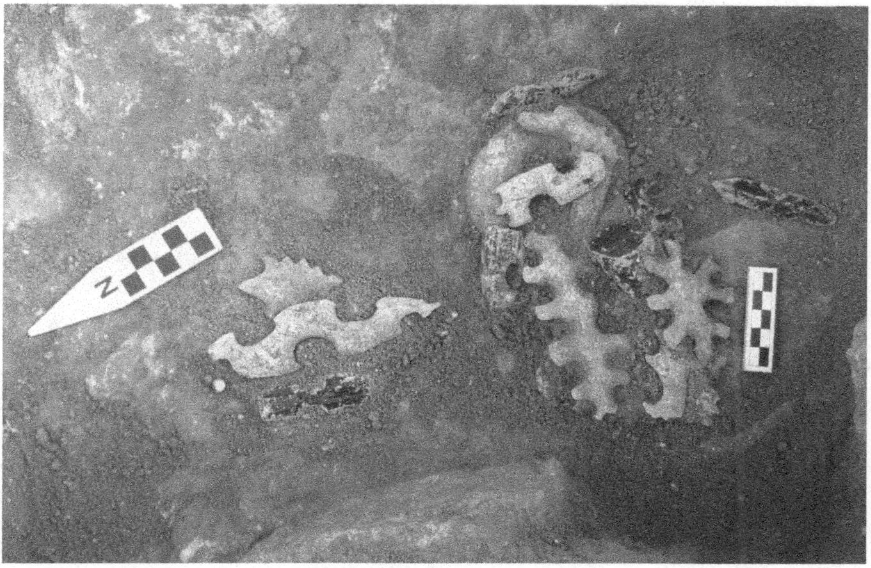

FIGURE 8.7. *Position of the silhouettes in Xunantunich Cache A3-1.*

JUXTAPOSITION OF THE CEREMONIAL CHERTS: SYMBOLIC, TECHNOLOGICAL, AND MATERIAL ATTRIBUTES

From a distance each group of squiggly silhouette forms seem to resemble one another. They are roughly the same size and tend to confuse our western sensibilities with their complexity, which has led some archaeologists to lump them together as lightning symbols. In fact, Karl Taube (Agurcia Fasquelle et al. 2016) has recently argued that as a general class of item, they are meant to resemble fulgurites: the glass tubes formed when lightning hits silica-rich earth. Upon closer inspection, however, details and patterns emerge.

Hruby (2006) suggests that while the Maya seem to have had a limited range of basic eccentric forms, they also conveyed more complex meanings by combining multiple forms in a single eccentric. Concepts borrowed from Maya epigraphy and iconography to describe these patterns are general *naturalistic* or *stylistic* format, and then the combination conventions of *conflation*, *superimposition*, and *infixation*. These conventions seem to support the notion that these objects are meant to be understood two-dimensionally, and not as three-dimensional sculpture, although in some cases we would argue that is not the case. *Superimposition* occurs when one form simply overlies another, as in this case from Altun Ha (figure 8.8). It is a combination of a mat or centipede form and a laurel-leaf form, one laid on top of the other but with both retaining their original full size and form (this example is quite large, at 40.4 cm).

Conflation is used to combine different forms, sometimes retaining some

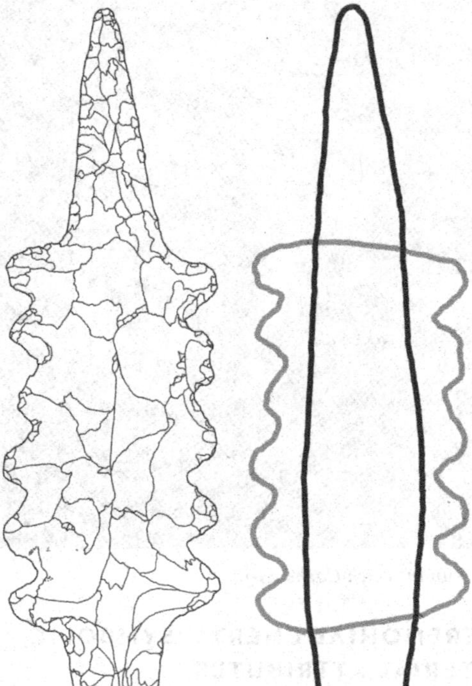

FIGURE 8.8. *An example of superimposition from a chert silhouette from British Honduras (40.4 cm long, Drawing by Hruby after the British Museum).*

aspects of two or more forms but clearly creating a blended "Frankenstein" of various characteristics to create a new form. We place serration in this category; for example, crescent forms may be smooth or serrated, creating two different possible meanings. Finally, *infixation* is not common but usually occurs with discoidal forms with negative space in the middle, although it may also be used more rarely with other forms. All of these combinations are found in the eccentrics of the caches presented here. In the following sections we compare the symbolic forms in each cache, and then the technology to elucidate any obvious patterns in the cached cherts.

COMPARISON OF SYMBOLIC FORMS

The Ucanal and Xunantunich caches feature silhouettes that individually do not seem to have much in common but placed in a broader framework share some similarities (figure 8.9, please see notation next to each example for analysis that follows). The A3-1 Cache from Xunantunich actually fits into a recognizable, pan-Maya format of nine cherts with many similarities to Late and Terminal Classic Tikal caches. The format is not identical to any of the Eccentric Flint Classes 1–4 identified by Coe and Moholy-Nagy (2008) but rather shares some affinities with each, both technologically and symbolically. Shared forms include sun and moon silhouettes that refer directly to these celestial bodies but also to the gods that represent them (Moholy-Nagy and Coe 2008; Coe Types EF1 and EF2, respectively; EF, i.e., eccentric flint). They may be serrated or smooth, although there is good evidence to suggest that there are subtle or even overt differences in symbolic meaning between the two. The serrations may reflect an additional meaning of *yax*, meaning green-blue or "new," instead of "sharpness" or "chipped stone," since Early Classic iconographic representations of *yax* are identified by these notches or serrations (Houston 2014).

The A3-1 examples include a serrated version of the crescent and a sun form not found at Tikal, which includes a quatrefoil infixed sign that may represent a cave opening but more likely reflects the internal petaloid nature of solar imagery. The negative space in this silhouette also may be an elaborate reference to the "doughnut" or "ring" shape common at Tikal (Coe Type EF4), which may itself refer to yet another aspect of the sun. One of almost the same size, shape, and technology was found at Altar de Sacrificios (Willey 1972, 187).

Another refence to the moon may be seen in what appears to be a conflation of a crescent form and a serrated variety of the laurel-leaf form (Coe Types EF2 and EF11, respectively in figure 8.9; Artifact A4 in figure 8.10). This blending of forms may be referring to combinations of different deities or celestial bodies in different positions or states in the sky. It should be noted that a number of caches at Xunantunich and Naranjo feature this serrated laurel-leaf form and often have multiple varieties of the lunar crescent in the same cache (Carreño S. 2013; Salas, personal communication, 2019). There is also a clear Type 8 form in the cache as well (figure 8.10, Artifact A2), likely signifying the *pop*, or mat, sign, which overlaps with centipede symbolism (Hruby 2007) and may be a reference to the Milky Way. What appears to be a serrated projectile point may be a stylized version of Coe's Type EF3, which he interpreted as a *Spondylus* shell but has not been demonstrated iconographically as such. A typical and easily identifiable scorpion form (figure 8.10, Artifact A5)—a likely reference to the Scorpius constellation—was included (Coe Type EF5) as was a serpent form (figure 8.10, Artifact A3; Coe Type EF12), which was probably meant to evoke lightning or one of the Maya embodiments of lightning: God K or K'awiil. There is also a possible serrated disc in A-3-1 (figure 8.10, Artifact A7), which Coe identified as a tick at Tikal but was not demonstrated iconographically. In this case, however, its lack of symmetry may place the Xunantunich example into the category of a

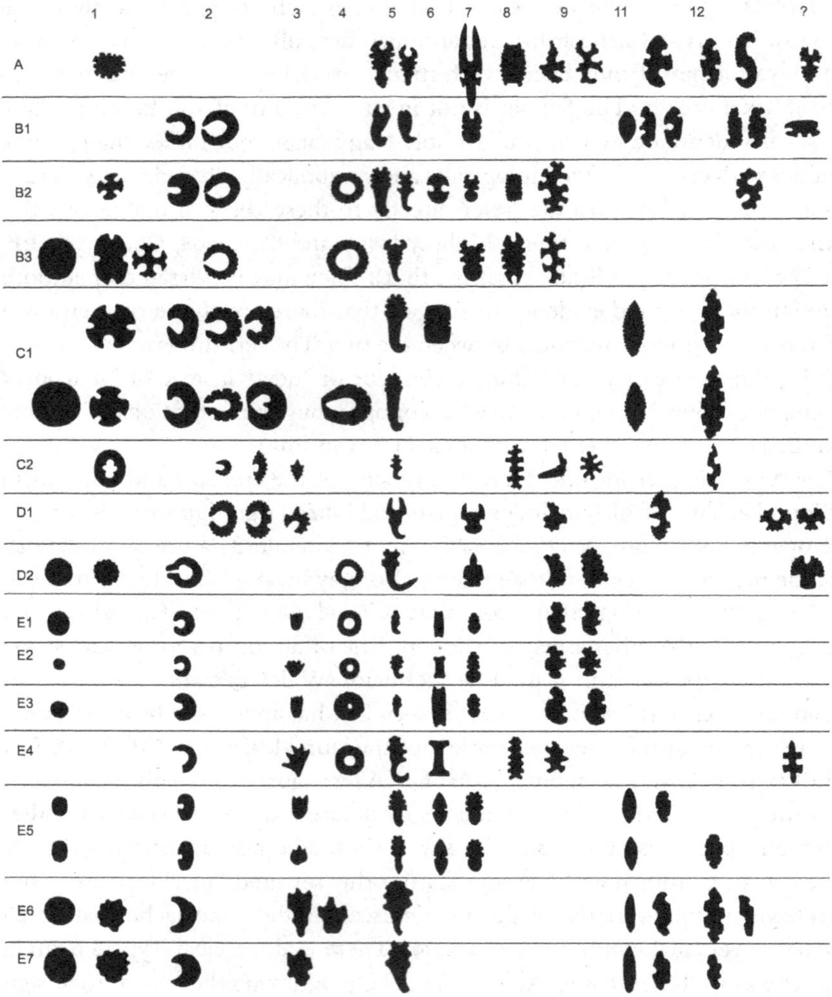

FIGURE 8.9. *A cross-site comparison of cached chert silhouettes from Yaxha (Row A), Naranjo (Row B), Xunantunich (Row C), Ucanal (Row D), and Tikal (Row E) caches, organized by column according to William Coe's eccentric types (i.e., EF; Drawing by Hruby partially after Moholy-Nagy and Coe 2008).*

deity portrait (Coe Type EF9). If this is true, then, in addition to another unique portrait in A-3-1, the presence of two portrait eccentrics aligns it with the Class 1 and 2 formats from Tikal.

In summary, Cache A-3-1 shows some affinities to the Tikal traditions (i.e., Classes), which were often carried out for over a century at a time, as well as other caches at Xunantunich and Naranjo. Tikal, however, boasts the greatest

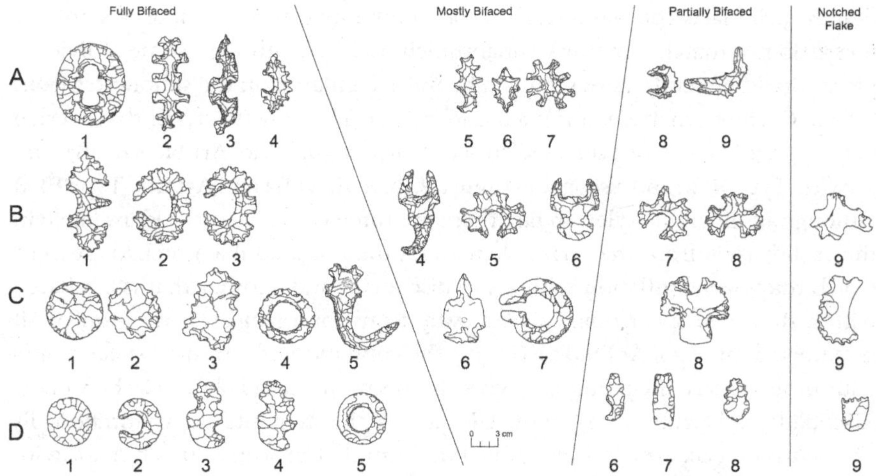

FIGURE 8.10. *A comparison of technological types (column 1: fully bifaced; column 2: mostly bifaced; column 3: partially bifaced; column 4: notched flake) from Xunantunich (row A: Cache A3-1), Ucanal (row B: Cache 1-1; Row C: Cache 20-1), and Tikal (row D: Cache 56).*

conservatism in chert eccentric forms of any site Hruby has studied. At Tikal knappers do not practice conflation, superimposition, or infixation of different forms, except inasmuch as they modify the basic laurel-leaf type EF11 to evoke more complicated silhouettes in EF Class 2, 3, and 4 caches (Moholy-Nagy and Coe 2008). This practice is more common in the notched core type obsidian eccentrics all over the Maya Lowlands, which are all, to a certain extent, based on the lenticular form of an exhausted blade core. Cache A-3-1 seems to follow suit, with caches found outside of Tikal and its subsidiary sites via the manufacture of greater creative and idiosyncratic forms based on infixation and conflation (see earlier). Importantly, Coe Type EF7 from Tikal, which Hruby identifies as a sacred bundle or knotted jewel with a three-dot device (see also Moholy-Nagy and Coe 2008), is not found at Xunantunich, or indeed many sites outside of Tikal (but see Willey 1973). Nevertheless, the suite of types found in Xunantunich A-3-1, and shared with many types from Classes 1 and 2 from Tikal, seem to confirm a Late Classic date. Naranjo silhouettes appear to be quite similar to those from Yaxha and Xunantunich but in at least one case adopt a modified, fine brown laurel-leaf eccentric more common at El Tigre, Campeche, Altar de Sacrificios, and more generally, Northern Belize.

The Ucanal Cache 1-1 is rather different. Since two of the silhouettes (figure 8.10, Artifacts B1 and B7) were not found in direct association with the main concentration of seven, it is possible that they were part of a different offering or that the original cache was disturbed somehow. Either way, the suite of symbolic forms do not match any from Xunantunich or Tikal, although some individual pieces do. In general, there seems to be more conflation in the silhouettes from Ucanal Cache 1-1 than Xunantunich Cache A-3-1. The only two types that overlap between the caches are a serrated crescent form (figure 8.10, Artifacts B2, B3, and A8; Coe Type EF2) and a scorpion (figure 8.10, Artifact B4 and A5; Coe Type EF5), although in different styles. In fact there are three crescent forms identifiable in the cache; aside from the serrated crescent, there is a double-notched crescent, which may be a conflation with a sun disc form, and another that is conflated with a stemmed spear point placed away from the main concentration of silhouettes (figure 8.5). At Piedras Negras the conflation of stemmed spear points with other forms, including crescents, is common, especially in Cache O-13-57 (Escobedo and Hruby 2002). In the Ucanal case the stem juts out asymmetrically. The scorpion eccentric is a different style from that of Xunantunich A-3-1, with a much more naturalistic rather than stylistic representation, including more defined pincers and abdomen. There appears to be another scorpion included through superposition, however, where the pincers and thorax of the scorpion are attached to a disc. Hence, there are two possible, but not certain, representations of the sun in Ucanal Offering 1-1.

One type, not directly represented at Tikal, is repeated in Ucanal 1-1. It consists of a trilobed form (figure 8.10, Artifacts B5 and B8) notched out of a serrated disc form, which is also found at Piedras Negras. It is possible that this is a reference to a solar disc, but more likely it is an alternate depiction of the three-dot jewel or knotted bundle device depicted as Coe Type EF7 at Tikal. Instead of chipping the knotted attachment, these Ucanal eccentrics seem only to depict the circular three-dot device. In the incised obsidians of Tikal the three-dot device is depicted positioned in the mouth of the Principal Bird Deity and sometimes with a penitential version of Chaak, the Maya storm god. Its overall meaning remains unclear and is not clearly related to any planetary body or constellation, but it may be related to penitential acts and imbibing psychoactive substances for ritual. There is one naturalistic form (figure 8.10, Artifact B9), which probably is a deity portrait of the Tikal Type EF9 and, finally, the most complex artifact (figure 8.10, Artifact B1), which seems to conflate a number of forms. At its foundation the eccentric is based on a serrated laurel-leaf form (Coe Type EF11 or perhaps a more stylized version of Coe Type EF8), but it features an infixed spear point stem, as well as large and small notches. In that sense, this eccentric shares some affinities with the crescent and serrated laurel-leaf eccentric from A-3-1.

To summarize, Ucanal Offering 1-1 as a whole does not fit any known patterns from Tikal, Piedras Negras, or Xunantunich. There were no obsidians in this cache, which may explain the various combinations of forms, perhaps to maintain some kind of numerology of symbols without creating separate items. This is just a hypothesis, however, which remains to be tested through comparisons with other caches. Importantly, the further away from Tikal one moves, the more arcane the symbol combinations become, which may have to do with the expression of local deities, creating local symbolic capital, or as-yet-unknown facets of local practices and traditions.

The final cache from Ucanal, Cache 20-1, conforms to a recognizable cache format, and one that is both clearly found at Tikal and Naranjo. Interestingly, while iconographic, monumental, and architectural styles across the Maya world changed significantly during the Terminal Classic, the eccentrics did not. Tikal provides the best example of this conservative aspect of Maya culture, since Coe's Class 1 caches retain the same forms as earlier periods and seem to become even more standardized over time. Although the size of the silhouettes is somewhat smaller than earlier examples, complete bifacial reduction seems more common at Tikal or Ucanal (see "Comparison of Technological and Material Attributes" section), and it most closely resembles Tikal Caches 44, 45, 56, 94, and 95. There are some differences, however, that seem to be related to style rather than symbolism. Likewise, Ucanal Cache 20-1 also shares some forms with Naranjo eccentric caches, most notably the combination of Coe Type EF1 or solar disc (figure 8.10, Artifacts C1 and C2), Type EF4 or ring form (figure 8.10, Artifact C4), Type EF2 or lunar crescent (figure 8.10, Artifact C7) and scorpion forms (figure 8.10, Artifact C5; Carreño S. 2013, fig. 5). The crescent, however, has two squared protrusions at the tips, which have not been seen elsewhere so far. There is also a serrated disc (figure 8.10, Artifact C2) or perhaps a more mildly chipped "tick" form Coe labeled Type EF14; however, it is likely that both of these forms simply represent another aspect of the sun or another shining deity. The ring form, common at Tikal as Type EF4, here is serrated as it is in at least two eccentric cherts from Naranjo (Carreño S. 2013; Miriam Salas, personal communication, 2019). All three of the circular forms from the Ucanal cache are virtually the same diameter within a matter of millimeters (figure 8.10). There are two Tikal Type EF9 portrait eccentrics (figure 8.10, Artifacts C3 and C9), as well as a rather simplified scorpion form with diminutive thoracic appendages (figure 8.10, Artifact C5; Coe Type EF5).

The two remaining forms do not fit neatly into Coe's Class 1 eccentric flint grouping. One is trilobed, which may be a more abstract, later version of the three-dot device discussed earlier. However, the notches are not radial in nature, create a primarily bilateral axis, and thus may be an unidentified form (figure 8.10, Artifact C8). The trilobed form is also present in Ucanal Cache #1-1 (figure 8.10, Artifacts B5 and B8), and differs from the common quadrilobed disc forms from caches from Xunantunich, Baking Pot, and Naranjo (Carreño S. 2013, fig. 5;

Miriam Salas, personal communication, 2019; Sullivan 2017, figs. 14, 21, 24, 25, 30). Finally, there is a pointed form (figure 8.10, Artifact C6) with bilateral symmetry, which may be an alternate version of the Tikal Type EF10. Of significance, however, Type EF10 seems to be related to early Late Classic Class 4 style caches, which also extend back in time to the Early Classic. Thus, Ucanal Cache 20-1 shares many forms with Late Classic lithic caches found elsewhere, but some of the shared types were executed in very different styles.

COMPARISON OF TECHNOLOGICAL AND MATERIAL ATTRIBUTES

Clearly there are some basic differences in symbolic forms and styles between all three caches, even if some of the symbols are quite similar to one another or at least repeat in various ways. In terms of basic technological types, production strategy, and raw material choices, however, the caches reveal further affinities (see figure 8.10). The first characteristic all three caches share is the selection of materials with a wide array of visual characteristics. Rather than automatically assume that the color of the material was of insignificance to the ancient Maya, we instead note both the surface colors of the chert as well as those that emerge in their translucency seen with backlighting. All three caches feature silhouettes made of gray, light brown, dark brown, light blue, and reddish colors. No matter the surface color, the pieces usually appear to be white to yellow to orange in color with back lighting. All of the materials are a chert-chalcedony mix, meaning that they feature characteristics of more opaque chert, as well as the clear icy appearance of chalcedony with intermittent dendrites within. These characteristics make the stone beautiful but also difficult to work, since chalcedony is more rich in silica gel, lacks opal, and is thus extremely tough. In the case of Ucanal Cache 20-1, it seems that the knappers selected and worked with inclusions within the material. While the ring forms and other eccentrics with enclosed negative spaces are generally agreed to have been created by selecting blanks with vugs, these pieces seem to indicate that inclusions were included in the overall planning as well. According to figure 8.10, Type EF10 (Artifact C6) and Type EF5 (Artifact C5) forms include a fossil inclusion and a limestone inclusion in the center mass of each piece. It should be noted that other sites do not follow this pattern. Material choice and reduction techniques change over the centuries at Tikal, while the basic symbols stay the same. Likewise, while Ucanal and Xunantunich are located along the same waterway, and there was undoubtedly communication between the two centers; also, chert resources were plentiful enough in the two regions to indicate that they had their own chert sources (Laporte and Mejía 2000).

Xunantunich A-3-1 silhouettes were made from flakes and blanks derived from five or six different nodules. This estimate is based on material similarities and could have been a lesser number if a single nodule featured a wide variety of colors within it. At least one of the pieces was heat modified. The serrated crescent was either made from a burned piece of found chert, a purposefully heat-treated flake, or was burned after it was produced but before it was interred in the cache. It was clearly chosen for its bright red color. The scorpion and the serrated laurel leaf are partially bifaced percussion flakes, probably removed from the same nodule. The Type EF3 eccentric and the serpent form appear to be the same material, which was a gray color. The Type EF3 eccentric was mostly bifaced and was likely made from a nodule reduction flake removed during the production of the serpent, which was completely bifaced. The more discoidal portrait (Type EF9) and the solar disc (Type EF1) followed the same pattern, where debitage from an earlier stage of production was retained to create another eccentric. The other portrait bears no material resemblance to the others and appears to be a one off as a modified flake technological type. Like the discoidal portrait, the mat-style eccentric features incredibly deep notches. Since these are relatively uncommon, it is likely that these pieces were finished by the same knapper. In terms of overall difficulty, the solar disc and serpent forms required the greatest skill to produce. The remaining eccentrics do not demonstrate that knappers of different skill levels were making the pieces but rather that the blanks and flakes used to make them may have been too thin to sufficiently reduce bifacially. However, the differences in style and finishing do indicate multiple hands making these items. The fact that materials from the same nodule were used indicates that at least some of these items were made during the same knapping event.

Ucanal Offering 1-1 represents a slightly more concerted effort using a more limited number of nodules. The serrated and notched crescent, the two tripartite silhouettes, and the portrait all have identical light gray color and white or cream translucency. The fact that the notched crescent (figure 8.10, Artifact B2) is fully bifaced, one tripartite disc is mostly bifaced (Artifact B5), another is partially bifaced (Artifact B8), and the Type EF9 portrait is a modified flake (Artifact B9) suggests that a single nodule was reduced with the primary goal of making the crescent and that the other pieces were made subsequently from the remaining debitage. In addition, the scorpion form (Artifact B4) is made of a similar color stone but with more dendrites within. It is unclear whether this fully bifaced example was derived from the same nodule, but it is unlikely given the obvious small size of the local nodules. The serrated crescent and the conflated scorpion and disc form appear to be made from the same nodule, and the fact that the crescent is fully bifaced and the conflated piece is mostly bifaced adds credence to that interpretation. The final two eccentrics, located away from the main concentration, also appear to have been made from the same nodule of whitish cream color and white translucency. The conflated laurel-leaf form (Artifact B1) was fully bifaced, and the conflated crescent form was partially bifaced (Artifact B7), again indicating a sequence of production based on a central form. All of these eccentrics are relatively consistent in terms of style and manufacturing technique, leaving open the possibility that they were made by the same team of knappers.

Ucanal Cache 20-1 features a wider array of materials than Cache 1-1. A complete material comparison could not be made, since two of the pieces were left uncleaned for future analysis. However, they appear to have been produced from at least four different nodules. The serrated disc (figure 8.10, Artifact C2) and ring (Artifact C4) seem to have been made out the of the same color material both on the surface and with back lighting. It would be a rare case of two fully bifaced silhouettes coming from the same nodule, but given their diminutive size it is possible. The fully bifaced scorpion (Artifact C5) and the trilobed partially bifaced (Artifact C8) eccentric appear to be from the same nodule, although the latter is a bit more rough textured. These are reddish-brown to maroon in color, and the concretions remaining in the matrix are of the same type. The portrait (Artifact C3), which appears to be a K'awiil silhouette, is singular in the cache and unique in color to any other eccentric analyzed by Hruby. It is a rainbow-colored chert-chalcedony with black dendrites and even displays some chatoyancy when placed under light. These colors do not remain with backlighting, however. Unlike the other portrait, which is simply a modified flake, this example is fully bifaced and made with a high degree of difficulty. The second portrait, although uncleaned, seems related to the ring eccentric and crescent material. However, both the ring and the crescent are fully bifaced. These are of a light cream to tan color. The Type EF14 silhouette (Artifact C6) is uncleaned and of an unknown material but was fully bifaced. As mentioned above, these cherts are notable for seven out of the nine being completely bifaced (table 7.1).

DISCUSSION AND CONCLUSIONS

According to the analyses presented here we can isolate a few loci where communities of practice may have expressed continuity: (1) symbolic type, although the expectations and requirements for different forms may have been required above and beyond the knappers; (2) technology and *chaînes opératoires*, that is, how each piece was made and how flakes and blanks were chosen for candidacy as a silhouette; and (3) what kinds of nodules, blanks, flakes, and colors were chosen based on their size and color. In this brief comparison we can say that from a synchronic perspective, knappers from Ucanal, Xunantunich and other Belize Valley sites, Naranjo, and Tikal all chose overlapping symbolic forms, but they or their patrons chose very different ways to express these symbolic forms and in different combinations. Even in the case of the most common types, such as the crescent form, no two examples had the same outline. These disjunctures between basic symbolic forms and execution suggest that different communities of practice may be linked by shared histories and ideas tied together through brokers and boundary objects but not necessarily characterized by the same communities of practices. Such phenomenon are seen elsewhere in the Maya area, such as in the anthropomorphic eccentrics from Copan and from Altar de Sacrificios that were executed using very different techniques and raw materials but evoked the same meaning of a K'awiil Maize God (Agurcia Fasquelle et al. 2016; Willey 1972). Such constellations of practice were not shared by all sites. For example, some sites in the Maya area, such as Calakmul, Palenque, and Caracol, did not participate in chert silhouette production and caching practices to any significant degree (Chase and Chase 2017, 198).

That different symbols in different combinations were also chosen at Ucanal and Xunantunich may indicate that they were emphasizing different deities and celestial bodies in their caches, perhaps as a reflection of their own local religious traditions. Comparing A-3-1 with other caches from Xunantunich shows that different symbolic forms were chosen for different caches but some forms were more common than others, such as lunar forms and quadrilobed discs (Sullivan 2017). While the sample from Ucanal is just too small to identify broader patterns with much confidence, they appear to more loosely align with Xunantunich, Naranjo, or Tikal. In contrast, Xunantunich and Naranjo appear to have more closely aligned with each other than with Ucanal, although the published samples from Naranjo are a similarly small number.

In addition to basic symbolic forms, however, the Ucanal and Xunantunich caches share other production techniques and embodied knowledge, although close-knit communities of practice may not have been necessarily responsible for such similarities. Knappers from both sites appear to have made an effort to use various material colors for each eccentric. While it is clear that color had tremendous symbolic import among the Maya, the idea that certain colors were chosen for specific sets of form types has not been demonstrated and would require a detailed study on that topic. In addition, it is clear in all three caches that the knappers who created the pieces were adept at direct percussion, at least two separate indirect percussion techniques, and pressure flaking. Alone these techniques would have been understood and practiced by most accomplished knappers in the ancient Maya world. Using them to create detailed notches, maintaining proportions, and creating negative enclosed spaces, however, is a different level of artistry—one controlled by specialists.

Diachronically speaking, it is possible that chosen symbols may change at Ucanal and Xunantunich over time, although less so at Tikal. At Tikal there was an incredible continuity in symbolic forms for, arguably, the longest period anywhere in the Maya world. If Hruby (chapter 1) is correct that the eccentric tradition was born of a syncretic union between Maya and Teotihuacan knapping and religious practices, then it is possible that the canonization of these forms began at Tikal. As Hruby (2002) has argued for the scribal tradition at Tikal, it may have perceived itself as a keeper of those traditions. Outside of Tikal, it is possible that knappers worked with more flexibility in design and execution of forms. Thus, while Xunantunich may feature continuity in basic symbolic forms, there was more variety in the execution of those signs in a semiotic sense. It should be noted here that cached obsidians at Tikal are a completely different story—a locus of greater creativity and experimentation.

One thing that is clear from each of these caches is that at least some of the objects were made at the same time. In other words, while some pieces may have been selected *ad hoc* to complete a set of eccentrics for a caching ritual, others were made together as part of a planned exercise. We would argue that this is the case for most eccentric caches based on pan-Maya comparative data but perhaps not all. At some sites, finished bifaces were imported long distance and then reduced, probably by local knappers (Meza Rodríguez 2008). This does not appear to be the case for the Ucanal and Xunantunich, where local materials were used. Part of the argument for this reconstruction is that some of the debitage from the production of larger eccentrics appears to have been used for making the smaller ones.

The implications of this pattern are important. First, many of the knapping techniques were probably not carried out by a single person. They were a team effort requiring at least two to three people. The same type of silhouette was sometimes executed in different styles, for example, the two tripartite eccentrics from Ucanal Cache 1-1, which might suggest that two different people were making them but from flakes of the same nodule. Thus, people in close proximity to one another, throughout the process, were likely working together toward the same goal. They must have shared the same kinds of esoteric production and religious knowledge. That some of the smaller eccentrics may be more crude than the larger pieces may also suggest a hierarchy of knowledge and practice, where perhaps apprentices were involved in the less-complicated procedures. These patterns suggest very tight, if quite restricted, communities of practice at each site. Furthermore, the process implies that the knappers were likely involved in the very early stages of production, as well as the final ones. This practice would contrast with those from other sites that imported finished bifaces that were then modified into silhouettes by specialists in religious forms (chapter 1). The group or team type of organization of production may reflect more general approaches to even utilitarian knapping, or a unique exchange occurring in this hyperspecialized setting. Only detailed analyses of workshops may resolve this question.

In summary, the caches from Xunantunich and Ucanal do not seem to indicate a form of constant or continuous information exchange between the elite knappers or elite-sponsored knappers of each site. While adhering to a basic pan-Maya symbol system is clear, the execution of these symbols appears unique to each site, Tikal included. In addition, the Ucanal examples date to the Terminal Classic, when the site began to engage in new expressions of political identity that forged ties less with dominant powers, such as Tikal and Naranjo. In general, however, Ucanal and Xunantunich may not have been constrained by the same conservative canons of knowledge production established at Tikal. The reasons for this are unclear but may have to do with more turbulent histories and population changes over time outside of the very large cities. In addition, the benefit of using a *chaîne opératoire* and community of practice approach to understanding these objects is clear. Trying to match a particular symbolic form between caches and sites is only a single facet of the study. Instead, one must consider all stages of production, from the earliest stages of nodule selection and decortication, to the final group activities and shared practices that brought about their final form, as a way to consider the social relations within and between communities of practice. It is difficult to imagine that taste was a major factor in how these objects were made, as would be the case for a painted polychrome pot or woven pattern, given how small they were and limited their distribution. We have to wonder how big of a role aesthetics played, if any, and whether there was simply a culturally acceptable way that they had to be made to function properly as divining tools (Hruby 2007, Hruby and Ware 2009). There also is an immediacy to these items. They were not used for long, in archaeological terms, but were meant after periods of one, five, ten, or twenty years to be buried and never seen again by human eyes.

REFERENCES

Agurcia Fasquelle, Ricardo, Payton Sheets, and Karl Andreas Taube. 2016. *Protecting Sacred Space: Rosalila's Eccentric Chert Cache at Copan and Eccentrics Among the Classic Maya*. Monograph 2. San Francisco: Precolumbia Mesoweb Press.

Audet, Carolyn M. 2006. "The Political Organization of the Belize Valley: Evidence from Baking Pot, Xunantunich, and Cahal Pech." PhD diss., Department of Anthropology, Vanderbilt University, Nashville.

Awe, Jaime J., Christophe Helmke, Diane Slocum, and Douglas Tilden. 2019. "Let's Talk of Graves, Eccentrics and Epitaphs: The Socio-Political Implications of Recent Discoveries at Xunantunich, Belize." *Research Reports in Belizean Archaeology* 16:57–74.

Awe, Jaime J., Christophe Helmke, Diane Slocum, and Douglas Tilden. 2020. "Ally, Client, or Outpost? Evaluating the Relationship Between Xunantunich and Naranjo in the Late Classic Period." *Ancient Mesoamerica* 31(3):494–506.

Blanton, Jeffrey. 1976. "The Role of Symbiosis in Adaptation and Sociocultural Change in the Valley of Mexico." In *The Valley of Mexico: Studies in Pre-Hispanic Ecology and Society*, edited by Eric R. Wolf, 181–202. Albuquerque: University of New Mexico Press.

Bourdieu, Pierre. 1977. *Outline of a Theory of Practice*. Translated by Richard Nice. Cambridge: Cambridge University Press.

Carreño S., Hugo René. 2013. "Ideas, tecnología y simbolismo: Lado social en la producción artesanal de la lítica en los sitios arqueológicos Yaxha-Nakum-Naranjo." In *XXVI Simposio de Investigaciones Arqueológicas en Guatemala, 2012*, edited by Bárbara Arroyo and Luis Méndez Salinas, 655–671. Guatemala City: Museo Nacional de Arqueología y Etnología.

Chase, Arlen F., and Diane Z. Chase. 2017. "E Groups and the Rise of Complexity in the Southeastern Maya Lowlands." In *Maya E Groups: Calendars, Astronomy, and Urbanism in the Maya Lowlands*, edited by David Freidel, Arlen Chase, Anne Dowd, and J. Murdock, 31–71. Gainesville: University Press of Florida.

Clark, John E. 1991. "Modern Lacandon Lithic Technology and Blade Workshops." In *Maya Stone Tools: Selected Papers from the Second Maya Lithic Conference*, edited by Thomas Hester and Harry J. Shafer, 251–265. Madison, WI: Prehistory Press.

Clark, John E., and Stephen D. Houston. 1998. "Craft Specialization, Gender, and Personhood Among the Post-Conquest Maya of Yucatan, Mexico." In *Craft and Social Identity*, edited by Cathy L. Costin and Rita P. Wright, 31–46. Archaeological Papers of the American Anthropological Association, 8. Arlington, VA: American Anthropological Association.

Coe, William R. 1959. *Piedras Negras Archaeology: Artifacts, Caches, and Burials*. Museum Monographs. Philadelphia: University Museum, University of Pennsylvania.

Crown, Patricia L. 2016. *The Pueblo Bonito Mounds of Chaco Canyon: Material Culture and Fauna*. Albuquerque: University of New Mexico Press.

Cruz Gómez, Carlos, and Jose Luis Garrido. 2016. "Excavaciones en el Grupo J del sitio arqueológico Ucanal, Operación 1." In *Proyecto Arqueológico Ucanal, 2da Temporada de Campo, 2016*, edited by Christina T. Halperin and Jose Luis Garrido, 23–44. Report submitted to the Instituto de Antropología e Historia (IDAEH), Ministerio de Cultura y Deportes, Dirección General del Patrimonio Cultural y Natural, Guatemala City.

Escobedo, Héctor L., and Zachary X. Hruby. 2002. "El contexto y producción del Escondite O-13-57 de Piedras Negras, Guatemala." In *XV Simposio de Investigaciones Arqueológicas en Guatemala, 2001*, Vol. 2, edited by Juan Pedro Laporte, Héctor Escobedo, and Bárbara Arroyo, 927–935. Guatemala City: Museo Nacional de Arqueología y Etnología.

Freidel, David A. 1979. "Culture Areas and Interaction Spheres: Contrasting Approaches to the Emergence of Civilization in the Maya Lowlands." *American Antiquity* 44(1):36–54.

Gosselain, Olivier P. 2008. "Mother Bella Was Not a Bella: Inherited and Transformed Traditions in Southwestern Niger." In *Cultural Transmission and Material Culture: Breaking Down Boundaries*, edited by Miriam T. Stark, Brenda J. Bowser, and Lee Horne, 150–177. Tucson: University of Arizona Press.

Gosselain, Olivier. 2012. "D'une histoire à l'autre: Retour sur une théorie des liens entre langues et techniques en Afrique." In *La préhistoire des autres: Perspectives archéologiques et anthropologiques*, edited by Nathan Schlanger and Anne-Christine Taylor, 83–100. Paris: La Découverte, l'Institut national de recherches archéologiques préventives et le musée du quai Branly.

Halperin, Christina T., and Jose Garrido López. 2019. "Architectural Aesthetics, Orientations, and Reuse at the Terminal Classic Maya Site of Ucanal, Petén, Guatemala." *Journal of Field Archaeology* 45(1):46–66.

Halperin, Christina T., and Jose Luis Garrido, eds. 2016. Proyecto Arqueológico Ucanal, 2da Temporada de Campo, 2016. Report submitted to the Instituto de Antropología e Historia (IDAEH), Ministerio de Cultura y Deportes, Dirección General del Patrimonio Cultural y Natural, Guatemala City.

Halperin, Christina T., Jose Garrido López, Miriam E. Salas Pol, and Jean-Baptiste LeMoine. 2020. "Convergence Zone Politics at the Archaeological Site of Ucanal, Petén, Guatemala." *Ancient Mesoamerica* 31:476–493.

Halperin, Christina T., and Simon Martin. 2020. "Ucanal Stela 29 and the Cosmopolitanism of Terminal Classic Maya Stone Monuments." *Latin American Antiquity* 31(4):817–837. https://doi.org/10.1017/laq.2020.70.

Helmke, Christophe, and Jaime J. Awe. 2016a. "Death Becomes Her: An Analysis of Panel 3, Xunantunich, Belize." *PARI Journal* 16(4):1–14.

Helmke, Christophe, and Jaime J. Awe. 2016b. "Sharper than a Serpent's Tooth: A Tale of the Snake-Head Dynasty as Recounted on Xunantunich Panel 4." *PARI Journal* 17(2):1–22.

Houston, Stephen D. 2014. *The Life Within: Classic Maya and the Matter of Permanence.* New Haven, CT: Yale University Press.

Houston, Stephen D., ed. 2021. *A Maya Universe in Stone*. Los Angeles: Getty.

Hruby, Zachary. 2002. "Evidence for Linguistic Conservatism in the Hieroglyphic Script of the Central Maya Lowlands." *Mayab* 15:49–59.

Hruby, Zachary X. 2006. "The Organization of Chipped-Stone Economies at Piedras Negras, Guatemala." PhD diss., University of California, Riverside.

Hruby, Zachary X. 2007. "Ritualized Chipped-Stone Production at Piedras Negras, Guatemala." In *Rethinking Craft Specialization in Complex Societies: Archaeological Analysis of the Social Meaning of Production*, edited by Zachary Hruby and Rowan K. Flad, 68–87. Archaeological Papers of the American Anthropological Association 17. Arlington, VA: American Anthropological Association.

Hruby, Zachary X, and Gene Ware. 2009. "Painted Lithic Artifacts from Piedras Negras, Guatemala." In *Maya Archaeology*, vol. 1, edited by Charles Golden, Stephen Houston, and Joel Skidmore, 76–85. San Francisco: Mesoweb Press.

Inomata, Takeshi. 2001. "The Power and Ideology of Artistic Creation: Elite Craft Specialists in Classic Maya Society." *Current Anthropology* 42(3):321–349.

Jordan, Jillian, Julie A. Hoggarth, and Jaime J. Awe. 2020. "Pottery, Practice, and Place: A Communities of Practice Approach to Commoner Interaction in the Late to Terminal Classic Belize River Valley." *Journal of Anthropological Archaeology*. http://doi.org/10.1016/j.jaa.2020.101148.

Joyce, Rosemary A. 2017. *Painted Pottery of Honduras: Object Lives and Itineraries*. Leiden, Netherlands: Brill.

Laporte, Juan Pedro, and Héctor Eduardo Mejía. 2000. *Atlas Arqueológico de Guatemala, Reporte 14: Registro de Sitios Arqueológicos del Sureste de Petén*, Reporte 14. Ministerio de Cultura y Deportes, Instituto de Antropología e Historia, Kreditanstalt für Wiederaufbau, Universidad de San Carlos de Guatemala, Escuela de Historia, Guatemala.

Lave, Jean. 2008. "Epilogue: Situated Learning and Changing Practice." In *Community, Economic Creativity, and Organization*, edited by Ash Amin and Joanne Roberts, 283–296. Oxford: Oxford University Press.

Lave, Jean, and Etienne Wenger. 1991. *Situated Learning: Legitimate Peripheral Participation*. Cambridge: Cambridge University Press.

LeCount, Lisa, and Jason Yaeger. 2010. "Provincial Politics and Current Models of the Maya State." In *Classic Maya Provincial Politics: Xunantunich and Its Hinterlands*, edited by Lisa J. LeCount and Jason Yaeger, 20–45. Tucson: University of Arizona Press.

López López, María Fernanda. 2020. "Excavaciones en la Estructura K–2 del Grupo K del sitio Ucanal (Operación 20B)." In *Proyecto Arqueológico Ucanal: 5ta Temporada de Campo, 2019*, edited by Christina T. Halperin and José Luis Garrido, 107–132. Report submitted to the Instituto de Antropología e Historia (IDAEH), Ministerio de Cultura y Deportes, Dirección General del Patrimonio Cultural y Natural, Guatemala City.

Martin, Simon. 2017. "The Caracol Hieroglyphic Stairway." Maya Decipherment. https://decipherment.wordpress.com/2017/01/20/the-caracol-hieroglyphic-stairway/.

McAnany, Patricia. 2010. *Ancestral Maya Economies in Archaeological Perspective*. Cambridge: Cambridge University Press.

Meza Rodríguez, Carolina. 2008. "Los Instrumentos de Pedernal (tuk 'tuunich) en El Tigre, Campeche: Procesos de explotación y consumo." Tesis licenciatura, archive, INAH, Mexico City.

Mills, Barbara J., Jeffery J. Clark, Matthew Peeples, W. R. Haas Jr., John M. Roberts Jr., J. Brett Hill, Deborah L. Huntley, Lewis Borck, Ronald L. Breiger, Aaron Clauset, and M. Steven Shackley. 2013. "Transformation of Social Networks in the Late Prehispanic US Southwest." *Proceedings of the National Academy of Sciences* 110(15):5785–5790.

Moholy-Nagy, Hattula. 2008. *The Artifacts of Tikal: Ornamental and Ceremonial Artifacts and Unworked Material, Tikal Report 27A*. Philadelphia: University of Pennsylvania Museum of Archaeology and Anthropology.

Moholy-Nagy, Hattula, and William R. Coe. 2008. "Appendix 11: Typology and Description of Flint and Obsidian Eccentrics and Incised Obsidians." In *The Artifacts of Tikal:*

Ornamental and Ceremonial Artifacts and Unworked Material, Tikal Report 27A. Philadelphia: University of Pennsylvania Museum of Archaeology and Anthropology. Accessed February 1, 2018. https://doi.org/10.6067/XCV8CJ8D43.

Perea, Marta L., and Camille Dubois Francoeur. 2020. "Excavaciones del Group J sitio Ucanal: Sur de Estructura J-2, Operación 1B." In *Proyecto Arqueológico Ucanal, Quinta Temporada de Campo, 2019*, edited by Christina T. Halperin and José L. Garrido, 23–46. Guatemala City: Report submitted to the Instituto de Antropología e Historia (IDAEH), Ministerio de Cultura y Deportes, Dirección General del Patrimonio Cultural y Natural.

Roddick, Andrew P. 2016. "Scalar Relations: A Juxtaposition of Craft Learning in the Lake Titicaca Basin." In *Knowledge in Motion: Making Communities and Constellations of Practice Across Time and Place*, edited by Andrew P. Roddick and Ann B. Stahl, 126–154. Tucson: University of Arizona Press.

Roddick, Andrew P., and John W. Janusek. 2018. "Moving Between Homes: Landscape, Mobility, and Political Action in the Titicaca Basin." In *Powerful Places in the Ancient Andes*, edited by Justin Jennings and Edward R. Swenson, 287–322. Albuquerque: University of New Mexico Press.

Roddick, Andrew P., and Ann B. Stahl. 2016. " 'Knowledge in Motion': Making Communities and Constellations of Practice Across Time and Place." In *Knowledge in Motion: Making Communities and Constellations of Practice Across Time and Place*, edited by Andrew P. Roddick and Ann B. Stahl, 3–35. Tucson: University of Arizona Press.

Sassaman, Kenneth E., and Wictoria Rudolphi. 2001. "Communities of Practice in the Early Pottery Traditions of the American Southeast." *Journal of Anthropological Research* 57(4):407–425.

Shafer, Harry J., and Thomas R. Hester. 1991. "Lithic Craft Specialization and Product Distribution at the Maya Site of Colha, Belize." *World Archaeology* 23:79–97.

Slocum, Diane L. 2018. "Contextualizing Xunantunich in the Late Classic Upper Belize Valley Through Investigations of Structure A9." MA thesis, Northern Arizona University, Flagstaff.

Sullivan, Kelsey J. 2017. "Caching It In: Local Patterns in Ancient Maya Ritual Caches of Eccentric Lithics Within the Belize Valley." MA thesis, Northern Arizona University, Flagstaff.

Wenger, Etienne. 1998. *Communities of Practice: Learning, Meaning, and Identity*. Cambridge: Cambridge University Press.

Willey, R. Gordon. 1972. *The Artifacts of Altar of Sacrifices, Papers of the Peabody Museum* 64(1). Cambridge: Peabody Museum.

9

Of Water Lilies and Lightning

Technology and Symbolism of Late Classic Lithics Caches in Northwestern Belize

JOSHUA J. KWOKA

> *Since a great many are quite symmetrical, the term "eccentric" is, of course, inappropriate. But it is retained because it is a traditional one applied in Maya archaeology to impractical exotic flints of a variety of forms found in votive deposits and very infrequently burials.*
>
> William R. Coe, *Piedras Negras Archaeology: Artifacts, Caches, and Burials* (1959)

By the late nineteenth century, descriptions of ancient Maya lithics had begun to populate the pages of British antiquarian journals. Excavations conducted by colonial authorities in Belize, then British Honduras, generated a steady flow of artifacts for export to England. Due to chance inclusion in ship ballast, Maya lithics had even been recovered from the banks of the Thames (Rice 1909). Many of these objects were recognized as analogous to Old World tools, such as axes or lance points. However, it was not these artifacts—those of foreign manufacture yet familiar design—that captivated British audiences but rather the perceived exoticism of a group of artifacts that were dubbed eccentric flints (Franks 1877;

https://doi.org/10.5876/9781646427819.c009

Lane Fox 1871; St. George Gray 1916). Introduced by Thomas Gann (1918, 68, 87–88), the term "eccentric" was applied to lithic artifacts that were viewed as peculiar owing to their intricate anthropomorphic, zoomorphic, or seemingly abstract designs. Furthermore, archaeologists working in Belize recognized that eccentrics recovered from ritual deposits were often accompanied by utilitarian stone tools, and so the term served a classificatory function as well (Gann 1918; Thomas Joyce et al. 1928). Paradoxically, adoption of the term was rapid (e.g., Heye 1925; Joyce et al. 1928; Kidder 1947), and it remains in use despite early protestations concerning its validity. As illustrated by the chapter epigraph, William Coe's (1959) critique was grounded in a rather technical understanding of the term that invokes centrality and symmetry in design principles, but it is clear that Gann's usage was colloquial in referring to unconventional shapes. Joyce (1932, xvii) emphasized that obsidian was also utilized in eccentric production.

Dissatisfaction with the term remains, though critiques have grown in sophistication concomitant with the development of Maya lithic studies. While I acknowledge these cogent arguments for alternatives (e.g., Zachary Hruby, chapter 1; Alejandro Pastrana and Hruby, chapter 2; Lucas R. M. Johnson et al., chapter 10), the term is retained here to designate a subset of artifacts. Eccentrics recovered from primary depositional contexts such as caches or burials would certainly qualify as ceremonial or ritual implements. Nevertheless, this functional designation does not allow these items to be distinguished from utilitarian tool forms that are often found in accompaniment. This leads to the question of whether there is an inherent property of eccentrics that justifies the creation of a distinct type. Employment of a monothetic approach to classification based on either function or morphology provides affirmation although, as illustrated in the sections that follow, these attributes are intertwined. Both eccentrics and utilitarian implements are imbued with symbolic force when included in a ritual deposit, but for the latter, it is the context that prompts the transition from utilitarian to ritual function. Herein lies the reasoning behind the term's persistence, as the forms of eccentrics preclude their use in quotidian tasks (Coe 1959, 16; Kidder 1947, 4). "Eccentric" morphologies provide a further point of distinction and indicate that lithics contained within ritual deposits carried differential symbolic loads. Counts, arrangements, and raw material types of cached objects invoke symbolic meanings, yet the designs of eccentrics exhibit a higher degree of specificity in referencing supernatural beings, social roles, places, and acts (Kwoka et al. 2019). Lithicists have made significant advances in deciphering these symbols, and so the classification of eccentrics should be further refined when possible (e.g., Agurcia Fasquelle et al. 2016, 9–36; Carballo 2011, 131–154; Hruby 2006, 289–309; 2007; Hruby and Ware 2009; Kwoka and Shackley 2019; Kwoka et al. 2019; Meadows 2001, 2003; Moholy-Nagy and Coe 2008; Stocker and

Spence 1973; Sugiyama 2005, 135–140). At the very least, "eccentric" possesses a flow that "impractical exotic flints" does not.

In contrast to early descriptive studies, recent years have witnessed important advances in the study of cached lithics. Technological analyses that seek to reconstruct production processes are increasing in frequency. Though often implicit, artifact life-cycles (Kopytoff 1986; Schiffer 1987) are approached through an organization of technology framework (Carr and Bradbury 2011; cf. Nelson 1991) that spans raw material acquisition to social variables. These positive developments will generate baseline data necessary for broader comparative studies, which are currently few in number (though, see chapters in this volume). The lack of individualistic cache studies is understandable; trained specialists are few and the object of study complex. Nevertheless, it is through comparison that continuity and innovation in technological and ritual practices emerge, and novel lines of inquiry may be identified. For example, the case studies presented below date to the Late Classic period (600–850 CE), an era that coincides with a shift in ritual practice across the Central and Southern Lowlands. Earlier ritual deposits often included a broad array of materials, but it was during this period that lithics served as the primary elements of many caches (e.g., Agurcia Fasquelle et al. 2016; Aoyama 2006; Coe 1959; Iannone and Conlon 1993; Pendergast 2006; Stemp et al. 2018). Even so, raw material type is the only consistent variable among the examples discussed in this chapter. The processes underlying this change in caching behavior are unclear, but comparative studies documenting the social transmission of technical knowledge via communities of practice will play a crucial role (see chapters 1, 5, 6, and 8; Kwoka 2014, 2017; Starzmann 2013). Indeed, further investment in the social domain of ritually deposited lithics will provide significant returns, as few contexts provide such rare affordances in terms of intentionality, materiality, and ideology.

CASE STUDIES: LATE CLASSIC LITHICS CACHES FROM THE THREE RIVERS REGION

The remainder of this text presents analyses of three lithics caches—variously comprised of eccentrics and utilitarian tool forms—recovered from two sites in northwestern Belize. In doing so, many of the core themes outlined in the introductory chapter (1) and the preceding paragraph are engaged. As a prelude, the reader is provided with a brief introduction to the study sites and descriptions of the caches. One of the caches was recovered from the site of Blue Creek, and the other two from Tz'unun. Both are located within the Belizean portion of the Three Rivers Region (figure 9.1), an area that encompasses the tributaries of the Río Hondo—Booth's River, Río Azul, and Río Bravo. Despite being separated by only 25 km, the sites are characterized by different environments and exhibit significant variation in artifact and architectural styles. In terms of physiographic

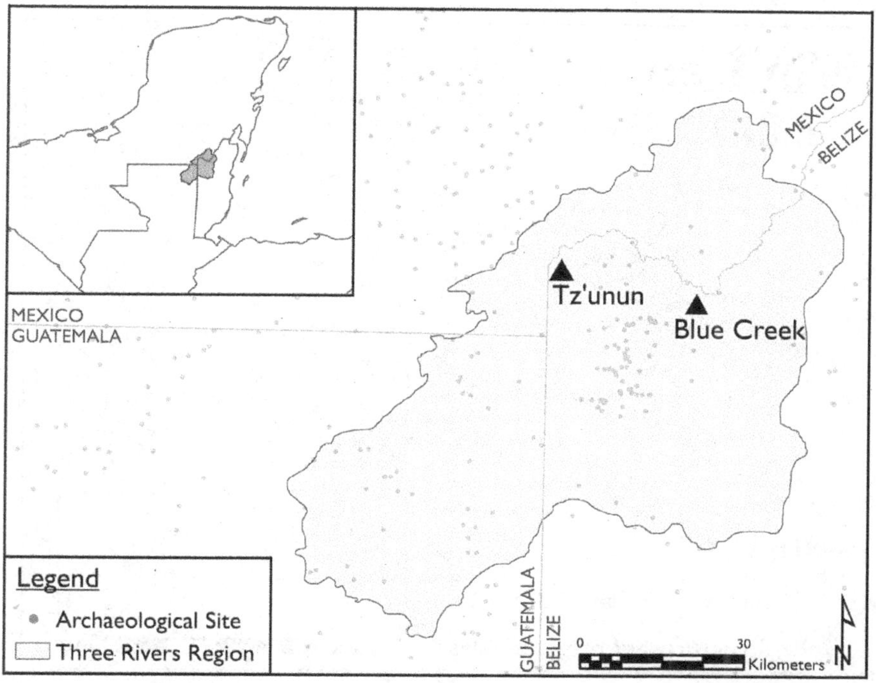

FIGURE 9.1. *Map of the Three Rivers Region showing location of case study sites.*

regions (Dunning et al. 2003, 15–16), Blue Creek sits along the edge of an escarpment that forms the eastern border of the Río Bravo Terrace Uplands. Tz'unun is located within the Azucar Lowlands to the west and occupies a prominence along the eastern margin of the Bajo Alacranes, a large seasonal wetland.

K'in T'an, Blue Creek

Founded during the Late Preclassic period (400 BCE–250 CE), Blue Creek became the seat of an independent polity sometime during the Terminal Preclassic period (150–250 CE). The city's rise in wealth and influence were attributable to an abundance of productive agricultural land and its position along the Río Hondo, a strategic trade route linking the interior Central Lowlands with the Caribbean coast. At its Early Classic (250–600 CE) peak, Blue Creek was surrounded by a network of satellite settlements occupying an area perhaps as large as 100 km^2 (Guderjan 2007, 2016). The site core had grown to include two central plazas bounded by monumental architecture, with multiple large residential groups situated along the periphery. Near the opening of the sixth century CE, over 900 jade artifacts, the majority broken and perhaps ritually "killed," were deposited within Structure 4, one of the temples at the site. Following this

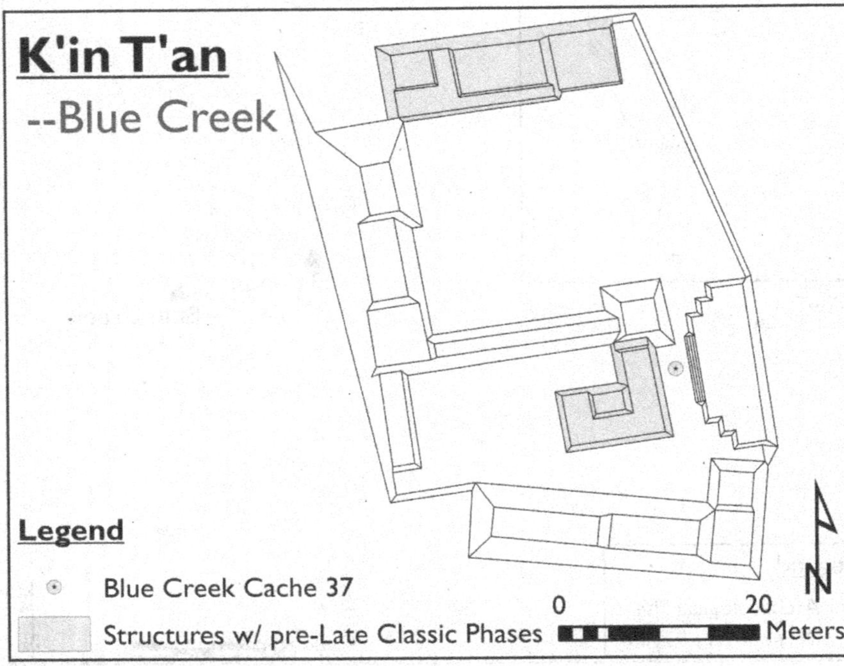

FIGURE 9.2. *Location of Cache 37 within K'in T'an* plazuela *group, Blue Creek.*

event, which Guderjan (1998, 2007) interprets as marking the city's loss of independence, the jade that had previously been abundant ceased to be imported. Although its influence was diminished, a construction boom within the site's residential groups provides evidence of continued prosperity until the beginning of the ninth century, after which Blue Creek was abandoned.

Evidence for the Late Classic flurry of construction activity is provided by the Structure 37 *plazuela* (figure 9.2), as only two structures contained earlier construction phases. This complex forms part of the larger K'in T'an settlement cluster located 600 m northwest of the site core. Comprised of multiple structures bounding one or more patios, *plazuela* groups represent extended family households (Thompson 1931, 233). Cache 37 (figure 9.3) was recovered from a narrow passage within the patio. Placement within the subfloor ballast of the final floor indicates that it was deposited as part of this construction event (Guderjan et al. 2003).

Blue Creek Cache 37

Excavated in 1999 (Hanratty 2002), Cache 37 (hereafter BC37) represents a complex arrangement of lithic raw materials and artifact types (figure 9.4). A large tridentate eccentric served as the focal point, as indicated by its central placement.

FIGURE 9.3. *Blue Creek Cache 37 in situ.*

Surrounding the tridentate were twenty-one stemmed bifaces deposited in three distinct groups. Group membership numbered 6 in the south, 7 in the northeast, and 8 in the northwest. Bifaces within the southern and northeastern groups exhibited conformity in orientation, while the northwestern group showed no patterning. The final act of this ritual event involved the placement of a large laurel-leaf biface in a position that straddled the southern biface group and the tridentate. Also present were twenty-one complete, late-series prismatic blades made of El Chayal obsidian; however, available records do not describe their placement, and they are not visible in photographs of the cache in situ.

Structure 21, Tz'unun

Tz'unun was discovered by the author in 2016, with excavation beginning in 2018, and so the site's history and position within the region are poorly understood. Following Thomas Guderjan's (1991) rank order system, a site score of 27.5 marks it as a statistical outlier (Garrison 2007, table 6.1) and thus one of the larger sites within the Three Rivers Region. Tz'unun's size may be attributable to its strategic position along the margin of the Bajo Alacranes, an expansive feature that would have functioned as a natural barrier. Multiple lines of evidence suggest this *bajo* may have been a frontier zone separating the hostile Tikal and Calakmul dynasties during the sixth and seventh centuries (Adams 1999, 144–145; Helmke et al. 2019; Stuart 2017; Šprajc 2004). The earliest evidence

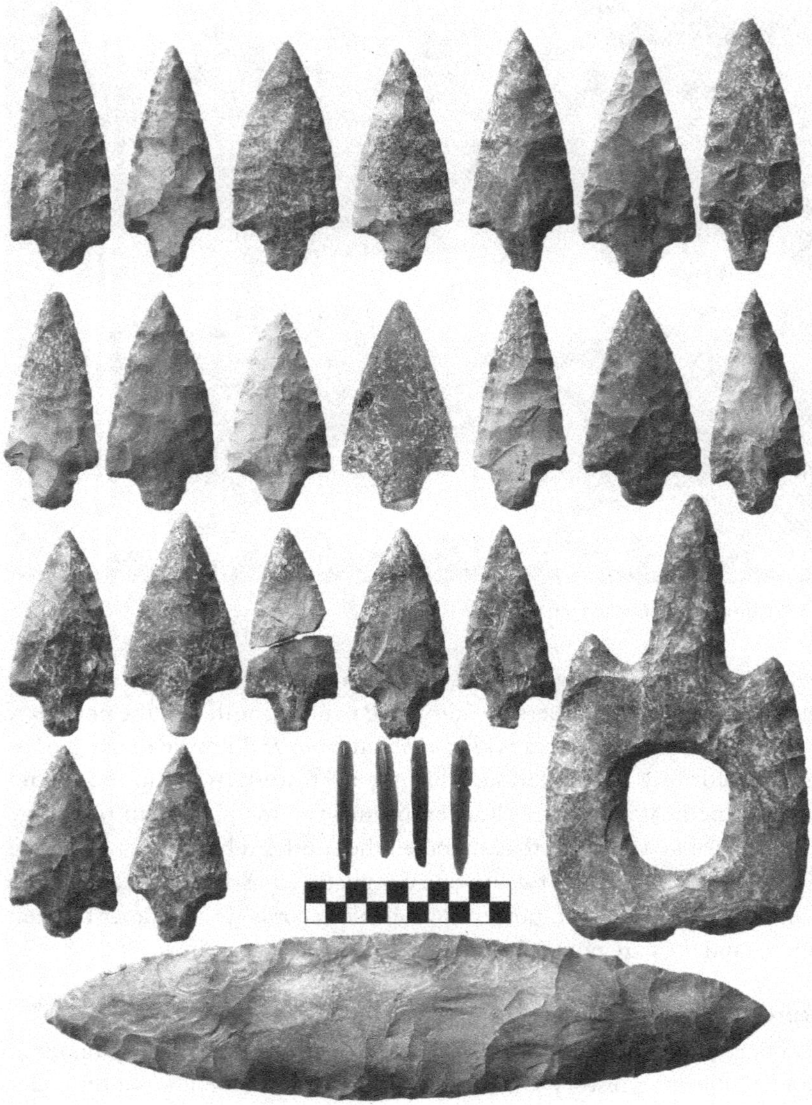

FIGURE 9.4. *Contents of Blue Creek Cache 37, including 21 stemmed bifaces, tridentate eccentric, large laurel-leaf biface, and 4 of the 21 obsidian late-series prismatic blades. Artifacts depicted at the same scale (cm).*

for occupation of Tz'unun is in the form of late Middle Preclassic (ca. 600 BCE) ceramics recovered from the ballcourt alleyway and plaza of the Cenote variant E Group. Final construction phases of all excavated contexts have produced ceramics consistent with the Late Classic–Terminal Classic transition, indicating

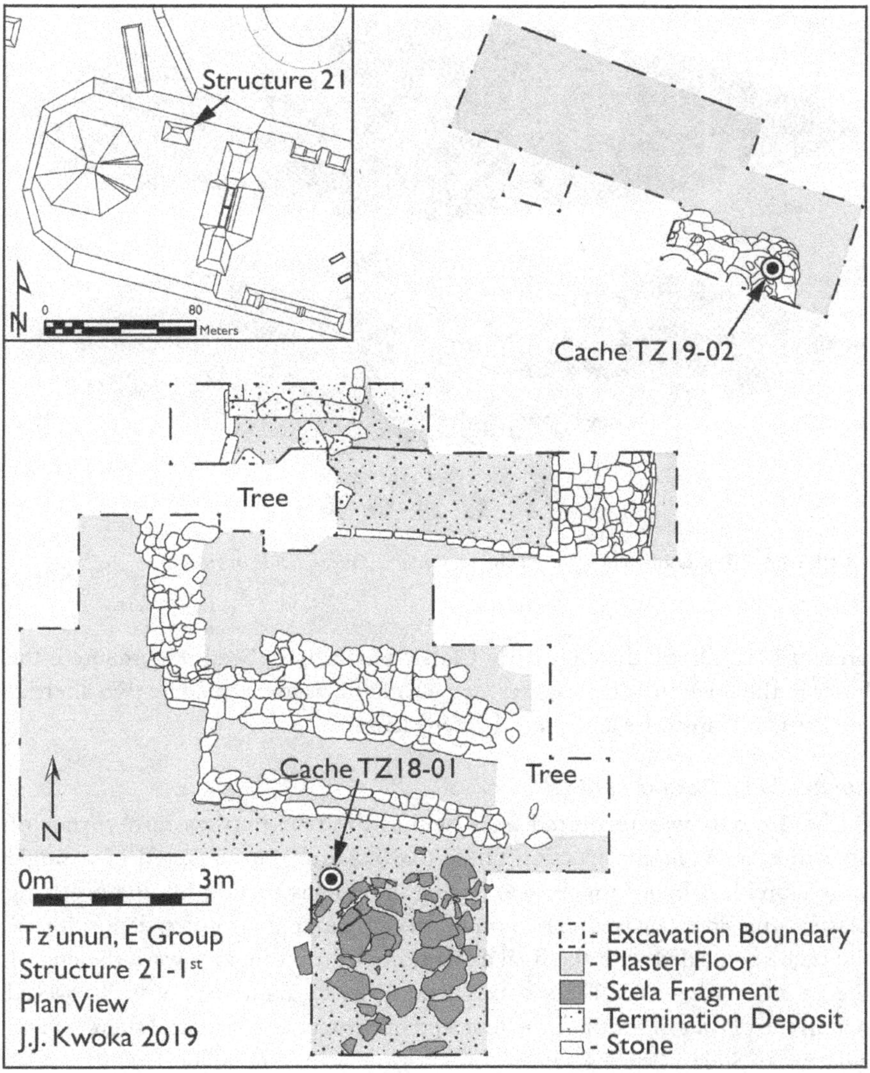

FIGURE 9.5. *Location of Tz'unun Structure 21, Cache TZ18-01 and Cache TZ19-02.*

abandonment sometime during the ninth century. Recovery of a side-notched Postclassic arrow point and Chen Mul–style *incensario* fragments from within the E Group demonstrates that the site remained an important place postabandonment. Structure 21 is a relatively small building that bounds the northern edge of the E Group plaza (figure 9.5). Of Tz'unun's three known stelae, two were set in front of the structure, and a third was repurposed as the first step of the building's penultimate construction phase. Excavations are ongoing, but at

FIGURE 9.6. *Large laurel-leaf biface, Tz'unun Cache TZ19-02. Scale in cm.*

present Late Classic through Early Classic phases have been documented, the latter in the form of a Tzakol 1 (ca. 300 CE) tomb accompanied by two alternating layers of capstones and chert debitage lenses.

Caches TZ19-02 and TZ18-01

Cache TZ19-02 was recovered from within the rear northeastern corner of Structure 21's final construction phase (Kwoka 2020). It consisted of a single, large laurel-leaf biface (figure 9.6) resting upon construction fill, the overlying plaster long since decayed. Based on the accompanying ceramics, deposition of the cache coincided with the final expansion of Structure 21 toward the end of the Late Classic period. The second cache, TZ18-01 (figure 9.7), was discovered within the plaza 1 m south of the first step of Structure 21's central staircase and adjacent to Stela 2 (Kwoka 2019).

Though its contents are similar to sub-stelae caches (Moholy-Nagy and Coe 2008), it was deposited as part of a larger event centered on the summit of Structure 21 and accompanying Stelae 1 and 2. Both areas were covered with thick deposits of Late and Terminal Classic sherds weighing a total of 166 kg ($N = 16{,}830$). TZ18-01 was placed on the surface of the final plaza floor, and consisted of fourteen stacked lithic implements. To facilitate the stacking process, a small stone block was employed as a side support, and sherds were utilized as wedges to support and level individual artifacts. Of the fourteen cache artifacts (figure 9.8), thirteen were bifaces, one of which had been further modified into

FIGURE 9.7. *Tz'unun Cache TZ18-01 in situ. Note side support block left of cache's center and ceramic wedges along the cache's west side, extending from south end of cache to north edge of side support block. Fragment of collapsed Stela 2 visible in top right of photograph.*

the form of an undulating eccentric, or a Type EF-12 following Coe's typology (Moholy-Nagy and Coe 2008, 9). A unifacial trident eccentric, representing a variant of Coe's Type EF-10, was also included (Moholy-Nagy and Coe 2008, 8). As discussed in this chapter, the author views tridents and tridentates as distinct types. Cache deposition proceeded with placement of the undulating eccentric followed by the remaining bifaces, with the unifacial eccentric capping the deposit.

COMPARATIVE ANALYSIS

An organization of technology framework serves as the point of embarkation for the following comparative analysis. Specific lines of comparison include raw material properties and acquisition, technology, number of artisans, artisan skill, and production context. Evidence for the use of suboptimal materials in the

FIGURE 9.8. *Contents of Tz'unun Cache TZ18-01. Artifacts arranged in reverse order of deposition: 1 = last artifact deposited / top of lithic stack; 14 = first artifact deposited / bottom of lithic stack. Scale in cm.*

production of expedient flake and formal tools (e.g., ovoid and general utility bifaces) is ubiquitous in the Maya Lowlands. However, quality parameters are less pliable for intricately shaped eccentrics or large bifaces that push size boundaries. In addition to setting limitations on artifact size and design, raw material characteristics are informative of provenance, location of manufacture, and the process through which artifacts were obtained for caching. Technological analyses that elucidate production techniques and sequence models are essential to identifying broader patterns that may be reflective of social units. Such analyses also generate data that can be employed to assess the number of artisans that were involved in the production of cached artifacts, their skill level, and how their efforts were organized. This section concludes with a consideration of cache function and symbolism.

Raw Material Properties and Acquisition

The Blue Creek cache artifacts are heavily patinated, but excavation damage to the laurel-leaf biface, tridentate, and two of the stemmed bifaces revealed a homogenous fine-textured chert, with colors spanning a variety of banded brown hues. Internal variation in color was sufficient to prohibit determination of how many raw material packages were represented. By all measures, this chert is high quality and consistent with material from the chert-bearing zone of northern Belize (Hester and Shafer 1984; Wright et al. 1959). Only five of the artifacts are cortex bearing, with coverage less than 1 cm^2 on a tip of the laurel-leaf biface, the tridentate's central prong, and three biface stems. A lack of cortical artifacts and the length of the laurel-leaf biface (i.e., 35 cm) indicate that large packages of high-quality material were available to the artisans who produced the Cache 37 artifacts. The large laurel-leaf biface from TZ19-02 was also produced from a homogenous fine-grained chert but was darker in color, with hues of brown differentiated by banding and mottling. As with the Blue Creek implements, the raw material package for this biface was obtained from the chert-bearing zone.

Raw material properties of the TZ18-01 artifacts present a stark contrast to those just described. From the perspective of a minimum analytical-nodule approach (Larson and Kornfeld 1997), twelve distinct chert packages can be identified. Two sets of artifacts (figures 9.8.1 and 9.8.9; 9.8.5 and 9.8.6) are similar in color and texture, and may have derived from the same nodules. Six of the bifaces exhibit internal heterogeneity in particle size, with textures ranging from fine to coarse. When combined with the presence of inclusions and vugs, chert quality can be characterized as medium to poor. Available raw material packages appear to have been small. None of the artifacts exceeds 11 cm in length, and three of the bifaces retain cortex on both tips.

In comparing the raw material properties of the three caches, an obvious contrast is provenance. The contents of TZ18-01 were produced from locally available nodular chert, and the other two from northern Belize chert bearing zone materials. Colha stemmed macroblades and ovoid bifaces have been recovered from surface contexts at Tz'unun (Kwoka and Telepak 2017), thus demonstrating that the latter fell within the former's consumer sphere. However, there is no evidence that unmodified or preformed chert packages were exported outside of the chert-bearing zone. Furthermore, no lithic workshops have been identified at Blue Creek nor Tz'unun, indicating that the contents of BC37 and TZ19-02 were produced in northern Belize and then imported to their respective sites. Another important point of distinction is raw material quality and, in the present context, the resultant implications for the assessment of production skill. This issue is explored in the following sections.

Technology

A detailed technological analysis of the BC37 stemmed bifaces has been published elsewhere (Kwoka et al. 2019). A summary is presented here, along with an expanded discussion of the tridentate eccentric. Eight of the stemmed bifaces retain a large platform on the distal end of the stem, indicating that they were produced from macroflake blanks. Reduction proceeded with the creation of an ovoid biface utilizing percussion, and concluded with the production of a tapered stem through the removal of flakes from the proximal corners of the biface at an approximately 45-degree angle. Based on flake scar sequences, three distinct stem production techniques were identified. With the first technique ($N = 12$), the hafting element was produced by detaching flakes from a single face. For the other two techniques, flakes were detached from only the left ($N = 7$) or right ($N = 2$) margin regardless of which face is viewed. This pattern suggests that after creating the first side of the stem, whether right or left, the biface was flipped, and then flakes were detached from the same side to complete the stem. The left/right only techniques may correspond to laterality, or handedness, while the single face technique likely represents the use of ulnar and radial flexion to orient the biface during stem production, rather than ambidexterity. Studies of contemporary and prehistoric lithic production have shown that methods of supporting and manipulating objective pieces are idiosyncratic, even when knappers are producing the same tool form (Uomini 2005; Whittaker 1987, 1994). Thus, it is probable that the stem production techniques described here reflect the efforts of three different artisans.

In contrast to the stemmed bifaces, the laurel-leaf biface and tridentate were produced from individual nodules. Neither artifact's longitudinal profile exhibits the characteristic curvature of macroflake blanks. James Stemp and colleagues (2014, 145) report that a series of Belizean bifacial "macroliths," the largest of which measures 96.5 × 15.5 × 6.5 cm, were produced from macroblade blanks. These would be exceptional cases, as the use of such large blanks, whether macroflakes or macroblades, has not been documented elsewhere. For comparison, macroflakes recovered from Colha's biface workshops rarely exceeded 30 × 20 cm (Shafer and Hester 1991, 85). Though approximate in size to the BC37 biface (35 × 8 × 2 cm; 649 g), the dimensions provided are for unmodified macroflakes. Due to the subtractive nature of lithic technology, any resultant biface would be significantly smaller. Percussion was the primary technique utilized to produce the laurel-leaf biface; pressure flake scars are few and unpatterned.

Creation of the tridentate's most unique feature—the perforation—required a chert nodule with particular characteristics—specifically, a soft carbonate inclusion that extended through the nodule's interior. In order to identify suitable candidates, nodule testing through removal of decortication flakes would have been necessary. A curvilinear shape indicates that production proceeded with the creation of an ovoid bifacial preform utilizing direct percussion. Initial shaping of the central prong may also have employed direct percussion, but there is clear evidence that certain areas were shaped through indirect percussion. Two element isolation notches (*sensu* Titmus and Woods 2003, 135) were created to define the short lateral prongs. The constricted dimensions of the notches would prohibit flake detachment via direct contact with a percussor. A series of 3–4 cm wide conchoidal flake scars, partially obscured by later pressure removals, emanate from the notches. The large size of these flakes and the force required to detach them exceed the known capabilities of pressure flaking (Cotterell and Kamminga 1987; Titmus and Woods 2003). Creation of the perforation likely occurred after the exterior margins and prongs had been shaped, in order to preserve mass, absorb shock, and thus minimize the risk of production failure through breakage (Titmus and Woods 2003, 136). A technique was employed that appears to have developed in northern Belize, whereby the carbonate inclusion was removed through drilling. Evidence of this method, consisting of inclusion remnants along perforation margins, has been documented at sites such as Altun Ha (Pendergast 1990), Colha (Shafer and Hester 1991), Lamanai (Meadows 2001, 2003), and Blue Creek (Kwoka et al. 2019, fig. 5a). No such remnants are present on the Blue Creek tridentate, due to extensive modification of the perforation margin. As with the element isolation notches described here, a narrow circumference and large/wide flake scars emanating from the margin demonstrate that final shaping of the perforation must have employed an indirect percussion technique.

Based on the size of the TZ19-02 laurel-leaf biface (35 × 9 × 0.75 cm; 335 g) and absence of longitudinal curvature, it also was produced from a nodule. Percussion was the primary manufacturing technique, but in contrast to the large Blue Creek biface, the lateral margins exhibit extensive pressure flake scars. These scars are unpatterned and centered on ridges resulting from earlier percussion flakes. Maximum pressure flake scar dimensions are 2 × 1 cm, though the majority are half this size, perhaps indicating that the primary function of pressure removals was to regularize the biface's lateral margins.

Whereas eight of the Blue Creek stemmed bifaces retained the original macroflake platforms, the artifacts from TZ18-01 were manufactured from small nodules. A lone exception is provided by the trident eccentric. This artifact was produced from a hard hammer decortication flake, the platform of which is present on the tapered end. Shaping of the eccentric occurred through bimarginal retouch. Larger, more invasive detachments from the ventral surface removed the flake's bulb of percussion. Notching was used to define the three protrusions, but the type of percussion technique employed is unclear. Many of the conditions that would necessitate the use of indirect percussion, such as those described for the Blue Creek tridentate, are present. Space between the protrusions is narrow (2–3 cm), and the eccentric's thickness required significant force to detach flakes at steep edge angles. Of all the cache artifacts, only the undulating bifacial eccentric exhibits evidence of pressure flaking. In contrast to the large laurel-leaf bifaces, pressure was used as a form of invasive unimarginal retouch to produce the notches. Pressure flake scar patterning indicates use of the flipping method described for the Blue Creek stemmed bifaces. Identification of specific handedness (i.e., right or left) is not possible in the absence of an orienting feature (e.g., stem).

Artisans, Skill, and Production Context

BC37 provides robust evidence for the work of multiple artisans in the form of three distinct stem production techniques; all other evidence is circumstantial. For example, the large laurel-leaf bifaces and tridentate are substantial artifacts. Indirect percussing utilizing a freehand method of supporting the objective piece would not be sufficient to prevent a production failure via stress fracture. In a series of anthropomorphic-eccentric replication experiments, Gene Titmus and James Woods (2003, 138; see also John Clark, chapter 4) employed a single-person method of cradling the objective piece between the legs within a piece of leather that functioned to support, cushion, and immobilize the objective piece. That the authors were successful in producing replicas demonstrates this technique's effectiveness. Payson Sheets (Agurcia Fasquelle et al. 2016, 72) presents an alternative model resulting from his analysis of the large laurel-leaf biface from Copán's Rosalila cache. He proposes that a two-person indirect percussion technique was employed that consisted of a punch operator and another person dedicated solely to ensuring proper support (cf. Clark, chapter 4). Regarding the Blue Creek tridentate, access to the perforation would have been difficult using the cradling support technique, perhaps indicating use of the two-person method. The bifaces from TZ18-01 exhibit significant variation; in other words, they are unstandardized. This variation in biface morphology could reflect the work of multiple artisans, but, again, this is a matter of conjecture.

Of the approaches to assessing knapping skill (Bamforth and Finlay 2008, 5, table 1), the use of a coefficient of variation (CV) as a quantitative measure of artifact standardization is an oft-employed method (e.g., Clark 2003; Costin 1991; Finlay 2008). The underlying argument is that an elevated skill level is required to produce standardized forms. Unlike ceramic production, which provides opportunities for error correction prior to firing, mistakes made during lithic reduction can only be rectified through removing further mass. As such, lithic artifacts with CV values of 10–15 percent are considered standardized (Eerkens 2000, 666–667). CV values were calculated for the Blue Creek stemmed bifaces, and only stem thickness (17.07%) exceeded the upper limits of this range (Kwoka et al. 2019, 357, table 1). If standardized production is correlated with knapping

TABLE 9.1. Size characteristics of oversized laurel-leaf bifaces

Site	*Length (cm)*	*Width (cm)*	*Thickness (cm)*	*Weight (g)*	*Source*
Blue Creek, Cache 37	35	8	2	649	Kwoka et al. (2019, 357)
Tz'unun, Cache 19-02	35	9	0.75	335	Kwoka (2020, 14)
Copán, Rosalila Cache	36	9.6	1.2	517	Agurcia Fasquelle et al. (2016, 72)
Buenos Aires Biface	41.7*	8.5	1.27	540*	Clark, chapter 4

* Clark's biface is broken and missing approximately 6 cm of one tip. Length and weight figures are estimates for the complete biface.

ability, then the bifaces resulted from skillful labor. Qualitative measures of knapping skill include the frequency of flake termination errors. Few step or hinge terminations are present on the two large laurel-leaf bifaces and tridentate, and, where present, they exist as isolated errors (i.e., no stacking of step fractures). With the tridentate, only flakes detached from the perforation margin resulted in hinge terminations. An exquisite piece of artisanship, the TZ19-02 laurel-leaf biface provides the clearest demonstration of skill. The majority of percussion-thinning flake scars extend beyond the biface's centerline, with those present on both midsection faces measuring 7–8 cm long. As significant force would have been required to detach relatively narrow, long flakes that spanned up to 90 percent of the biface's width, the knapper(s) must have possessed expert knowledge of platform preparation. When compared to other oversized bifaces, the Tz'unun example is notable for its thinness (table 9.1). Application of the same measures to TZ18-01 suggests the bifaces represent the products of less-skilled artisans. Small and unstandardized, all bifaces exhibit abundant step and hinge terminations. However, such an assessment fails to account for the relation between raw material properties and production limitations. Due to the suboptimal raw material quality, it is doubtful that even an exceptionally skilled knapper would be able to produce oversized bifaces and eccentrics from the Tz'unun nodular chert.

The data presented thus far allow inferences to be made regarding production contexts. Artifacts from TZ19-02 and BC37 were manufactured from nonlocal chert, indicating an origin of manufacture somewhere within the northern Belize chert-bearing zone. Both quantitative and qualitative measures are indicative of a high level of knapping skill, thus suggesting production by specialists. In comparison to other areas of the Maya Lowlands, a large number of lithic workshops have been identified in northern Belize, some of which produced eccentrics

(Shafer and Hester 1983). The Blue Creek artifacts were produced in the same workshop by three artisans. All are similar in color and texture, the stemmed bifaces are standardized, and three stem production techniques were identified. Comparison of the Blue Creek and Tz'unun oversized laurel-leaf bifaces supports an argument for different workshops. Both were made from high-quality chert, but there are notable differences in color and texture. There is also evidence of differential biface-thinning abilities and the use of pressure flaking to regularize lateral margins. Within northwestern Belize, household lithic reduction focused on the production of expedient flake tools and formal tool maintenance (e.g., Trachman 2009). As such, it is probable that the contents of TZ18-01 were manufactured in one of the regional workshops (e.g., Barrett 2011; Houk and Zaro 2015), or a workshop within Tz'unun that has not yet been identified.

Cache Function and Symbolism

BV37 and TZ19-02 are consistent with dedication/foundation offerings that functioned to ensoul, or animate, places (for ethnographic analogs, see Stross 1998; Vogt 1998). As such, these caches were placed in conjunction with a construction event, such as the raising of a plaza (Blue Creek) or expansion of a building (Tz'unun). TZ18-01 represents the antithesis of this practice, whereby the life force of a building was released (Schele and Freidel 1990, 437–438). After placement on the plaza floor adjacent to Stela 2, both the cache and Structure 21 were covered with a thick lens of ceramic sherds. Ceramic refits indicate that at least a portion of the vessels were smashed in place and then scattered as part of a reverential termination event (Pagliaro et al. 2003). Existing data suggest that the termination of Structure 21 occurred as part of the broader abandonment of Tz'unun.

Blue Creek Cache 37

Since TZ19-02 is comprised of a single artifact, this discussion focuses on the other two offerings. Beginning with BC37, it is doubtful that the inclusion of an equivalent number of prismatic blades and stemmed bifaces (21), as well as the sequential grouping of bifaces (6, 7, and 8) were coincidental. The numerological significance of this patterning is unclear, but a calendrical association is possible (see Aveni 2011). Placement of the stemmed bifaces into distinct groups resembles the three-stone hearth (Taube 1998) and may have served to mark the deposit as an axis mundi. Reference to Maya art provides substantial insight into the tridentate eccentric. Before proceeding, it is important to note that depictions of tridentates exhibit morphological variation, some with prongs of equal length and others with a large central prong accompanied by smaller lateral prongs (e.g., Kerr vessel numbers K5509 and K8351). Based on the accompanying scenes, both variants were functionally equivalent. Depictions of these objects are found on a variety of polychrome vessels and monuments dating to the

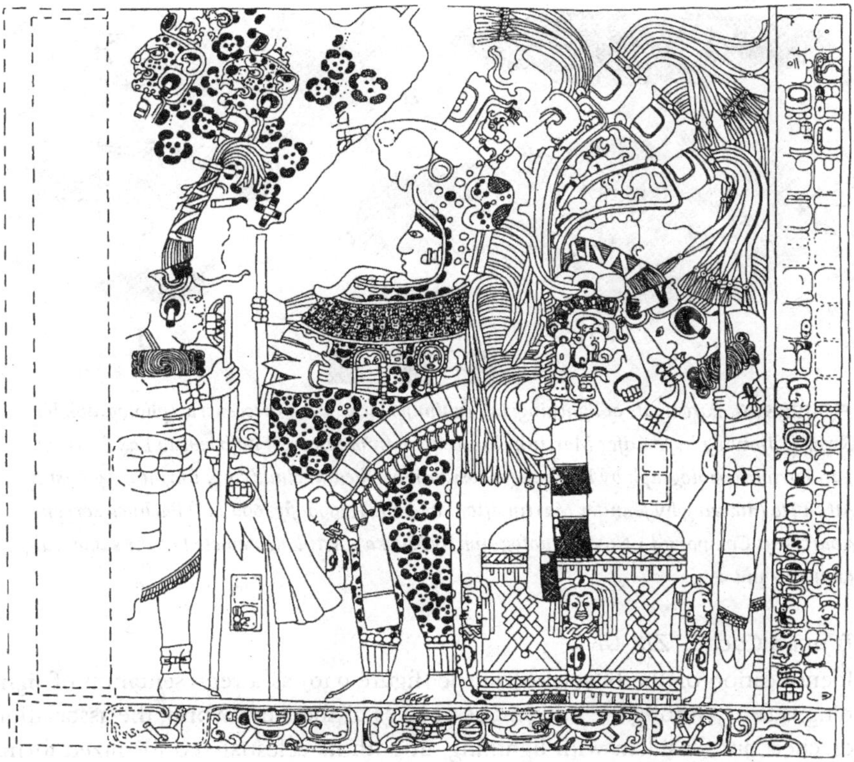

FIGURE 9.9. *Lintel 2 of Tikal Temple 3. Note tridentate eccentrics held by Dark Sun and two attendants.*

seventh to ninth centuries CE (Kwoka et al. 2019, 358, table 2). In the case of the former, the death gods and associated *way* employ tridentates in sacrificial rites involving decapitation and dismemberment. On monuments, they are wielded by rulers and accompanying courtiers—depicted in the guise of the Jaguar God of the Underworld—conducting ritual burning events (figure 9.9).

Andrew Scherer and Stephen Houston (2018, 122) identify these individuals as *ch'ajoom*, or "person of incense," a Late Classic title for ritual specialists associated with the burning of human remains, particularly those of infants and adolescents. Mirroring the contents of BC37, Tikal Altar 5 shows Jasaw Chan K'awiil and a lord of Maasal bearing the markings of *ch'ajoom* and conducting a burning ritual; one holds a tridentate and the other an oversized laurel-leaf biface. This same toolset also appears on a polychrome vessel (K2286) portraying three *way* of the Akan complex. Thus, the tridentate and large laurel-leaf biface invoke acts of ritual sacrifice and the burning of human remains, ritual practices central to the institution of divine kinship.

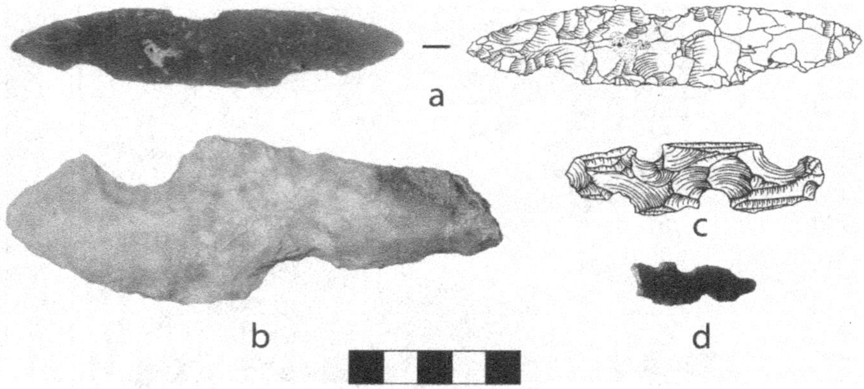

FIGURE 9.10. *Portrayals of lightning: (a) Tz'unun Cache TZ18-01eccentric (photograph by Kwoka; drawing by Jennifer Martin); (b) eccentric from reservoir adjacent to Las Pinturas, San Bartolo (photograph by Kwoka); (c) obsidian polyhedral blade core, Burial E-54/9 1st B, Altun Ha (drawing by Jennifer Martin after Pendergast 1990, fig. 69d); (d) Pachuca serpent, Apartment Compound 1:N6W3 (photograph by Kwoka). Artifacts depicted at the same scale (cm).9.10 and 9.11.*

Tz'unun Cache TZ18-01

Identification of the bifacial eccentric (figure 9.10) as a representation of lightning is firm. Among the indigenous peoples of the New World, the association of chert and obsidian with lightning, in both its celestial and fossilized forms (i.e., fulgurites), has a broad spatiotemporal distribution (Agurcia Fasquelle et al. 2016, 16–19).

Serpents also share in this symbolism owing their undulating form of locomotion to lightning (see figure 9.10d). Drawing on Thompson's (1950, 72) discussion of the *Imix* glyph, Coe (Moholy-Nagy and Coe 2008, 8) suggests that trident eccentrics may represent water lilies, a view supported by iconographic data. In Maya art, water lilies are depicted as a trident that represents the sepals of the flower's calyx, the fourth of which is hidden due to the profile view (figure 9.11). Furthermore, the calyx can be depicted in isolation (i.e., pre-anthesis) or accompanied by a flower (McDonald and Stross 2012). *Naab'*, the word for water lily, is homonymous with a number of aquatic terms, such as "lake," "river," and "swamp" (Schele and Freidel 1990, 417), and can signify "plaza" as well (417). Although the connection is speculative, tridentates resemble the water stacks of the *ha'al* glyph (figure 9.11d), which translates as "watery" or "rainy" (Kettunen and Helmke 2013).

Identification of the eccentrics as lightning-serpent and water lily symbols provides important clues to the meaning of the deposit. Of the fourteen cache artifacts, thirteen are bifaces. This latter number has particular significance in

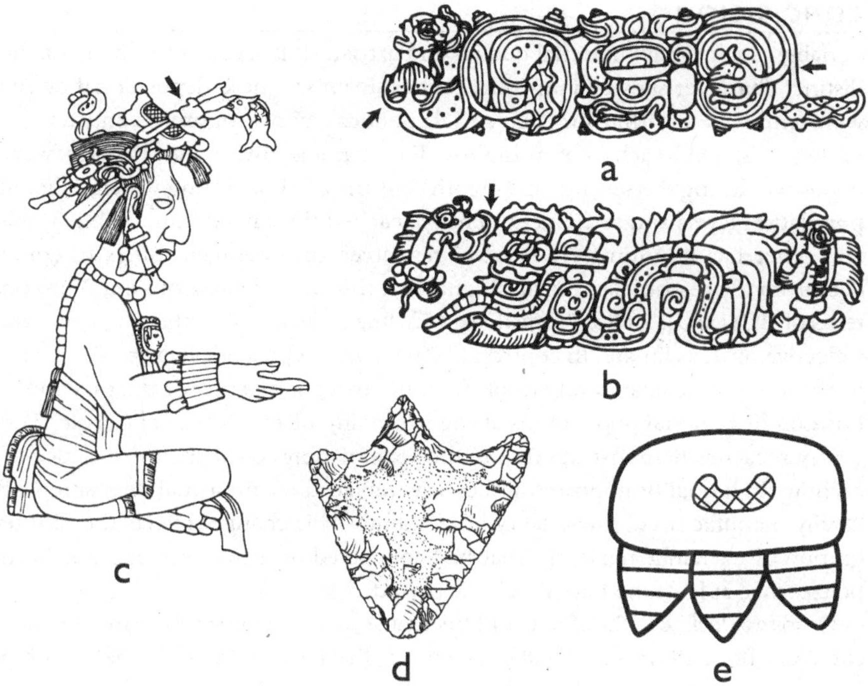

FIGURE 9.11. *Portrayals of water lilies, with arrows indicating calyx: (a) Water Lily Fish Monster (drawing by Linda Schele); (b) Water Lily Monster (drawing by Linda Schele); (c) K'inich Janaab' Pakal I wearing water lily headdress (drawing by Linda Schele); (d) ha'al glyph (drawing by Kwoka after Stone and Zender 2011, 163); (e) eccentric from Tz'unun Cache TZ18-01 (drawing by Jennifer Martin).*

relation to a deity known as Water Lily Serpent. The supernatural patron of the number 13, for which it serves as the head variant, was also depicted as the personified *tun*, or year sign (Miller and Taube 1993, 184). Identified by Nicholas Hellmuth (1987), Water Lily Serpent symbolizes open, still water, with additional references to "earth, abundance, the underworld, and the primordial sea" (Ishihara-Brito et al. 2006, 218). To this list, Robertson (2011) adds rain, running water, and blood sacrifice. The cache's location within a plaza strengthens this connection, as David Freidel and colleagues (1993, 139) argue that the Maya viewed these spaces as primordial seas. During seasonal storms, the impermeable plaster coating of plaza floors would have produced short-lived, shallow pools. Water Lily Serpent features prominently in elite regalia and architectural design, an association that may have emerged from the efforts of Maya rulers to control an essential life-sustaining resource (Ishihara-Brito 2006).

CONCLUSION

Variables such as raw material quality and artisan skill account for many of the distinct characteristics of individual caches. Even so, these depositional events were embedded within broader social, political, and economic contexts—all of which shaped cache compositions. For example, BC37 and TZ19-02 were deposited during the seventh and eighth centuries, a time during which regional population levels and settlement density reached their maximum. While these and other factors ultimately stressed resources and presaged things to come, data from the region's largest sites signal that this was a prosperous era. The contents of BC37 and TZ19-02—prestige goods imported from northern Belize—are reflective of this climate. In contrast, TZ18-01 was deposited circa 850 CE in conjunction with the abandonment of Tz'unun. Its contents mirror the precarious position of regional populations at the beginning of the Terminal Classic. The Tz'unun caches demonstrate that in the span of only 100 years, the practice of caching high-quality imported lithics was replaced by the ritual deposition of locally manufactured, low-quality implements. This change reflects the disintegration of exchange networks that had connected northwestern and northern Belize since at least the Late Preclassic period.

Despite a decline in the quality of lithic implements, cache symbolism remained complex. Both BC37 and TZ18-01 reference the institution of divine kingship, though this connection is more direct, and perhaps less legible at Blue Creek. In this context, legibility refers to audience size in terms of the proportion of a population that would have understood the symbolic associations of cached lithics. It is unclear whether rituals involving the sacrifice and burning of human remains were intimate or public spectacles. Nonetheless, depictions of *ch'ajoom* and their accompanying ritual implements, such as those contained in BC37, are infrequent, and even fewer tridentates have been recovered through excavation. In comparison, lightning eccentrics are ubiquitous in Mesoamerica and have been found in a variety of contexts, suggesting this symbol was legible to a broad segment of the populace. If the identification of tridents as water lily symbols proves correct, then these eccentrics would also have had a high degree of legibility. At Tikal, which provides a substantial and robust dataset, lightning and water lily eccentrics account for 39.3 percent of all cached lithics (Moholy-Nagy and Coe 2008, 8–9).

In closing, replication experiments, past and present, continue to provide the foundation upon which other analyses rest by illuminating both commonalities and idiosyncrasies in production techniques. Viewing these data through intrasite and regional comparisons of lithic technologies opens a path to the identification of salient social groups; the transmission of technical knowledge occurs within the context of learning communities. While acknowledging the record of excellence in elucidating how things were made, much work remains to be done in relation to the social domain of lithic production. Caches comprised solely of stone implements will often fall within the purview of lithic specialists due to the division of archaeological labor. This provides an opportunity to not only apply and refine existing strengths but also expand our analyses to nontraditional areas, such as artifact and cache symbolism. The chapters in this volume embody this ethos, and signal the vibrant state of Mesoamerican lithic studies.

REFERENCES

Adams, Richard E. W. 1999. *Río Azul: An Ancient Maya City*. Norman: University of Oklahoma Press.

Agurcia Fasquelle, Ricardo, Payson Sheets, and Karl A. Taube. 2016. *Protecting Sacred Space: Rosalila's Eccentric Chert Cache at Copan and Eccentrics Among the Classic Maya*. San Francisco: Precolumbian Mesoweb Press.

Aoyama, Kazuo. 2006. "Political and Socioeconomic Implications of Classic Maya Lithic Artifacts from the Main Plaza of Aguateca, Guatemala." *Journal de la Société des Américanistes* 92(92-1 and 92-2):7–40.

Aveni, Anthony F. 2011. "Maya Numerology." *Cambridge Archaeological Journal* 21(2):187–216.

Bamforth, Douglas B., and Nyree Finlay. 2008. "Introduction: Archaeological Approaches to Lithic Production and Craft Learning." *Journal of Archaeological Method and Theory* 15(1):1–27.

Barrett, Jason W. 2011. "Ancient Maya Exploitation of Non-Renewable Resources in the Eastern Maya Lowlands." In *The Technology of Maya Civilization: Political Economy and Beyond in Lithic Studies*, edited by Zachary X. Hruby, Geoffrey E. Braswell, and Oswaldo Chinchilla Mazariegos, 57–68. Sheffield: Equinox, UK.

Carballo, David M. 2011. *Obsidian and the Teotihuacan State: Weaponry and Ritual Production at the Moon Pyramid*. University of Pittsburgh Memoirs in Latin American Archaeology No. 21. Center for Comparative Archaeology. Pittsburgh and Mexico City: University of Pittsburgh, and Instituto de Investigaciones Antropológicas, Universidad Nacional Autónoma de México.

Carr, Philip J., and Andrew J. Bradbury. 2011. "Learning from Lithics: A Perspective on the Foundation and Future of the Organization of Technology." *PaleoAnthropology* 2011:305–319.

Clark, John E. 2003. "Craft Specialization as an Archaeological Category." *Research in Economic Anthropology* 16:267–294.

Coe, William R. 1959. *Piedras Negras Archaeology: Artifacts, Caches, and Burials*. Museum Monographs. Philadelphia: University Museum, University of Pennsylvania.

Costin, Cathy Lynne. 1991. "Craft Specialization: Issues in Defining, Documenting, and Explaining the Organization of Production." *Archaeological Method and Theory* 3:1–56.

Cotterell, Brian, and Johan Kamminga. 1987. "The Formation of Flakes." *American Antiquity* 52(4):675–708.

Dunning, Nicholas, John G. Jones, Timothy Beach, and Sheryl Luzzadder-Beach. 2003. "Physiography, Habitats, and Landscapes of the Three Rivers Region." In *Heterarchy, Political Economy, and the Ancient Maya: The Three Rivers Region of the East-Central Yucatán Peninsula*, edited by Vernon L. Scarborough, Fred Valdez Jr., and Nicholas Dunning, 14–24. Tucson: University of Arizona Press.

Eerkens, Jelmer W. 2000. "Practice Makes Within 5% of Perfect: Visual Perception, Motor Skills, and Memory in Artifact Variation." *Current Anthropology* 41(4):663–668.

Finlay, Nyree. 2008. "Blank Concerns: Issues of Skill and Consistency in the Replication of Scottish Later Mesolithic Blades." *Journal of Archaeological Method and Theory* 15:68–90.

Franks, A. W. 1877. "Stone Implements from Honduras." *Journal of the Anthropological Institute of Great Britain and Ireland* 6:37–41.

Freidel, David, Linda Schele, and Joy Parker. 1993. *Maya Cosmos: Three Thousand Years on the Shaman's Path*. New York: William Morrow.

Gann, Thomas W. F. 1918. *The Maya Indians of Southern Yucatan and Northern British Honduras*. Bureau of American Ethnology Bulletins 64. Washington, DC: Smithsonian Institution.

Garrison, Thomas G. 2007. "Ancient Maya Territories, Adaptive Regions, and Alliances: Contextualizing the San Bartolo-Xultun Intersite Survey." PhD diss., Harvard University, Cambridge, MA.

Guderjan, Thomas H. 1991. "Aspects of Maya Settlement in the Río Bravo Area." In *Maya Settlement in Northwestern Belize: The 1988 and 1990 Seasons of the Río Bravo Archaeological Project*, edited by Thomas H. Guderjan, 103–110. Culver City, CA: Labyrinthos.

Guderjan, Thomas H. 1998. "The Blue Creek Jade Cache: Early Classic Ritual in Northwestern Belize." In *The Sowing and the Dawning: Termination, Dedication, and Transformation in the Archaeological and Ethnographic Record of Mesoamerica*, edited by Shirley Boteler. Mock, 101–112. Albuquerque: University of New Mexico Press.

Guderjan, Thomas H. 2007. *The Nature of an Ancient Maya City: Resources, Interaction, and Power at Blue Creek, Belize*. Tuscaloosa: University of Alabama Press.

Guderjan, Thomas H. 2016. "Implementing Multiple Research Strategies: A Case Study from Blue Creek, Belize." In *The Ancient Maya City of Blue Creek, Belize: Wealth, Social Organization and Ritual*, edited by Thomas H. Guderjan, 5–14. BAR International Series 2796. Oxford: Archaeopress.

Guderjan, Thomas H., Robert J. Lichtenstein, and C. Colleen Hanratty. 2003. "Elite Residences at Blue Creek, Belize." In *Maya Palaces and Elite Residence: An Interdisciplinary Approach*, edited by Jessica Joyce Christie, 13–45. Austin: University of Texas Press.

Hanratty, C. Colleen. 2002. "Excavations in the Structure 37 Plazuela." In *The Blue Creek Project: Working Papers form the 1998 and 1999 Field Seasons*, edited by Thomas H. Guderjan and Robert J. Lichtenstein, 73–80. Fort Worth, TX: Maya Research Program.

Helmke, Christophe, Gail Hammond, Thomas Guderjan, Pieta Greaves, and Colleen Hanratty. 2019 "Sighting a Royal Vehicle: Observations on the Graffiti of Tulix Mul, Belize." *PARI Journal* 19(4):10–30.

Hellmuth, Nicholas. 1987. *Monster und Menschen in der Maya-Kunst: Eine Ikonographie der alten Religionen Mexikos und Guatemalas*. Graz, Austria: Academische Druk- u. Verlagsanstalt.

Hester, Thomas R., and Harry J. Shafer. 1984. "Exploitation of Chert Resources by the Ancient Maya of Northern Belize, Central America." *Mines and Quarries* 16(2):157–173.

Heye, George G. 1925. "Eccentric Chipped Objects from British Honduras." Museum of the American Indian, Heye Foundation. *Indian Notes* 2:99–102.

Houk, Brett A., and Gregory Zaro. 2015. "Lithic Production and Domestic Economy in an Ancient Maya Neighborhood at Chan Chich, Belize." *Research Reports in Belizean Archaeology* 12:127–134.

Hruby, Zachary X. 2006. "The Organization of Chipped-Stone Economies at Piedras Negras, Guatemala." PhD diss., University of California, Riverside.

Hruby, Zachary X. 2007. "Ritualized Chipped-Stone Production at Piedras Negras, Guatemala." In *Rethinking Craft Specialization in Complex Societies: Archaeological Analysis of the Social Meaning of Production*, edited by Zachary Hruby and Rowan K. Flad, 68–87. Archaeological Papers of the American Anthropological Association 17. Arlington, VA: American Anthropological Association.

Hruby, Zachary X, and Gene Ware. 2009. "Painted Lithic Artifacts from Piedras Negras, Guatemala." In *Maya Archaeology*, vol. 1, edited by Charles Golden, Stephen Houston, and Joel Skidmore, 76–85. San Francisco: Mesoweb Press.

Iannone, Gyles, and James M. Conlon. 1993. "Elites, Eccentrics, and Empowerment in the Maya Area: Implications for the Interpretation of a Peripheral Settlement Cluster near Cahal Pech, Cayo District, Belize." *Papers from the Institute of Archaeology* 4:77–89.

Ishihara-Brito, Reiko, Karl A. Taube, and Jaime J. Awe. 2006. "The Water Lily Serpent Stucco Masks at Caracol, Belize." *Research Reports in Belizean Archaeology* 3:211–222.

Joyce, Thomas A. 1932. "Presidential Address: The 'Eccentric Flints' of Central America." *Journal of the Royal Anthropological Institute of Great Britain and Ireland* 62:xvii–xxvi.

Joyce, Thomas A., Thomas Gann, E. L. Gruning, and R. C. E. Long. 1928. "Report on the British Museum Expedition to British Honduras, 1928." *Journal of the Royal Anthropological Institute of Great Britain and Ireland* 258:323–350.

Kettunen, Harri, and Christophe Helmke. 2013. "Water in Maya Imagery and Writing." *Contributions in New World Archaeology* 5:17–38.

Kidder, Alfred V. 1947. *The Artifacts of Uaxactun, Guatemala*. Publication 576. Washington, DC: Carnegie Institution of Washington.

Kopytoff, Igor. 1986. "The Cultural Biography of Things: Commoditization as Process." In *The Social Life of Things: Commodities in Cultural Perspective*, edited by Arjun Appadurai, 64–91. Cambridge: Cambridge University Press.

Kwoka, Joshua J. 2014. "Ideological Presentism and the Study of Ancient Technology: Preclassic Lithic Production at San Bartolo, Guatemala." PhD diss., State University of New York at Buffalo.

Kwoka, Joshua J. 2017. "Finding Value in the Mundane: Chert Features and Communities of Practice at San Bartolo, Guatemala." In *The Value of Things: Commodities in the Maya Region from Prehistoric to Contemporary*, edited by Jennifer P. Mathews and Thomas H. Guderjan, 49–66. Tucson: University of Arizona Press.

Kwoka, Joshua J. 2019. "Exploration of Tz'unun's E Group: Structure 21 and Associated Monuments." In *The 26th Annual Report of the Blue Creek Archaeological Project*, edited by Thomas H. Guderjan and C. Colleen Hanratty, 1–28. Tyler, TX: Maya Research Program.

Kwoka, Joshua J. 2020. "Tz'unun Structure 21: Results of the 2019 Field Season." In *The 27th Annual Report of the Blue Creek Archaeological Project*, edited by Thomas H. Guderjan and C. Colleen Hanratty, 1–45. Tyler, TX: Maya Research Program.

Kwoka, Joshua J., and Justin Telepak. 2017. "Maya Research Program Regional Reconnaissance: 2016." In *The 25th Annual Report of the Blue Creek Archaeological Project*, edited by Thomas H. Guderjan and C. Colleen Hanratty, 125–137. Tyler: Maya Research Program and the Center for Social Science Research, University of Texas at Tyler.

Kwoka, Joshua J., H. Colleen Hanratty, and Thomas H. Guderjan. 2019. "From Crafting to Caching: Technological and Iconographic Analyses of Blue Creek Cache 37." *Journal of Field Archaeology* 43(6):352–366.

Kwoka, Joshua J., and M. Steven Shackley. 2019. "Technological Analysis and Source Provenance of Obsidian Artifacts from a Sun Pyramid Substructure Cache, Teotihuacan, Mexico." *Latin American Antiquity* 30:205–210.

Lane Fox, A. H. 1871. "A Flint Implement Brought from Honduras." *Proceedings of the Society of Antiquaries of London* 5:93–95.

Larson, Mary Lou, and Marcel Kornfeld. 1997. "Chipped Stone Nodules: Theory, Method, and Examples." *Lithic Technology* 22:4–18.

McDonald, J. Andrew, and Brian Stross. 2012. "Water Lily and Cosmic Serpent: Equivalent Conduits of the Maya Spirit Realm." *Journal of Ethnobiology* 32:74–107.

Meadows, Richard. 2001. "Crafting K'awil: A Comparative Analysis of Maya Symbolic Flaked Stone Assemblages from Three Sites in Northern Belize." PhD diss., University of Texas, Austin.

Meadows, Richard. 2003. "From Eccentric Lithics to Material Symbols: Aspects of Ancient Maya Cultural Production in Northern Belize." *Mono y Conejo* 1:15–24.

Miller, Mary, and Karl Taube. 1993. *An Illustrated Dictionary of the Gods and Symbols of Ancient Mexico and the Maya*. London: Thames and Hudson.

Moholy-Nagy, Hattula, and William R. Coe. 2008. "Appendix 11: Typology and Description of Flint and Obsidian Eccentrics and Incised Obsidians." In *The Artifacts of Tikal: Ornamental and Ceremonial Artifacts and Unworked Material, Tikal Report 27A*. Philadelphia: University of Pennsylvania Museum of Archaeology and Anthropology. Accessed February 1, 2018. https://doi.org/10.6067/XCV8CJ8D43.

Nelson, Margaret C. 1991. "The Study of Technological Organization." *Archaeological Method and Theory* 3:57–100.

Pagliaro, Jonathan B., James F. Garber, and Travis W. Stanton. 2003. "Evaluating the Archaeological Signatures of Maya Ritual and Conflict." In *Ancient Mesoamerican Warfare*, edited by M. Kathryn Brown and Travis W. Stanton, 75–89. Walnut Creek, CA: AltaMira Press.

Pendergast, David M. 1990. *Excavations at Altun Ha, Belize, 1964–1970*. Vol. 3. Toronto: Royal Ontario Museum.

Rice, R. Garraway. 1909. "Notes on a Flint in Human Shape Found in the Thames." *Proceedings of the Society of Antiquaries of London* 22:359–360.

Robertson, Merle Greene. 2011. "The Celestial God of Number 13." *PARI Journal* 12(1):1–6.

Schele, Linda, and David Freidel. 1990. *A Forest of Kings: The Untold Story of the Ancient Maya*. New York: William Morrow and Company.

Scherer, Andrew K., and Stephen Houston. 2018. "Blood, Fire, Death: Covenants and Crises among the Classic Maya." In *Smoke, Flames, and the Human Body in Mesoamerican Ritual Practice*, edited by Vera Tiesler and Andrew K. Scherer, 109–150. Washington, DC: Dumbarton Oaks.

Schiffer, Michael B. 1987. *Formation Processes of the Archaeological Record*. Albuquerque: University of New Mexico Press.

Shafer, Harry J., and Thomas R. Hester. 1983. "Ancient Maya Chert Workshops in Northern Belize, Central America." *American Antiquity* 48(3):519–543.

Shafer, Harry J., and Thomas R. Hester. 1991. "Lithic Craft Specialization and Product Distribution at the Maya Site of Colha, Belize." *World Archaeology* 23:79–97.

Šprajc, Ivan. 2004. "Maya Sites and Monuments in SE Campeche, Mexico." *Journal of Field Archaeology* 29(3–4):385–407.

St. George Gray, H. 1916. "On a Chipped Flint Implement Found in British Honduras." *Royal Anthropological Institute of Great Britain and Ireland* 16:154–155.

Starzmann, Maria Theresia. 2013. "Spontaneity and Habitus: Stone Tool Production in Communities of Practice at Fıstıklı Höyük." In *Interpreting the Late Neolithic of Upper Mesopotamia*, edited by Oliver Nieuwenhuyse, Reinhard Bernbeck, Peter M. M. G. Akkermans, and Jana Rogasch, 161–169. Turnhout, Belgium: Brepols Publishers.

Stemp, James W., Jaime T. Awe, and Christophe Helmke. 2014. "The Macrolith of El Chiquero, Belize." *Mexicon* 36(5):145–150.

Stocker, Terrance L., and Michael W. Spence. 1973. "Trilobal Eccentrics at Teotihuacan and Tula." *American Antiquity* 38(2):195–199.

Stone, Andrea, and Mark Zender. 2011. *Reading Maya Art: A Hieroglyphic Guide to Ancient Maya Painting and Sculpture*. New York: Thames and Hudson.

Stross, Brian. 1998. "Seven Ingredients in Mesoamerican Ensoulment: Dedication and Termination in Tenejapa." In *The Sowing and the Dawning: Termination, Dedication, and Transformation in the Archaeological and Ethnographic Record of Mesoamerica*, edited by Shirley Boteler. Mock, 31–40. Albuquerque: University of New Mexico Press.

Stuart, David. 2017. "An Intriguing Date on the Tz'unun Panel." *Maya Decipherment: Ideas on Ancient Maya Writing and Iconography*. Accessed December 1, 2018. https://decipherment.wordpress.com/2017/01/20/an-intriguing-date-on-the-tzunun-panel/.

Sugiyama, Saburo. 2005. *Human Sacrifice, Militarism, and Rulership: Materialization of State Ideology at the Feathered Serpent Pyramid, Teotihuacan*. Cambridge: Cambridge University Press.

Taube, Karl A. 1998. "The Jade Hearth: Centrality, Rulership, and the Classic Maya Temple." In *Function and Meaning in Classic Maya Architecture*, edited by Stephen D. Houston, 427–478. Washington, DC: Dumbarton Oaks.

Thompson, J. Eric S. 1931. *Archaeological Investigations in the Southern Cayo District, British Honduras*. Field Museum of Natural History Anthropology Series 17(3). Chicago: Field Museum.

Thompson, J. Eric S. 1950. *Maya Hieroglyphic Writing: Introduction*. Publication 589. Washington, DC: Carnegie Institution of Washington.

Titmus, Gene L., and James C. Woods. 2003. "The Maya Eccentric: Evidence for the Use of the Indirect Percussion Technique in Mesoamerica from Preliminary Experiments Concerning Their Manufacture." In *Mesoamerican Lithic Technology: Experimentation and Interpretation*, edited by Kenneth G. Hirth, 133–146. Salt Lake City: University of Utah Press.

Trachman, Rissa M. 2009. "Ancient Maya Social Organization: A Mutli-Scale Analysis from the Dos Hombres Hinterland, Northwestern Belize." *Research Reports in Belizean Archaeology* 6:109–117.

Uomini, Natalie. 2005. *A Study of Individual Variation in Bimanual Knapping Configurations, and the Effects of Handedness of Knapping Products*. Technical Report of Experiments Undertaken July 2005. Lejre, Denmark: Lejre Historical-Archaeological Research Centre.

Vogt, Evon Z. 1998. "Zinacanteco Dedication and Termination Rituals." In *The Sowing and the Dawning: Termination, Dedication, and Transformation in the Archaeological and Ethnographic Record of Mesoamerica*, edited by Shirley Boteler Mock, 21–30. Albuquerque: University of New Mexico Press.

Whittaker, John C. 1987. "Individual Variation as an Approach to Economic Organization: Projectile Points at Grasshopper Pueblo, Arizona." *Journal of Field Archaeology* 14:-465–479.

Whittaker, John C. 1994. *Flintknapping: Making and Understanding Stone Tools*. Austin: University of Texas Press.

Wright, A. C., D. H. Romney, R. H. Arbuckle, and V. E. Vial. 1959. *Land Use in British Honduras*. Colonial Research Publication 24. London: HMSO.

PART THREE

Studies of Obsidian Artifacts

10

An Interpretative Framework and Description of Ritualized Obsidian from Caracol, Belize

LUCAS R. M. JOHNSON,
ARLEN F. CHASE, AND DIANE Z. CHASE

Obsidian was one of the most commonly ritualized materials in Precolumbian Mesoamerica, which is particularly true for the Lowland Maya city of Caracol, Belize, from the Late Preclassic through Terminal Classic periods (ca. 300 BCE–900 CE). While our focus is on the technological and distributional analysis of "ritualized obsidian," we offer an overview of important definitions and theories that may help to infer a level of meaning to the selective use of obsidian over other materials. To be sure, symbolic meaning may be inferred apart from archaeological context by interpreting stylized forms and placing a given form within the structuring constraints of religion or a cosmological order (Agurcia Fasquelle et al. 2016). This practice is commonplace among analyses of obsidian "eccentrics," where an abstract symbol or symbolic sign can have cosmological reference and not be considered constrained by its material properties and the agency (i.e., choices) of crafters (see Levine and Carballo 2014). While we do not argue that there are overarching religious or cosmological connotations structuring residential rituals at Caracol and elsewhere, we choose to not use the term "eccentric," in that it may continually mystify the social and physical *making* of

https://doi.org/10.5876/9781646427819.c010

rituals. Here we situate rituals as actions and as materialized processes that manifest in historically contingent places and times (Bell 1992; Fogelin 2007; Swenson 2015). The focus of many rituals at Caracol occurred in domestic spaces; therefore, we consider how residential rituals were produced in materially, spatially, and temporally patterned ways, all of which serve to (re-)create and (re)affirm social ties. Understanding these data may then dialectically inform a shared cosmology or religion across a broad cityscape and beyond.

SOME USEFUL ANALYTICAL CONCEPTS

An analysis of flaked stone typically includes some transformative connotations (Sellet 1993). Some call these referential schemes "object biographies" (Appadurai 1986; Kopytoff 1986), life histories (LaMotta and Schiffer 2001), artifact genealogies (Clark 2007), or the like (Hoskins 1998). These concepts examine the relationships between people and materials over both short and long temporal scales. For example, cores may be reduced in minutes or hours at workshops, whereas some flaked stone objects may experience longer and more detailed "itineraries" by being recycled or intentionally curated and removed from circulation to then be deployed in human projects many years or even decades later (Joyce and Gillespie 2015). If enough data are present, flaked stone artifact analysis may reveal a complex transformative, fragmented, and social history (L. M. Johnson 2016). Inferring various itineraries can reveal the multiple ways things evoke meaning, both symbolic and as a more direct reference to transformative events in the lives of materials and people (Haskell 2015). Transformative material practice may be especially apparent for objects used in rituals, which often retain physical markers of quotidian production. For example, an obsidian blade core shaped into a serpent or lightning bolt shape still retains vertical negative blades scars, indicating an intensive level of sophisticated pressure flaking and core management prior to its final more symbolic or iconic form. These physical traces would have been apparent to a large socioeconomic cross-section of Caracol society, especially as blade use was as common as blade production (but typically using locally available chert).

Obsidian objects can have meaning in multiple ways as they move in and out of contexts over time or are "bundled" (Keane 2005). As used by Webb Keane (2005), the "bundle" is not a physical assemblage of objects but rather refers to the multiple qualities a specific object may possess. Qualities include its material properties (color, texture, durability, sharpness) as well as its accumulated histories (i.e., itineraries) and future potential. Certain qualities may be more important at different times as an object shifts between contexts. It has been argued that for the Maya and Mesoamerica more broadly, obsidian was a cosmologically significant material and obsidian blades were both especially dangerous during rituals and essential for daily living (Darras 2014; Garcia Cook and Merino

Carrion 2005; Graulich 205; Stemp and Awe 2014). As is outlined in this chapter, archaeological data from Caracol indicates that obsidian was available to most of the population and was utilized in many rituals. Furthermore, obsidian was a foreign material for much of the Maya region and acquired only through complex regional/local economic and political negotiation (L. M. Johnson 2016; Meierhoff et al. 2012; Moholy-Nagy et al. 2013; Nazaroff et al. 2010). Additionally, those who crafted obsidian into highly standardized blades were skilled and this production knowledge was likely restricted (Hruby 2007). Obsidian was therefore viewed with a deep social, historical, and regional weight. The very transformation from raw core to blade may have been perceived as a ritualistic act (Hruby 2007). Provided this level of depth, obsidian in any form can be viewed as an index of regional connections and skilled human action apart from any stylized (eccentric) form (L M. Johnson 2016, 354). A stylized form, if present, may be an icon of something that exists in nature. For example, many shaped obsidian artifacts resemble scorpions, snakes, lightning bolts, and even people; thus, through a specific knapping process an impression of reality is created. Icons can be viewed as symbols too, but symbols may take on other more arbitrarily assigned meanings that are dependent on specific contexts far removed from the present day (cf. Joyce 2007, 114).

Materials with unique physical properties and historical importance, including cosmological and economic, are often given a heightened emphasis in society. For the Maya, ritual contexts offer a window into how such an emphasis may have been displayed. Ritualization, as we use it here, refers to a process whereby elements of everyday life can be given added meaning (Bradley 2003). In this case we argue obsidian was commonly ritualized in Classic period households throughout the city of Caracol. Research at Caracol has examined residential rituals anchored at eastern mausoleum structures, a common feature of over 70 percent of all household groups at the site (A. Chase and D. Chase 2008, 2013, 2014a, 3; D. Chase and A. Chase 1998, 2011). These structures provide significant information surrounding group histories that include the burying of specific individuals, architectural remodeling, and materializing time through repeated caching events often associated with prior inhumations (D. Chase and A. Chase 2024). A cursory analysis of caches from these structures underscores the broad use of ceramic urns, often with modeled deity faces (i.e., face caches), that can contain shell, coral, stingray spines, jadeite, and obsidian (A. Chase and D. Chase 2013). These caches are analyzed to consider how certain materials were "ritualized" as part of cyclical and timed residential events (A. Chase and D. Chase 2013) and are clearly set apart from quotidian tasks.

Therefore, in following Bradley (2003, 12), who defines ritualization as "both a way of acting which reveals some of the dominant concerns of society, and a process by which certain parts of life are selected and provided with an added emphasis," we can argue that "ritualized objects" are those elements that emphasize historical and social importance and do not necessarily need to represent iconic forms or provide symbolic inference (see also L. M. Johnson 2016, 49). Within this framework, obsidian as a commonly used quotidian material can take on added emphasis in particular moments or events during household rituals. At Caracol, specifically, we can see that the Maya may have enacted specific events with the inclusion of other obsidian types (e.g., blade cores and blade-core objects) to emphasize those materials essential for quotidian tasks in daily life (e.g., obsidian blades). As explained in the section "The Contexts and Scale of Obsidian at Caracol," specific technological types of obsidian artifacts correlate with visibly identifiable contexts that are the material traces of specific activities.

Ultimately, we approach one aspect of ritual life at Caracol, Belize, through a technological and contextual analysis of ritualized obsidian objects. These items are typically termed "obsidian eccentrics," although "ritualized obsidian" more specifically enables a usable and socially interpretive framework based on archaeological context that relies less on symbolism for interpretation. Analyzing the obsidian contextually and technologically at Caracol emphasizes historically important materials and associated crafting, regional and local exchange, and depositional practices, especially those that include marking and materializing temporalities at eastern mausolea. While we describe some obsidian as being ritualized, as opposed to being quotidian, our primary purpose here is to describe the morphology, chronology, and spatial extent of ritualized obsidian at Caracol, Belize. Toward this end, we present information on some thirty-five years of consecutive and standardized excavation data since 1985 (A. Chase and D. Chase 1987; D. Chase and A. Chase 1994, 2017) to demonstrate the patterned ways obsidian was and was not often crafted into ritualized forms that were then deployed in important cyclical events. The cyclical events that often included ritualized obsidian took place among varied sizes of house groups situated over a vast settlement area far beyond the elite city center, thus indicating the use of obsidian was not an exclusively elite or royal practice. Furthermore, many of the ritualized obsidians from Caracol cannot necessarily be interpreted through an overtly symbolic approach, because a significant proportion of the obsidian recovered from ritual caches are the unmodified byproducts of core shaping/maintenance for blade manufacture.

OVERVIEW OF CARACOL

The city of Caracol, Belize, occupies a large karstic plateau between two rivers and is connected to the geology of the Petén. At Caracol's height around 650 CE, the city encompassed nearly 200 sq. km (see figure 10.1; D. Chase and A. Chase 2017, 185) and was home to more than 100,000 people situated within almost 9,000 residential groups. Its archaeological record spans 600 BCE to 900 CE (table 10.1).

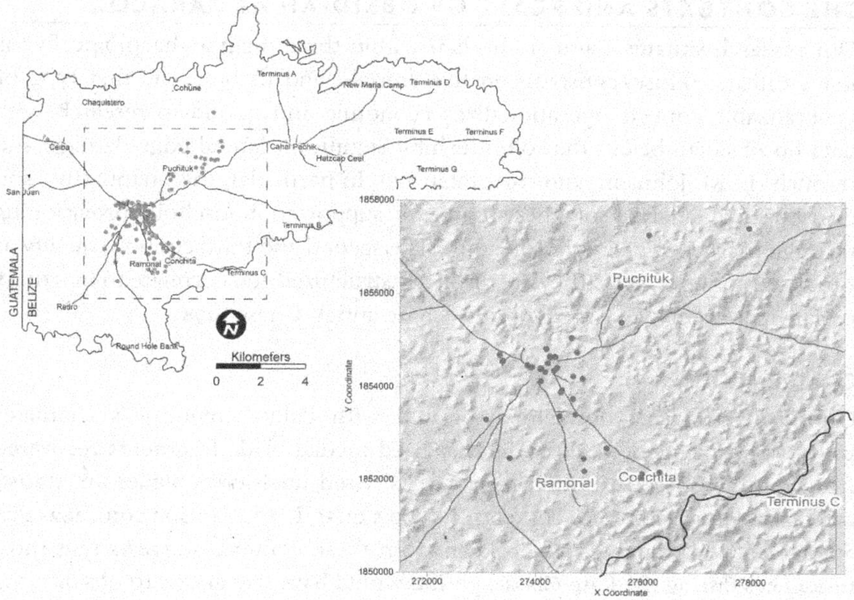

FIGURE 10.1. *Overview of Caracol with causeways and terracing extent (after D. Chase and A. Chase 2017, fig. 1) showing the distribution of archaeological investigations at the site represented as dots, and the distribution of seventy-eight caches with obsidian within a 15 m hill-shade background. Image by Lucas R. M. Johnson and Adrian S. Z. Chase.*

During the Late Classic and possibly earlier, markets served most of the population with quotidian (household domestic items), prestige (items thought to be associated with status and associated with the elite; see A. Chase and D. Chase 1992), and ritual-related (items utilized in terms of religion and belief systems) goods (A. Chase et al. 2015; D. Chase and A. Chase 2014, 2020). Recent technological and distributional analysis of obsidian (L. M. Johnson 2016) has reinforced the assertion that Caracol had multiple marketplace locales that had overlapping provisioning radii (A. S. Z. Chase 2016). Analyses of both burial and cache contexts has shown that historically early ceremonies took place within the city's central monumental architecture (A. Chase and D. Chase 1995, 2017a). Later, during the Late Classic period, similar rituals spread throughout the residential settlement, indicating a shared ritual identity (A. Chase and D. Chase 2009; D. Chase and A. Chase 2004). Domestic rituals, mostly associated with eastern-related mausolea, had similar cache and burial assemblages yet varied in the degree of elaboration, a factor likely associated with wealth or status (A. S. Z. Chase 2017; A. Chase and D. Chase 2013; D. Chase and A. Chase 2011).

THE CONTEXTS AND SCALE OF OBSIDIAN AT CARACOL

Our research situates Caracol obsidian within three main archaeologically visible contexts: refuse/construction fills, burials, and caches. The first type of generalizable context operationalizes economic and quotidian research with data on obsidian blades that often exhibit significant lateral edge damage and retouch (L. M. Johnson 2016, 290, table 7-5). In particular, the quantitative and distributional analysis of these contexts supported household provisioning through markets (L. M. Johnson 2016). The second and third contexts qualify as ritualized contexts, since they are distinctly structured and correlate to mortuary or calendar events (D. Chase 1994; D. Chase and A. Chase 1998).

Quotidian Obsidian

Quotidian obsidian—obsidian associated with daily activities—is characterized by mostly moderately to heavily used medial blade fragments recovered from refuse and construction fill contexts. Used final-series blades are statistically correlated with these contexts ($p = 90.5 \pm 1.21$; L. M. Johnson 2016, 292–293). Spatial statistical analysis of obsidian from these contexts suggests that most residences during the Late Classic period would have had access to obsidian via a market distribution system within overlapping provisioning areas (D. Chase and A. Chase 2014, 213; L. M. Johnson 2016, 269, fig. 6.10). This relatively unrestricted access of a nonlocal material by an economically diverse cross-section of Caracol society has significant implications for whether obsidian may have been used for more ritualistic activities. In other words, if residences of different economic scale—a "middle-class" (e.g., A. Chase and D. Chase 1992; D. Chase and A. Chase 1992)—had access to obsidian for everyday activities, then it is extremely likely that they would have also used this material during domestic rituals. As the analysis showed, this is indeed true; yet there is a technological correlation between various archaeologically discernable contexts (e.g., blade cores from residential caches).

Burials with Obsidian

A total of 124 burials with obsidian were analyzed as of the 2015 season. This included three above-tomb obsidian assemblages, which totaled nearly 15,000 pieces combined. Complete or near complete final-series blades statistically correlate with burial contexts whether they be simple crypts or vaulted tombs from across a broad settlement sample of residences ($p = 90.1 \pm 4.5$, L. M. Johnson 2016, 320). On the other hand, the three epicentral above-tomb contexts include a range of obsidian artifacts representing a reconstructible blade-core reduction sequence where refits are common. Like other sites, these above-tomb assemblages are proxies for workshop behavior and have some ceremonial significance (L. M. Johnson 2016; Moholy-Nagy 1997; Trachman 2002; Trachman and Titmus 2003).

TABLE 10.1. Summary of cached obsidian by temporal period and technological type as of 2018.

Period	*Date Range*	*N = from Caches as of 2018*	*% of Total*	*Caches with Obsidian*	*% of Total*
Middle Preclassic	Before ca. 300 BCE	0	0	0	0
Late Preclassic 1	ca. 300–50 BCE	0	0	0	0
Late Preclassic 2	50 BCE–200 CE	1	<1	1	1
Early Classic 1	200–350 CE	19	3	2	3
Early Classic 2	350–550 CE	79	13	6	8
Late Classic 1	550–700 CE	240	38	21	27
Late Classic 2	700–800 CE	244	39	40	51
Terminal Classic	800–900 CE	43	7	8	10
Total		626	100	78	100

	Blade Production							Other			
	Macrodebitage	*Small Perc. Debitage*	*Initial-Series*	*Final-Series*	*Ind Blade*	*Rejuvenation Debitage*	*Exhausted Core*	*Flake*	*Shatter*	*Biface/Point*	*Pebble*
Middle Preclassic	0	0	0	0	0	0	0	0	0	0	0
Late Preclassic 1	0	0	0	0	0	0	0	0	0	0	0
Late Preclassic 2	0	0	0	0	0	0	0	0	0	0	1
Early Classic 1	1	0	0	1	0	1	2	0	0	0	0
Early Classic 2	1	0	1	3	1	1	4	0	0	2	0
Late Classic 1	3	1	3	19	0	12	11	3	0	0	0
Late Classic 2	14	2	4	28	0	10	16	5	0	1	0
Terminal Classic	3	1	2	4	0	5	4	0	1	0	0
Total number of caches by type	22	4	10	55	1	29	37	8	1	3	1

While these above-tomb layering events are clear examples of ceremonial use of obsidian across a broad region of the Maya area, the Caracol examples are not discussed here, because they are associated with city-center elites and are not representative of a shared ritual practice across the Caracol cityscape. Residential caches, on the other hand, offer significant insights into shared practice.

Caches with Obsidian

Analyses of obsidian from caches include seventy-eight special deposits (SD), or approximately 23 percent of all investigated cache contexts as of 2018 (see figure 10.1 for a distribution of these deposits). This sample is an increase from sixty-one analyzed as of 2015 (Johnson 2016) but does not include obsidian in stela/altar caches noted by A. Chase and D. Chase (1987, 4; see also L. M. Johnson 2016, 93), because these were not available for analysis. Caches as reported here are from residential architecture; however, it would include other types of caches if available. As previously reported, a more diverse suite of reduction stages and artifact types correlate with a caching context and are clearly distinct from the other two contexts described earlier (Johnson 2016, 233, fig. 5.34). Caches showed the strongest correlation with more varied technological types: macrocore shaping, $p = 27.9 \pm 9.59$; percussion debitage, $p = 4.9 \pm 4.61$; rejuvenation debitage, $p = 21.3 \pm 8.75$; blades (initial-series, $p = 13.1 \pm 7.21$; final-series, $p = 67.2 \pm 10.04$), and both modified and unmodified exhausted blade cores ($p = 52.5 \pm 10.68$). In other words, cache deposits, in contrast to the other two contexts (burials and refuse/construction fills), account for the overwhelming abundance of those steps associated with creating and maintaining cores as well as destroyed or otherwise shaped exhausted blade cores. We focus on caches with obsidian to demonstrate how obsidian was used to materialize rituals associated with temporal cycles, likely incorporating remembrance of familiar ancestors.

Caches with obsidian, while seemingly concentrated around the Caracol's epicenter, are also distributed more than 3 km distant, in many cases suggesting that this activity was not exclusively an elite activity. Temporal analysis indicates early caching in the epicenter and then a distribution throughout the settlement over time (see the section "Temporal and Spatial Trends of Ritualized Obsidian"). For a sense of scale, the current distribution of analyzed caches with obsidian covers 5,362 × 4,871 sq. m, or approximately 26 sq. km, suggesting about 3 caches per sq. km (or 12.8 per sq. km if using the total number of caches as of the 2018 field season, $n = 335$). The estimated number of structures—arranged in *plazuela* groups with three or more structures per group (averaging three to four structures arranged in NS/EW directions around a central plaza)—within the terraced area is about 27,000 or a structure density of approximately 135 per sq. km or 42 *plazuela* groups per sq. km (A. Chase et al. 2019). Here the median *plazuela* count equals 8,404. These two generalized data points suggest that at a minimum 30 percent of *plazuela* groups may have caches (12.8 caches distributed among 42 *plazuelas*) and that about 22 percent (or $n = 3$) of those caches will contain obsidian. Taken even further, these data suggest that if there are three caches with obsidian per each sq. km within a 200 sq. km area, then we might expect at least 600 such caches ($p = 7.0 \pm 0.36$). These broad distribution assertions paired with what is already known from Caracol, through standardized systematic investigations, indicate a shared ritual practice and a significant economic network that supplemented the provisioning of ritual materials.

Again, obsidian blade cores and blade-core objects are statistically correlated with these types of contexts, while other non-blade artifacts also have a strong correlation (L. M. Johnson 2016, 307). Non-blade artifacts—those other than pressure blades—include macro- and other percussion debitage from shaping macrocores and a range of proximal, lateral, and distal rejuvenation debris. Ritual cache contexts include obsidian, occasionally assembled with other materials such as jadeite and shell, in vessels with modeled facial features, either human or animal-like, otherwise known as "face caches" (A. Chase and D. Chase 1994, 2010; D. Chase and A. Chase 1998).

Investigations within the eastern structure of the Northeast (NE) Acropolis, an elite residence just adjacent to Caana, recovered two significant face caches. One, SDC117B-6, nicknamed "Mongo" (D. Chase and A. Chase 1998, fig. 7; 2010, fig. 2g), was located just beneath the first step of a buried construction; it contained fragments of human bone, shell, malachite, and jadeite, and eighteen ritualized obsidian objects (figure 10.2). The obsidian assemblage within the urn consisted of larger overshot or plunging fragments of obsidian, worked and unworked blades, worked and unworked platform preparation debitage, and modified exhausted blade cores. Blade cores were percussion flaked into bifacial forms, "S" forms, and other unilateral or bilaterally notched forms. The indirect percussion notching effectively removed proximal pressure platforms and disabled distal rejuvenation. Distal rejuvenation acts to reshape the distal end of core, which regularizes the overall shape for more effective blade removal. Some of the obsidian exhibited minor dorsal and ventral facial wear, suggesting that they had been curated sometime before their final deposition in the cache. Work during the 2007 field season just north of Caana at the GRB Group, multiple caches were recovered in association with the eastern mausoleum constructions (A. Chase and D. Chase 2007). One prominent cache was recovered that included two larger-than-normal face caches (A. Chase and D. Chase 2007, figs. 57–61). These face caches are associated with thirty-nine obsidian artifacts (see figure 10.2), a chert chunk, shell, coral, stingray spines, burned and unburned jadeite, faunal bone fragments, and a limestone bar. Refits can be made among the obsidian artifacts and show how proximal and lateral notching occurred sometime before the caching event. Within this cache, complete or partial refits are present in at least twelve with blade cores, suggesting nearby destruction of cores or in situ knapping of blade cores (see L. M. Johnson 2016, 313, table 7-10). Most of the obsidian artifacts were laterally notched exhausted blade cores forming an "E" shape. Others include proximal and distal notches on blade cores, bifacially or unifacially worked blade cores, and plunging overshots.

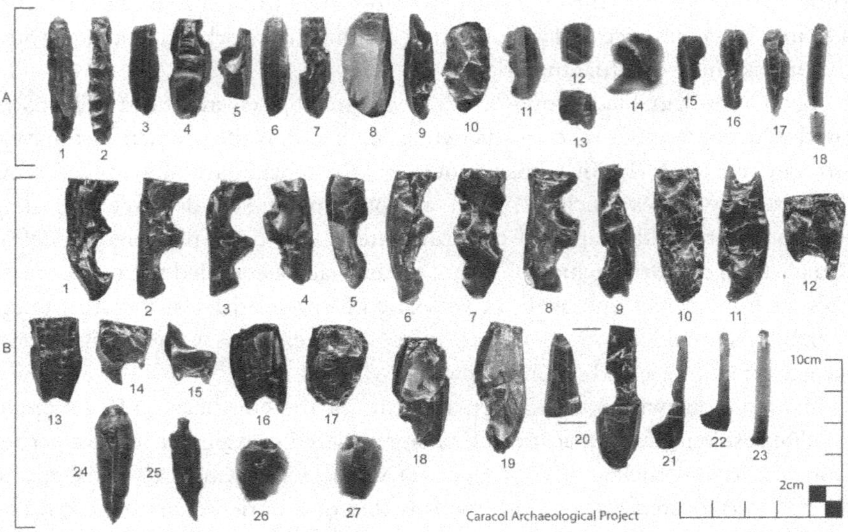

FIGURE 10.2. *(a) SDC117B-6, blade core: 1–13, 15–17; rejuvenation debitage: 14; final-series blade: 18. (b) SDC177D-9, blade core: 1–18, with refit: 18–20; plunging overshot: 21–22; percussion debitage: 26–27; initial-series blade: 24–25; final-series blade: 23.*

During investigations of other residential groups located southeast of Caana, excavations at the "Zumba Group" revealed another exemplary assemblage of caches associated with an eastern mausoleum (A. Chase and D. Chase 2012). Two caches are of note (figure 10.3). The first, SDC189B-1, is comprised of two partial cache vessels and a partial finger bowl (a typical ritual ceramic type at Caracol in which human fingers were placed in small lip-to-lip ceramic dishes or bowls; see A. Chase 1994, 166; A. Chase and D. Chase 1994, 59; D. Chase and A. Chase 1998, 308), as well as seven obsidian artifacts, slate fragments, and chert (i.e., one piece of debitage and a core tool) and quartzite debitage. The obsidian from this context consisted of two laterally modified exhausted blade cores, a core fragment that was symmetrically bilaterally notched to form a serrated object (not unlike a stingray spine), and percussion-related debitage. The second, SDC189B-6, also had two cache vessels and included jadeite chips, marine shell fragments, and eleven obsidian artifacts. These obsidian artifacts included multiple retouched blade cores (e.g., bilaterally notched and longitudinally split) and medial core fragments that indicate the presence of overshot or partial plunging terminations likely due to the very last attempts at removing pressure blades. A complete exhausted blade core is also present along with a medial core fragment with both ends exhibiting multidirectional percussion flaking.

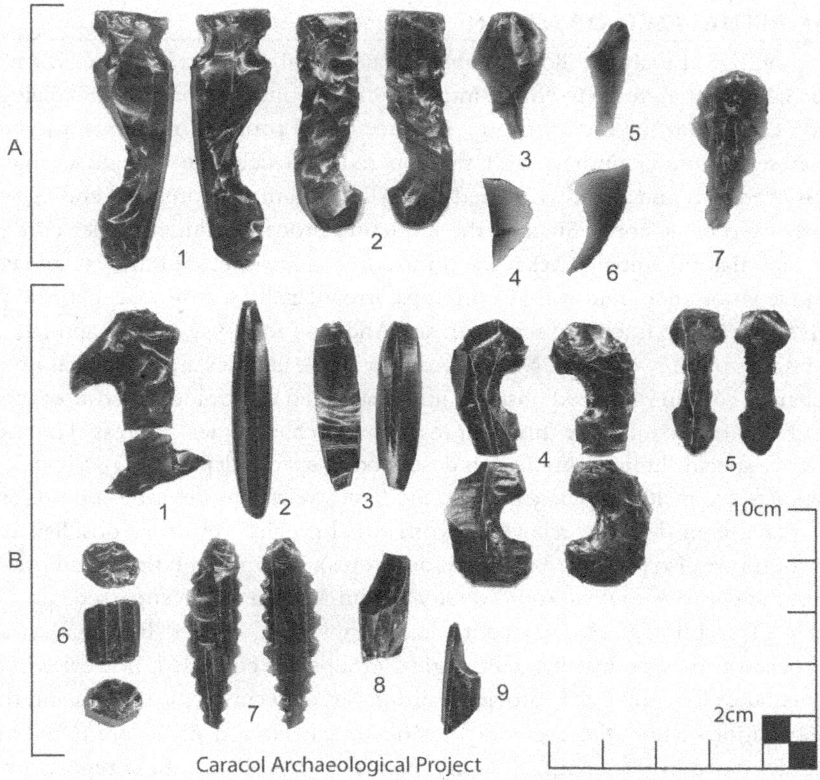

FIGURE 10.3. *(a) SDC189B-1, blade core: 1–2 (two views), 7; macrodebitage: 3, 5; flake: 4; rejuvenation debitage: 6. (b) SDC189B-6, blade core: 2, 6, 8, 9, 3 and 5 (two views); macrodebitage: 1 and 4; percussion blade: 7.*

These cache deposits do not offer an exhaustive representation of obsidian diversity from Caracol but introduce the need to focus on technological diversity and as well as on symbols. The examples given here illustrate blade cores that are both modified and unmodified; those that are modified are not necessarily iconic but are rather those technological stages associated with the rejuvenation and end stages of production. Some appear to have surface and lateral edge wear, indicating that they may have been curated, used, or otherwise ground to dull sharp edges. In the following section we present a suite of other caches that further reinforce the technological diversity among obsidian artifacts in many of Caracol's caches.

THE TECHNOLOGICAL CLASSIFICATIONS OF RITUALIZED OBSIDIAN

Technological analysis follows established analytical criteria (Hirth and Flenniken 2006; Hirth et al. 2006) to enable intrasite and regional comparison and dialogue (see Clark 2003 for a discussion). This section is roughly organized by reduction sequence, beginning with macropercussion debitage (i.e., macroflakes) removed to refine already roughed-out cylindrical macrocores and ending with exhausted blade cores. Steps in the reduction process include smaller core percussion flaking, initial-series blade removal, final-series blade removal, and core rejuvenation (e.g., removal of core tops, irregular distal core margins, and possible lateral step fractures; see Hirth and Andrews 2002, 3–4, figs. 1.1 and 1.2; and Johnson 2016, 145–150, figs. 5.1–5.6). Additionally, examples are given that do not separate certain ritualized obsidian into assembled technological groups; rather, artifacts are assembled within their respective archaeological context. Therefore, while a specific lithic technology is described by special deposit (SD), all obsidian artifacts are included. For example, the first few caches describe the presence of percussion debitage recovered from ritual caches; yet other obsidian technologies are also present (e.g., blades and cores). This presentation is intentional, especially as the context and diversity within a given locale supersede an overview of technological types contained within ritual caches. Information and association may be lost if technological groups are compiled; however, Lucas Martindale Johnson (2016) provides summaries of technological types and their distributions within the ancient city. Most of the cache deposits presented here are discussed in their entirety within annual archaeological field reports (available at www.caracol.org).

Core Shaping

Macrodebitage is typically larger and thicker than other percussion debitage, may retain portions of cortex, and usually exhibits multidirectional negative dorsal flake scars (see Hirth and Flenniken 2006; L. M. Johnson 2016, 159–161). Cache SDC204B-1 (figure 10.4), dated to the Late Classic 2 period and from a modest residential group comprised of about seven structures, includes macropercussion debitage and a single blade fragment arranged within a ceramic vessel (A. Chase and D. Chase 2014b, 203–204, figs. 148 and 149). A total of forty-eight pieces of macrodebitage have been recovered from twenty-two caches (table 10.1). Some of these are unworked ($n = 24$), while the others have unilateral or bilateral notching and could have functioned as some type of tool (Johnson 2016, 171, fig. 5.14). Deposit SDC177B-7 contained bilaterally notched macroflakes (A. Chase and D. Chase 2007).

Small percussion debitage that is generally 1.5 to 2.0 cm wide and only slightly longer is also associated with further refining the circular-plan form shape of

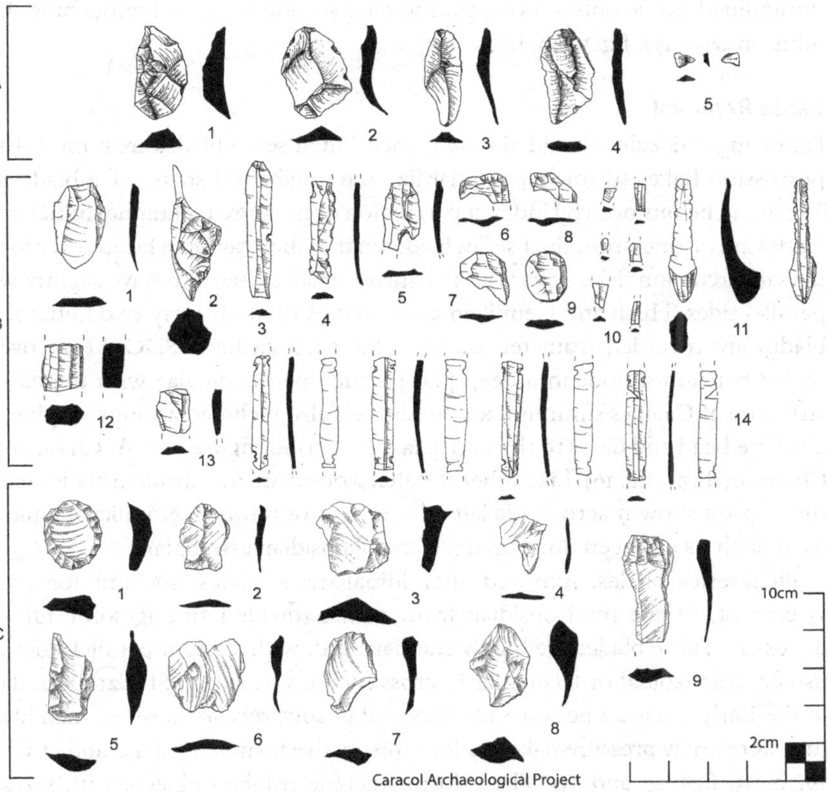

FIGURE 10.4. *(a) SDC204B-1, macrodebitage: 1–4; final-series blade: 5. (b) SDC1K-11, blade core: 2, 9, 12; macrodebitage: 1, 13; percussion debitage: 5, 6–8; plunging overshot: 11; initial-series blade: 4; final-series blade: 3, 10 (n = 6), 14 (n = 4). (c) SDC184B-4, macrodebitage: 6, 8–9; rejuvenation debitage: 1–5, 7.*

obsidian macrocores. These flakes exhibit unidirectional dorsal scarring and often have wide and flat multifaceted platforms that are "U" shaped in cross section (L. M. Johnson 2016, 167, fig. 5.12). Ventral eraillure scars are also common due to the force of impact. Two deposits that have cached small percussion debitage (figure 10.4) include SDC1K-11, recovered during investigations at the summit of Caana (A. Chase and D. Chase 2017b, 59, fig. 23), and SDC184B-4, recovered from a shrine room built into the western face of a buried eastern structure (A. Chase and D. Chase 2010, figs. 58–63). This cache dates to the Late Classic period and was directly above another special deposit, SDC184B-6, a crypt burial. Only 12 such flakes of small percussion debitage are known from all caches and are unequally divided among four caches. These flakes are usually unmodified, while only a few appear to have secondary lateral retouching (L. M. Johnson 2016, 171, fig. 5.14h–i).

Blade Removal

Following an idealized reduction sequence, initial-series blades are removed after percussion flakes to further standardize the polyhedral shape of a blade core. These can be removed by indirect percussion or pressure technique. Initial-series blades are distinct from final-series blades in that they can have faceted platforms; dorsal percussion flake scars with irregular dorsal arrises thus have slightly wavy parallel sides (Hirth and Flenniken 2006, 309). A total of thirty-two initial-series blades are recorded from ten caches. One such cache is SDC1K-6 recovered sealed beneath a floor in a deep plaza trench perpendicular with the eastern structure at Caana's summit B20; this deposit also included a stingray spine and a jadeite bead and dates to the Late Classic 2 period (figure 10.5; A. Chase and D. Chase 2017b, 52, fig. 16). Like other ritualized obsidian, the obsidian blades within this deposit show macroscopic lateral use-wear or retouch. No microscopic use-wear analysis has been done on the Caracol obsidian assemblage.

Final-series blades, removed after initial-series blades, account for 217, or 35 percent, of the total obsidian from caches (divided among some fifty-five deposits). These blades are highly standardized, with straight-parallel sides and usually trapezoidal or triangular in cross sections. A cache, SDC210B-10, dated to the Early Classic 2 period and excavated in 2017, recovered seven such blades that were finely pressure flaked to form distinctive forms (A. Chase and D. Chase 2017b, 83, figs. 47 and 53). These forms include trilobed objects with serrated edges, serpent-like forms with serrated edges, and two possible perforators (i.e., bloodletters) with anthropomorphic bases that are reminiscent of Charlie Chaplin figurines with outturned feet (see figure 10.5; Johnson 2018; Moholy-Nagy 2008). A range of artifacts that fit into the final-series blade technological type have an association with ritual caches. These include complete and partial unused blades, laterally notched blades resembling serpents or stingray spines, drilled blades, minor overshot pressure blades, and two blades exhibiting nacelle flakes (see figure 10.2; Hirth and Flenniken 2006, 309–310, photo A2; see also L. M. Johnson 2016, 186, fig. 5.19).

Rejuvenation

Generally, following extensive removal of final-series blades, cores can become plummet shaped, having smaller platform area or a restricted proximal portion and a wider distal diameter. Thus, core tops were removed, or cores may be otherwise sectioned to create a larger, wider-diameter pressure platform surface. Core tops, core sections, and platform preparation flakes may all result from a rejuvenation process (see Hirth et al. 2006, 72, fig. 3.8). Core tops may be faceted, ground,

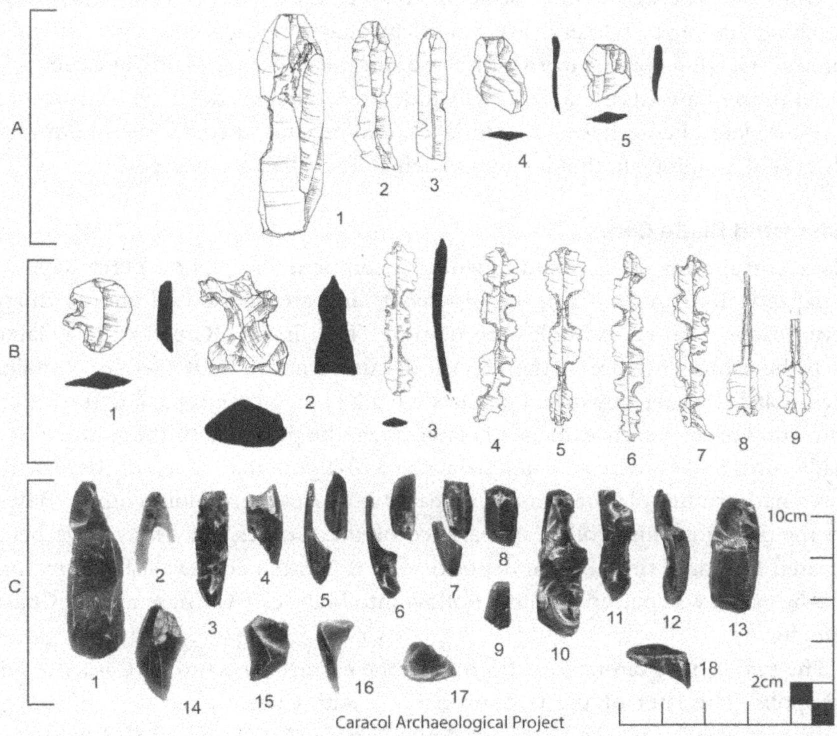

FIGURE 10.5. *(a) SDC1K-6, macrodebitage: 1; percussion debitage: 4–5; initial-series blade: 2; final-series blade: 3. (b) SDC210B-10, blade core: 2; macrodebitage: 2; final-series blade: 3–9. (c) SDC4E-9, blade core: 1, 3, 4, 5–13 (note refits in 5–7); macrodebitage: 14; percussion debitage: 2, 16; rejuvenation debitage: 17–18.*

striated, or cortical (Hirth et al. 2006, 73, fig. 3.1). Distal or lateral margins may also begin to have asymmetrical or irregular dimensions and may be removed to regularize a core's shape for productive pressure blade removal. Lateral rejuvenation flakes may also be removed, if step or hinge fractures occurred due to mistakes in the angle or strength in force. These types of debitage are lumped here but are described in further detail elsewhere (L. M. Johnson 2016). It is important to note that some core rejuvenation debitage as described here may reflect intentional acts to simply split exhausted cores (i.e., destroying cores). Some of the refits present in a few of Caracol's caches show that this did occur. In either case, however, the overall technological description (e.g., core sections) still applies, even if the process was not to rejuvenate cores per se.

Rejuvenation debitage accounts for 54 pieces (core sections n=16, core tops n = 7, distal orientation n = 2, platform preparation n = 17, miscellaneous n =

12) from 29 caches. Like other obsidian artifacts, these can be worked to exhibit notching that can be bilateral, asymmetrical, or symmetrical and create irregular shapes. SDC4E-9, a cache from within the summit of Structure B19 at Caana and dated to the Late Classic 2 period, included rejuvenation debitage of a striated core-top, lateral core flakes, macroflakes, and exhausted cores (see figure 10.5). This cache is unique in that it contains refits of small exhausted cores.

Exhausted Blade Cores

Blade cores, both worked and unworked, comprise the highest percentage of ritualized obsidian ($n = 247$, or 39 percent) and are distributed among thirty-seven caches. Three examples are of note. The first (SDC184B-5), associated with the same structure as the deposit discussed above (SDC184B-4), contains macroblades, other unworked percussion debitage, 2 core tops, a core section, and 6 unworked nearly exhausted blade cores; the presence of these unworked blade cores makes this cache unique at Caracol (figure 10.6). Each of these blade cores had ground platforms, and at least two cores were bidirectional. Based on the abrasions and rolled appearance of these cores, they may have been curated for some time before deposition in this ritual context. This Terminal Classic cache was placed within a hollow into bedrock (A. Chase and D. Chase 2010, fig. 65).

The two other examples are both from the eastern structure in Caracol's NE Acropolis. The later of the two, probably a Late Classic 1 cache (SDC205B-5), consists mostly of worked exhausted blade cores (A. Chase and D. Chase 2015, fig. 52). This context included bifacial and notched cores as well as the only scorpion-shaped artifact recovered at Caracol. The scorpion's pinchers were formed by deeply notching the core's platform, while the tail's curved shape utilizes a small overshot at the core's distal end. Lateral sides were serrated to give the appearance of insect legs. The earlier of the two caches from this locus, SDC205B-9, dated to the Early Classic 1 period (A. Chase and D. Chase 2015, fig. 56) and included two "S"-shaped notched cores and an "E" form. These two forms are common but not a regular pattern with all ritualized blade cores. As SDC205B-5 demonstrates, some cores are notched while others are not, suggesting a clear intentionality.

Furthermore, in terms of formalized styles, at least three common Maya silhouette or outline forms are present at Caracol: the "E," "S," and uniface/bifaces. These modified blade-core styles are usually created by removing one or both ends and laterally notching margins of completely or near completely exhausted blade cores. These forms first appear during the Early Classic 1 period and continue through the Late Classic 2 period (table 10.2); they are generally restricted to the downtown area, being associated with elite residential ritual practice.

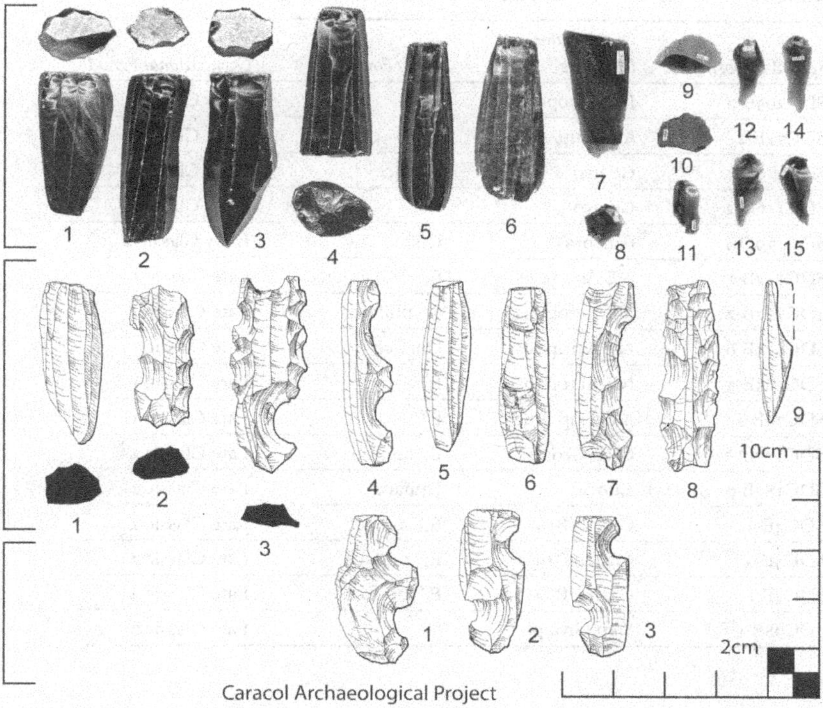

FIGURE 10.6. *(a) SDC184B-5, blade core: 1–7 (note 1–3 show proximal platforms and 4 shows distal end); percussion debitage: 13, 15; rejuvenation debitage: 8 (core section), 9–10 (core top), 11–12 (misc. debitage); initial-series blade: 14. (b) SDC205B-5, blade core: 1–7. (c) SDC205B-9, blade core: 1–3.*

Bifaces, Points, and Other Nonblade Objects

Other technological categories of obsidian artifacts are also present but are far fewer in number than those described so far. Bifaces or points are present but are not very common at Caracol. Only four examples are known, excluding those recovered during the 2010 field season from a cremation burial from the NE Acropolis (A. Chase and D. Chase 2011). Two obsidian bifaces made on larger macroblades were recovered from a ceramic box placed within the chest of a life-sized seated stucco figure in the A Group (SDC141C-2; D. Chase and A. Chase 2008, fig. 6; L. M. Johnson 2016, 226, fig. 5.32). The two other examples derived from Caana investigations (A. Chase and D. Chase 2017b, 57, fig. 21) and from smaller residential excavations (A. Chase and D. Chase 2008). Last, the earliest-known deposit with obsidian came from excavation within the Monterey Group. SDC118F-7, a Late Preclassic cache, contained, among other items, an

TABLE 10.2. List of "E," "S," and unifacial styles/forms by period

Special Deposit	*Group Name, Structure*	*Style/Form*	*Occupational Period*
SDC205B-9	NE Acropolis	"S"	Early Classic 1
SDC71E-2	A-Group, A2	"E"	Early Classic 1
SDC116C-1	Centro	Uniface	Early Classic 2
SDC116C-2	Centro	"S"	Early Classic 2
SDC179D-1	Culebras	Uniface	Early Classic 2
SDC117B-1	NE Acropolis	"E," "S," uniface	Late Classic 1
SDC117B-2	NE Acropolis	"S," uniface	Late Classic 1
SDC117B-6	NE Acropolis	Uniface	Late Classic 1
SDC205B-5	NE Acropolis	"E"	Late Classic 1
SDC70B-2	A-Group, A8	"E"	Late Classic 1
SDC177D-1	GRB Group	"E," uniface	Late Classic 2
SDC189B-6	Zumba	Uniface	Late Classic 2
SDC4E-3	Caana, B19	"E," uniface	Late Classic 2
SDC4E-9	Caana, B19	"E," uniface	Late Classic 2
SDC4F-1	Caana, B19	"E," uniface	Late Classic 2
SDC98B-1	Sam Group	"S"	Late Classic 2

obsidian Charlie Chaplin (A. Chase and D. Chase 2006, fig. 2). At a slightly later date in the same structure was another cache (SD 118F-6) associated with an obsidian pebble geochemically consistent with the San Luis source in Honduras (Aoyama et al. 1999; L. M. Johnson 2016, 225, fig. 5.30). Small black glassy pebbles are also reported from Tikal but were geochemically determined to be tektites (Hildebrand 2003).

TEMPORAL AND SPATIAL TRENDS OF RITUALIZED OBSIDIAN

Obsidian was ritualized throughout most represented time periods at Caracol (table 10.1). Broadly, the diversity and frequency in types increases, as does the frequency of use, into the Late Classic 2 period. This trend then changes, and the frequency diminishes in the Terminal Classic. The highest occurrence and diversity of types are during the Late Classic 1 period, but nearly the same phenomenon carries into the Late Classic 2 period. This increase in the ritualization of obsidian during the Classic period is contemporaneous with a broader distribution of the material and indicates that the epicentral elites set a precedent of some kind early in Caracol's history for including obsidian in ceremonies. The use of exhausted blade cores in these special deposits is at its highest during the Late Classic. The increase in typological diversity and frequency is followed by a subsequent decline in the Terminal Classic, in concert with an overall decline in caching practices in general (A. Chase and D. Chase 2009).

Further spatial and temporal analyses looked for potential differences in the practice of ritualizing obsidian across house groups of different statuses. A proxy for this status or wealth is house group size (L. M. Johnson 2016, 248–275). House groups were divided into four size classes: special use / pyramidal, large (≥6 structures), medium (4–5 structures), and small (≤3 structures). The temporal trends show that early in Caracol's occupation the ritualization of obsidian occurred in medium to special use / pyramidal groups, but that through time, as the city expanded, all types or sizes of house groups participated in domestic rituals using obsidian. Finally, this trend reversed to more elite, larger-sized house groups in Terminal Classic (table 10.1). This pattern supports a prior assertion by Arlen Chase and Diane Chase (2009) that varied status house groups obtained and used obsidian and a host of other standardized assemblages in domestic rituals. A. Chase and D. Chase (2009) refer to this phenomenon as "symbolic egalitarianism," which they argue was a bureaucratic or administrative strategy at Caracol used to create a shared identity through minimizing status and wealth differences, while at the same time increasing cooperation and collaboration. These and other data also support prior assertions that markets at the city were open systems that made nonlocal goods available to all status groups (A. Chase and D. Chase 2009; D. Chase and A. Chase 2014, 2017).

While some types or stylistic forms are few and spatially restricted to downtown Caracol, in general the ritualization of obsidian occurred everywhere throughout the sampled residential settlement and crosscuts at least four of the modeled districts (A. S. Z. Chase 2016). As stated earlier, the spatial distribution of ritualized obsidian is largely restricted to the site's epicenter during the Early Classic and is at its most spatially extensive during the Late Classic period.

DISCUSSION AND CONCLUSION: TOWARD THE RITUALIZATION AND MEANING OF STONE

Obsidian in any form was ritualized at Caracol; however, particular technological types and some stylistic forms of ritualized obsidian can be strongly correlated with contextual analysis, thus having the potential to inform us as to why obsidian was included in the materialization of rituals. Observations at Caracol using cached obsidian shows that a focus on "eccentricness" of the site's ritualized obsidian confounds the material processes of producing domestic ceremonies, especially because the "E," "S," and unifacial styles are only minimally represented—in contrast to a broad assortment of technological types largely associated with the early stages of production, rejuvenation processes, and the end stage of blade production (i.e., exhausted blade cores that were unworked, notched, or otherwise destroyed). The ritualization of these transformative stages may signal a heightened awareness or recognition of the importance of physical and social states. Because obsidian caches are generally associated with eastern mausolea, they may serve in some way to mark transitory stages of life and death.

Additionally, ritualized obsidian was not restricted to elites conducting more eventful ceremonies but rather was accessible to inhabitants in modest-sized house groups who lived a considerable distance from the downtown area and who carried out their own domestic ceremonies. Thus, cached obsidian may be linked to rituals that can serve to acknowledge obsidian's important role in daily life. Because it was so critical to a way of life, obsidian took on symbolic meaning as well. Yet, all things have multiple meanings and are not restricted to the symbolic sphere alone. In other words, the indexical association with broader economic and domestic practice recognizes the importance of obsidian during rituals. This perspective permits us to move away from an overtly symbolic approach in favor of a more balanced understanding that situates ritualized obsidian in terms more closely aligned to those agents caching the materials. At a minimum, it helps to further understand how a kind of identity and industry is shared and its importance materialized—through a collective notion of the order and place of things in space and time.

Finally, looking at the variation in artifact types and stylistic forms from the Caracol caches, it is evident that there is a greater emphasis on type over form. These types correspond with the core shaping, core rejuvenation, and the end stages of blade production. Perhaps the Maya recognized life cycles of materials as well as those of living individuals (Hruby and Rich 2014, 183). The various forms (i.e., the "S" and "E" form) are stylistic or symbolic elaborations on a broader theme of ritualizing exhausted blade cores. This association and the economic significance of obsidian suggest that specific ritualized obsidian, as seen from the vantage point of Caracol, emphasized the recognition of an essential quotidian industry and a social emphasis of physical transitory states of production from its very beginning to its "end-of-life" stage. The importance of obsidian was celebrated with stylized forms, such as the "S," "E," and, in a few cases, a scorpion, snake, stingray spine, or stylized perforator. But in making many of these forms, they are simultaneously disabling the lives of obsidian blade cores as the materials moved along their varied itineraries (L. M. Johnson 2016). These economic and sociophysical transitory associations, concomitant with cyclical calendric events (A. Chase and D. Chase 2013), may have been further "bundled" (Keene 2005), recognizing that obsidian always had multiple meanings.

Acknowledgments. This work was possible with the support and guidance from Steven Brandt, Augusto Oyuela-Caycedo, and Susan Milbrath from the University of Florida and Arlen and Diane Chase of the University of Houston.

Artifacts were loaned for this research with the help and permissions of Melissa Badillo, Silvia Batty, John Morris, and Jamie Awe of the National Institute of Culture and History (NICH) in Belize. A special thanks is necessary to Lisa Johnson, who read earlier versions of this chapter and is always available for supportive discussions. Funds for this research were provided through the National Science Foundation (#1556260) and a travel grant from the University of Florida Graduate School. Finally, this research is possible through the hundreds of students and local Belizeans who worked to carefully excavate and document the materials and contexts presented herein and elsewhere.

Abbreviations. SDC210B-10, Special Deposit Caracol 210 Suboperation B, the 10th special deposit in Suboperation 210.

REFERENCES

Agurcia Fasquelle, Ricardo, Payson Sheets, and Karl A. Taube. 2016. *Protecting Sacred Space: Rosalila's Eccentric Chert Cache at Copan and Eccentrics Among the Classic Maya.* San Francisco: Precolumbian Mesoweb Press.

Aoyama, Kazuo, Toshiharo Tashiro, and Michael D. Glascock. 1999. "A Pre-Columbian Obsidian Source in San Luis, Honduras: Implications for the Relationship Between Late Classic Maya Political Boundaries and the Boundaries of Obsidian Exchange Networks." *Ancient Mesoamerica* 10(2):237–249.

Appadurai, Arjun. 1986. "Introduction: Commodities and the Politics of Value." In *The Social Life of Things: Commodities in Cultural Perspective*, edited by A. Appadurai, 3–63. Cambridge: Cambridge Press.

Bell, Catherine. 1992. *Ritual Theory and Ritual Practice*. New York: Oxford University Press.

Bradley, Richard. 2003. "A Life Less Ordinary: The Ritualization of the Domestic Sphere in Later Prehistoric Europe." *Cambridge Archaeological Journal* 13(1):5–23.

Chase, Adrian S. Z. 2016. "Districting and Urban Services at Caracol, Belize: Intrasite Boundaries in an Evolving Maya Cityscape." *Research Reports in Belizean Archaeology* 13:15–28.

Chase, Adrian S. Z. 2017. "Residential Inequality Among the Ancient Maya: Operationalizing Household Architectural Volume at Caracol, Belize." *Research Reports in Belizean Archaeology* 14:31–39.

Chase, Arlen F. 1994. "A Contextual Approach to the Ceramics of Caracol, Belize." In *Studies in the Archaeology of Caracol, Belize*, Monograph 7, edited by Diane Chase and Arlen Chase, 157–182. San Francisco: Pre-Columbian Art Research Institute.

Chase, Arlen F., and Diane Z. Chase. 1987. *Investigations at the Classic Maya City of Caracol, Belize: 1985–1987*. Monograph 3. San Francisco: Pre-Columbian Art Research Institute.

Chase, Arlen F., and Diane Z. Chase. 1992. "Mesoamerican Elites: Assumptions, Definitions, and Models." In *Mesoamerican Elites: An Archaeological Assessment*, edited by Diane Chase and Arlen Chase, 3–17. Norman: University of Oklahoma Press.

Chase, Arlen F., and Diane Z. Chase. 1994. "Maya Veneration of the Dead at Caracol, Belize." In *Seventh Palenque Round Table, 1989*, edited by Merle Robertson and Virginia Fields, 55–62. San Francisco: Pre-Columbian Art Research Institute.

Chase, Arlen F., and Diane Z. Chase. 1995. "External Impetus, Internal Synthesis, and Standardization: E Group Assemblages and the Crystallization of Classic Maya Society in the Southern Lowlands." In *The Emergence of Lowland Maya Civilization: The Transition from the Preclassic to Early Classic, Acta Mesoamericana*. Vol. 8, edited by Nikolai Grube, 87–101. Berlin, Germany: Markt Schwaben Verlag A.

Chase, Arlen F., and Diane Z. Chase. 2006. "Before the Boom: Caracol's Preclassic Era." *Research Reports in Belizean Archaeology* 3:41–67.

Chase, Arlen F., and Diane Z. Chase. 2007. *Late Classic Ritual Variation in a Maya Community: Continued Investigations of Structures in and near Caracol's Epicenter*. Report prepared for the Belize Institute of Archaeology, Belmopan.

Chase, Arlen F., and Diane Z. Chase. 2008. *Household Composition and Ritual Patterning. Continued Investigation of Residential Groups near Caracol's Epicenter: 2008 Field Report of the Caracol Archaeological Project*. Report prepared for the Belize Institute of Archaeology, Belmopan.

Chase, Arlen F., and Diane Z. Chase. 2009. "Symbolic Egalitarianism and Homogenized Distributions in the Archaeological Record at Caracol, Belize: Method, Theory, and Complexity." *Research Reports in Belizean Archaeology* 6:15–24.

Chase, Arlen F., and Diane Z. Chase. 2010. "The Context of Ritual: Examining the Archaeological Record at Caracol, Belize." *Research Reports in Belizean Archaeology* 7:3–15.

Chase, Arlen F., and Diane Z. Chase. 2011. "Status and Power: Caracol, Teotihuacan, and the Early Classic Maya World." *Research Reports in Belizean Archaeology* 8:3–18.

Chase, Arlen F., and Diane Z. Chase. 2013. "Temporal Cycles in the Archaeology of Maya Residential Groups from Caracol, Belize." *Research Reports in Belizean Archaeology* 10:13–24.

Chase, Arlen F., and Diane Z. Chase. 2014a. "Ancient Maya Houses, Households, and Residential Groups at Caracol, Belize." *Research Reports in Belizean Archaeology* 11:43–53.

Chase, Arlen F., and Diane Z. Chase. 2014b. *Ancient Social Integration in a Maya Neighborhood: Investigations of Adjacent Residential Complexes near Caracol's Epicenter. Caracol Archaeological Project Investigations for 2014. A Continuation of the 2012 and 2013 Research Focus*. Report prepared for the Belize Institute of Archaeology, Belmopan.

Chase, Arlen F., and Diane Z. Chase. 2015. *Investigating Early Long-Distance Interaction in Caracol's Epicenter: Caracol Archaeological Project Investigations for 2015*. Report prepared for the Belize Institute of Archaeology, Belmopan.

Chase, Arlen F., and Diane Z. Chase. 2017a. "E Groups and the Rise of Complexity in the Southeastern Maya Lowlands." In *Maya E Groups: Calendars, Astronomy, and Urbanism in the Maya Lowlands*, edited by David Freidel, Arlen Chase, Anne Dowd, and J. Murdock, 31–71. Gainesville: University Press of Florida.

Chase, Arlen F., and Diane Z. Chase. 2017b. "Interpreting Ancient Maya Society through Residential Groups: Investigating Early Long-Distance Interaction in Caracol's Epicenter: Caracol Archaeological Project Investigations for 2017." Report prepared for the Belize Institute of Archaeology, Belmopan.

Chase, Arlen F., Diane Z. Chase, and Adrian S. Z. Chase. 2019. "A Decade Later: The Impact of Caracol LiDAR Surveys on Archaeological Interpretation in the Maya Area." Paper presented at XI Congreso Internacional de Mayistas, Chetumal, Mexico.

Chase, Arlen F., Diane Z. Chase, Richard Terry, Jacob M. Horlacher, and Adrian S. Z. Chase. 2015. "Markets Among the Ancient Maya: The Case of Caracol, Belize." In *The Ancient Maya Marketplace: The Archaeology of Transient Space*, edited by E. King, 226–250. Tucson: University of Arizona Press.

Chase, Diane Z. 1994. "Human Osteology, Pathology, and Demography as Represented in the Burials of Caracol, Belize." In *Studies in the Archaeology of Caracol, Belize, Monograph* 7, edited by Diane Chase and Arlen Chase, Pre-Columbian Art Research Institute, San Francisco, 123–138. San Francisco: Pre-Columbian Art Research Institute.

Chase, Diane Z., and Arlen F. Chase. 1992. "An Archaeological Assessment of Mesoamerican Elites." In *Mesoamerican Elites: An Archaeological Assessment*, edited by Diane Chase and Arlen Chase, 303–317. Norman: University of Oklahoma Press.

Chase, Diane Z., and Arlen F. Chase, eds. 1994. *Studies in the Archaeology of Caracol, Belize, Monograph* 7. San Francisco: Pre-Columbian Art Research Institute.

Chase, Diane Z., and Arlen F. Chase. 1998. "The Architectural Context of Caches, Burials, and Other Ritual Activities for the Classic Period Maya (as Reflected at Caracol, Belize)." In *Function and Meaning in Classic Maya Architecture*, edited by Stephen D. Houston, 299–332. Washington, DC: Dumbarton Oaks.

Chase, Diane Z., and Arlen F. Chase. 2004. "Archaeological Perspectives on Classic Maya Social Organization from Caracol, Belize." *Ancient Mesoamerica* 15:111–119.

Chase, Diane Z., and Arlen F. Chase. 2008. "Que no nos cuentan los jeroglíficos?: Arqueología e historia en Caracol, Belice." *Mayab* 20:-93–108.

Chase, Diane Z., and Arlen F. Chase. 2010. "Rituales mezclados: Analizando comportamientos públicos y privados en el registro arqueológico de Caracol." In *El ritual en el mundo maya: De lo privado a lo público, publicación* 9, edited by A. Ciudad Ruiz, M. J. Iglesias, and M. Sorroche, 107–128. Madrid: Sociedad de los Estudios Mayas, Grupo de Investigación Andalucía-América, CEPHIS-UNAM.

Chase, Diane Z., and Arlen F. Chase. 2011. "Ghosts Amid the Ruins: Analyzing Relationships Between the Living and the Dead Among the Ancient Maya at Caracol, Belize."

In *Living with the Dead: Mortuary Ritual in Mesoamerica*, edited by J. Fitzsimmons and I. Shimada, 78–101. Tucson: University of Arizona Press.

Chase, Diane Z., and Arlen F. Chase. 2014. "Ancient Maya Markets and the Economic Integration of Caracol, Belize." *Ancient Mesoamerica* 25:239–250.

Chase, Diane Z., and Arlen F. Chase. 2017. "Caracol, Belize, and Changing Perceptions of Ancient Maya Society." *Journal of Archaeological Research* 25:185–249.

Chase, Diane Z., and Arlen F. Chase. 2020. "The Ancient Maya Economic Landscape of Caracol, Belize." In *The Real Business of Ancient Maya Economies: From Farmers' Fields to Rulers' Realms*, edited by Marilyn Masson, David Freidel, and Arthur Demarest, 132–148. Gainesville: University Press of Florida.

Chase, Diane Z., and Arlen F. Chase. 2024. "The Materialization of Time in the Maya Archaeological Record: Examples from Caracol and Santa Rita Corozal, Belize." In *Crafting Time: Mythic History and Ritual Order in the Ancient Maya World*, edited by David Freidel, Arlen Chase, Anne Dowd, and J. Murdock, 363–386. Gainesville: University Press of Florida.

Clark, John E. 2003. "A Review of Twentieth-Century Mesoamerican Obsidian Studies." In *Mesoamerican Lithic Technology: Experimentation and Interpretation*, edited by K. Hirth, 15–54. Salt Lake City: University of Utah Press.

Clark, John E. 2007. In "Craft Specialization's Penumbra: Things, Persons, Action, Value, and Surplus." In *Rethinking Craft Specialization in Complex Societies: Archaeological Analysis of the Social Meaning of Production*, edited by Zachary Hruby and Rowan K. Flad, 20–36. Archaeological Papers of the American Anthropological Association 17. Arlington, VA: American Anthropological Association.

Darras, Veronique. 2014. "Ethnohistoric Evidence for Obsidian's Ritual and Symbolic Uses Among the Postclassic Tarascans." In *Obsidian Reflections: Symbolic Dimensions of Obsidian in Mesoamerica*, edited by Marc Levine and David Carballo, 45–74. Boulder: University Press of Colorado.

Fogelin, Lars. 2007. "The Archaeology of Religious Ritual." *Annual Review of Anthropology* 36:55–71.

García Cook, Ángel, and B. Leonor Merino Carrión. 2005. "Sobre tres elementos líticos con carácter ritual: Navajas de autosacrificio, navajas con cresta y cortadores o 'Tranchet.'" In *Reflexiones sobre la industria lítica*, edited by Leticia González Arratia and Lorena Mirambell, 299–323. Mexico City: INAH.

Graulich, Michel. 2005. "Autosacrifice in Ancient Mexico." *Estudios de Cultura Nahuatl* 36:301–329.

Haskell, David L. 2015. "Places to Go and Social Worlds to Continue: The Fractal Itinerary of Tarascan Obsidian Idols in Prehispanic Mexico." In *Things in Motion: Object Itineraries in Anthropological Practice*, edited by Rosemary A. Joyce and Susan D. Gillespie, 63–80. Santa Fe: SAR Press.

Hildebrand, Alan R. 2003. "Appendix H. Report on Tektites found at Tikal." In *Tikal Report No. 27, Part B, The Artifacts of Tikal: Utilitarian Artifacts and Unworked Material,* edited by H. Moholy-Nagy, 99–101. Philadelphia: University of Pennsylvania Museum Press.

Hirth, Kenneth G., and Bradford Andrews. 2002. "Pathways to Prismatic Blades: Sources of Variation in Mesoamerican Lithic Technology." In *Pathways to Prismatic Blades: A Study in Mesoamerican Obsidian Core-Blade Technology*, Monograph 45, edited by K. Hirth and B. Andrews, 1–14. Los Angles: Cotsen Institute of Archaeology Press at UCLA.

Hirth, Kenneth G., and J. Jeffrey Flenniken. 2006. "The Analytical Categories for Xochicalco Lithic Terminology, Appendix A." In *Obsidian Craft Production in Ancient Central Mexico: Archaeological Research at Xochicalco*, edited by K. Hirth, 301–314. Salt Lake City: University of Utah Press.

Hirth, Kenneth G., Bradford Andrews, and J. Jeffrey Flenniken. 2006. "A Technological Analysis of Xochicalco Obsidian Prismatic Blade Production." In *Obsidian Craft Production in Ancient Central Mexico: Archaeological Research at Xochicalco*, edited by K. Hirth, 63–95. Salt Lake City: University of Utah Press.

Hoskins, Janet. 1998. *Biographical Objects: How Things Tell the Stories of People's Lives*. New York: Routledge.

Hruby, Zachary X. 2007. "Ritualized Chipped-Stone Production at Piedras Negras, Guatemala." In *Rethinking Craft Specialization in Complex Societies: Archaeological Analysis of the Social Meaning of Production*, edited by Zachary Hruby and Rowan K. Flad, 68–87. Archaeological Papers of the American Anthropological Association 17. Arlington, VA: American Anthropological Association.

Hruby, Zachary X., and Michelle Rich. 2014. "Flint for the Dead: Ritual Deposition of Production Debitage from El Perú-Waka', Burial 39." In *Archaeology at El Perú-Waka': Ancient Maya Performances of Ritual, Memory, and Power*, edited by O. C. Navarro-Far and M. Rich, 167–183. Tucson: University of Arizona Press.

Johnson, Lisa M. 2018. "Little Helpers, Companions, and Ancestors: Miniature Figures in the Maya Lowland Region During the Late Preclassic and Classic Periods (AD 15–800)." *Latin American Antiquity* 29(3):616–622.

Johnson, Lucas Martindale. 2016. "Toward an Itinerary of Stone: An Investigation of Obsidian from Caracol, Belize." PhD diss., University of Florida, Gainesville.

Joyce, Rosemary A. 2007. "Figurines, Meaning, and Meaning-Making in Early Mesoamerica." In *Material Beginnings: A Global Prehistory of Figurative Representation*, edited by Colin Renfrew and Iain Morley, 107–116. Cambridge: McDonald Institute for Archaeological Research.

Joyce, Rosemary A., and Susan D. Gillespie. 2015. "Making Things Out of Objects That Move." In *Things in Motion: Object Itineraries in Anthropological Practice*, edited by Rosemary A. Joyce and Susan D. Gillespie, 3–20. Santa Fe: SAR Press.

Keane, Webb. 2005. "Signs Are Not the Garb of Meaning: On the Social Analysis of Material Things." In *Materiality: Politics, History, and Culture*, edited by Daniel Miller, 182–205. Durham, NC: Duke University Press.

Kopytoff, Igor. 1986. "The Cultural Biography of Things: Commoditization as Process." In *The Social Life of Things: Commodities in Cultural Perspective*, edited by Arjun Appadurai, 64–91. Cambridge: Cambridge University Press.

LaMotta, V. M., and M. B. Schiffer. 2001. "Behavioral Archaeology: Toward A New Synthesis." In *Archaeology Theory Today*, edited by Ian Hodder, 14–64. Malden, MA: Blackwell Publishers.

Levine, Marc N., and David M. Carballo. 2014. *Obsidian Reflections: Symbolic Dimensions of Obsidian in Mesoamerica*. Boulder: University Press of Colorado.

Meierhoff, James, Mark Golitko, and James D. Morris. 2012. "Obsidian Acquisition, Trade, and Regional Interaction at Chan." In *Chan an Ancient Maya Farming Community*, edited by C. Robin, 271–288. Gainesville: University of Florida Press.

Moholy-Nagy, Hattula. 1997. "Middens, Construction Fill, and Offerings: Evidence for the Organization of Classic Period Craft Production at Tikal, Guatemala." *Journal of Field Archaeology* 24:293–313.

Moholy-Nagy, Hattula. 2008. *The Artifacts of Tikal: Ornamental and Ceremonial Artifacts and Unworked Material, Tikal Report 27A*. Philadelphia: University of Pennsylvania Museum of Archaeology and Anthropology.

Moholy-Nagy, H., J. Meierhoff, M. Golitko, and C. Kestle. 2013. "An Analysis of pXRF Obsidian Source Attributions from Tikal, Guatemala." *Latin American Antiquity* 24(1):72–97.

Nazaroff, A. J., K. M. Prufer, and B. L. Drake. 2010. "Assessing the Applicability of Portable X-ray Fluorescence Spectrometry for Obsidian Provenance Research in the Maya Lowlands." *Journal of Archaeological Science* 37(4):885–895.

Sellet, Frederic. 1993. "Chaîne Opératoire: The Concept and Its Application." *Lithic Technology* 18(1/2):106–112.

Stemp, James W., and Jaime J. Awe. 2014. "Ritual Use of Obsidian from Maya Caves in Belize: A Functional and Symbolic Analysis." In *Obsidian Reflections: Symbolic Dimensions of Obsidian in Mesoamerica*, edited by M. Levine and D. Carballo, 223–254. Boulder: University Press of Colorado.

Swenson, Edward. 2015. "The Archaeology of Ritual." *Annual Review of Anthropology* 44:-329–345.

Trachman, Rissa M. 2002. "Early Classic Obsidian Core-Blade Production: An Example from the Site of Dos Hombres, Belize." In *Pathways to Prismatic Blades: A Study in Mesoamerican Obsidian Core-Blade Technology*, Monograph 45, edited by K. Hirth and B. Andrews, 105–120. Los Angeles: Cotsen Institute of Archaeology Press at UCLA.

Trachman, Rissa M., and Gene L. Titmus. 2003. "Pecked and Scored Initiations: Earl Classic Core-Blade Production in the Central Maya Lowlands." In *Mesoamerican Lithic Technology: Experimentation and Interpretation*, edited by K. Hirth, 108–119. Salt Lake City: University of Utah Press.

11

Obsidian Pressure-Blade Caches from the 8N-11 Structure 69C at Las Sepulturas, Honduras

JORGE RAMOS, ZACHARY HRUBY, AND XINWEI LI

The Institute of Archaeology of the Chinese Social Sciences Academy along with the Honduran Institute of Anthropology and History in 2015 launched an archaeological project at Group 8N-11, the northern portion of the area known as Las Sepulturas. Among the many new discoveries in the group were a series of architectural and plaza caches that contained obsidian. While chert and flint caches at Copan seem to focus on the Maya World's most complex effigy flints (see chapters 4, 5, and 6), obsidian caches feature mostly unmodified or lightly modified percussion or pressure blades. Although this pattern can be found at some sites, such as Palenque in the far Northwestern Lowlands, the most common form of cache obsidian is the eccentric or silhouette (see chapters 7 and 10), which has been found by the thousands throughout Northern Guatemala. Why or how Copan and Las Sepulturas chose not to participate in these traditions is the subject of a future study (but see Agurcia Fasquelle et al. 2016). This chapter focuses on how the ancient Copanecos expressed religious symbolism through blades, where they put them, how they made them, how they collected them, and perhaps most important, how they placed them within sacred spaces.

https://doi.org/10.5876/9781646427819.c011

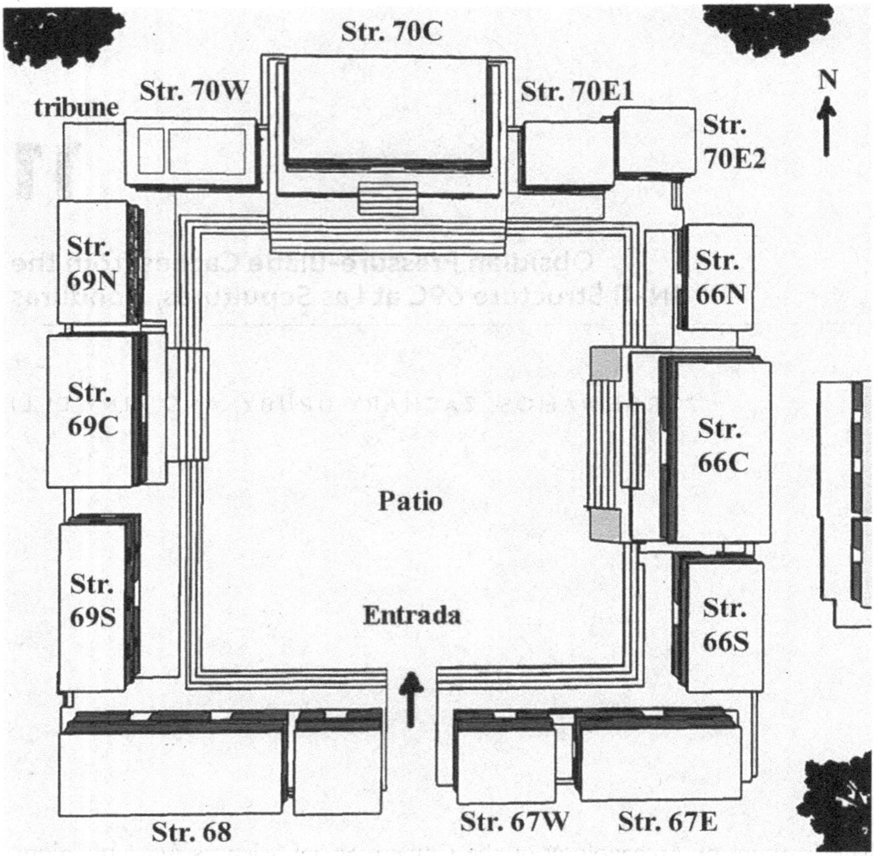

FIGURE 11.1. *Plan Map of Group 8N-11 with all structures labeled.*

Specifically, we focus on five pressure-blade caches (RC-2-101, RC-2-102, RC-2-105, RC-2-107, and RC-2-117) found on the centerline of Structure (Str.) 69C, the central and largest structure on the western platform of Group 8N-11 (figure 11.1). While they may seem rather diminutive compared to the massive Stromsvik Macroblade Cache from the Plaza of the Monuments in the Main Group (Hruby and Tokovinine n.d.), the Str. 69C caches gain much greater significance when considered in relation to the elite tomb that was placed before them and the possible stone workshop located just external to the 8N-11 Plaza. We argue that the Str. 69C pressure blades celebrate the production process by including blades taken as a series from one or two third-series (3s) blade cores (see Clark 1997 and Clark and Bryant 1997 for the typology). Production symbolism as part of the central theme of the caches is important for understanding one of the main economic and political functions of the group as a whole. For the sake

FIGURE 11.2. *Copan Valley with location of Group 8N-11 on the eastern side of urban core circled in white (upper left).*

of comparison, we also briefly mention two large double-notched blade caches (RC-1-219 and RC-1-224) found inside a terrace on the northwest corner of the group, in front of Structures 70W and 69N.

THE CONTEXT OF 8N-11

Group 8N-11 is a large residential compound located on the eastern end of the ancient road that once connected with the center of Copan. There were other important residential, economic living units along this *sacbe* (figure 11.2).

However, given: (1) this direct physical link with the epicenter of the city; (2) the fine quality and ideological symbolism of the architectural sculpture; (3) the size of the buildings; (4) the discovery of several tombs, including a very large crypt containing finely crafted jade and other prestige items; and (5) associated inscribed ceramics, it now seems likely that 8N-11 housed a member of the royal family. The sculpture of the group includes signatures of royal or dynastic connections to the Copan dynasty, such as the "skyband" bench discovered by David Webster and his team (1996), as well as the recently uncovered sculpture of Str. 70C and Str. 69N, which recalls the cult of Teotihuacan and the Copan dynastic founder, as well as the Maize God mythical cycle. All these elements along its location make 8N-11 of great importance to studies of cosmology, origins of urban complexity, political evolution, and religious symbolism during the Late Classic florescence of Copan's history.

Excavations at the group have revealed a detailed construction sequence, developed during a period of at least 300 years (ca. 550–850 CE), from late Acbi to Coner ceramic complexes. The last two phases of construction, the focus of our chapter, date to the end of the Late Classic period and the reign of the sixteenth king of Copan (763–820 CE). By this time the architectural layout had evolved into an enclosed compound that recalls the royal residential complex south of the Acropolis in the Principal Group. The formal southern entrance of Group 8N-11 bears a close resemblance to the eastern patio group of the Acropolis. Five superstructures in 8N-11 were decorated with mosaic sculptures—two on the east (Str. 66C and 66S), two on the west (Str. 69N and 69S), and one on the north (Str. 70C). Of these, Str. 69C and Str. 70W stand out from the others in terms of ritual obsidian deposits.

YUNQUES, PLATFORMS, AND STONEWORKING IN THE 8N-11 GROUP

While stoneworking was undoubtedly an important part of all household groups in the Copan pocket, the degree, kind, and style of stoneworking varied. We argue that 8N-11 was intensively involved in a wide array of stoneworking activities. The first pieces of evidence are what we have identified as *yunques*, or anvil stones, placed on the exteriors of Strs. 69C and 70C. The two identified *yunques* were made of massive river-rolled chert nodules, the cortex of which is a yellow-orange color. While cherts and jaspers are relatively common in the Copan River, stones of this size are not. In all likelihood these rocks would have had to be hauled by a group of people, from a few hundred meters along the river bar. Each boulder weighs between 200 and 500 lbs. and seems to have been positioned with the flattest surface facing up. They were then eased into position and leveled by wedging smaller cobblestones underneath.

We were able to identify these objects as anvil stones due to the wear patterns left on their superior surfaces. This wear consisted of extremely heavy hammer-marks and incipient cones in the surface of the chert—marks that could only be made with large hammerstones and cores. The *yunques* were almost certainly multifunctional in the sense that many different kinds of stone were worked on them in a variety of ways. Around the base of the *yunques* we collected dust, microdebitage, and small flakes of an array of materials, including obsidian, chert, volcanic tuff, rhyolite, orthoquartzite, quartz, and jade. We argue that this small debitage accumulated around the *yunques* because of the chert boulders used in stoneworking. Since the *yunques* flank the small platform, or "tribune," of Str. 70W, we propose that the platform, and this external portion of the group in general, was a stone working area.

The *yunques* were probably used for the earliest stages of manufacture. The hard rhyolite and quartzite stones found in the rivers were the raw materials Copanecos used to make their chopping tools. Since they did not have a highly developed system for making midsized to large bifaces, they relied on relatively crude choppers made from river cobbles to do the work. The cobbles would be first broken by placing them on the *yunque* and smashing them with another durable hammerstone. According to replication experiments, the working edge of the core could be flaked by placing the margin close to the surface of the *yunque*, and then struck by the hammer. The reason for such an elaborate flaking technique is that the cobbles are often quite round, which is a difficult format from which to start the reduction process, and also that the stone is so hard. In other words, the work would be far too damaging to the joints to carry out in the hand.

Even this heavy-handed labor, however, seems to have had some symbolic importance to the stoneworkers of 8N-11, since a cache of rhyolite and quartzite percussion flakes was placed on the back of the 70W Platform at the foot of its northern face in clear association to one of the burials of Stage I construction (Burial 14/1-3). The flakes were placed in a circular, stone-lined pit about 20 cm in diameter. Refit studies showed that some of the flakes came off in a sequence, but there was evidence of the reduction of at least five different cobbles in the cache.

Jade was worked in a very similar way in the Middle Motagua region of modern-day Guatemala. Massive jade *yunques* have been recovered there (Taube et al. 2011); they were used not only to chip the extremely tough jadeite but also to peck the jade bifaces into rounded celtiform axes. At 8N-11 we found no direct evidence of jade pecking, but we did identify a few pieces of jadeite microdebitage, suggesting early-stage jade working may have taken place on or around the *yunques*. Perhaps significantly, many jade items were found in the elite burials discovered in the group. Similar to Aguateca and Cancuen (Inomata 2001; Kovacevich 2011), crafters associated with royal household groups may have been responsible for the manufacture of their own inalienable goods.

Although obsidian tools may not have been produced on the *yunques*, it is possible that obsidian blades were made nearby and that obsidian tools were used to work other stones placed on or around the *yunques*. Shaping and carving architectural blocks were among the many possible functions of the anvil stones, and obsidian artifacts no doubt played a role in the finishing of the finest sculpture. A pure deposit of unused blade-production debitage was also found between 70W and 69N. The small dump of about 300 flakes, replete with consistent refits and the reduction of only about two cores, is one of just a few such dumps identified in the Copan Pocket (e.g., Mallory 1984; McNeil et al. 2022). Furthermore, one of the "steps" of the stairway up to this platform consists of two massive half-ton limestone slabs, which would have been useful in prismatic blade production. Actually, the step, which is more than 0.60 m high, may have been placed for the specific purpose of blade making. This limestone material does not occur in the geology of the immediate vicinity but on the northern hill at least 2 km from 8N-11. Therefore, this slab was moved over a large distance to be integrated into the architecture of Group 8N-11 to serve that function on the northwest corner in front of Str. 70W.

Finally, the 70W tribune also contained two massive obsidian blade caches (RC-1-219 and RC-1-224), which, although discussed briefly elsewhere in this chapter, may have marked the platform as an area of blade making. We believe that blade production was a major part of the socioeconomic identity of the people living in and around 8N-11 and that the blade caches discussed in this chapter tie directly into this ideology of blade production. Group 6N-1 is in the known area of Rastrojon, where Jorge Ramos along with Barbara and William Fash carried out research for several years (Fash et al. 2016; Ramos et al. 2011); it seems to have participated in some of the same socioeconomic activities, including obsidian blade production.

Although we do not have all the data, it is important to note that Str. 10L-29, the royal residential compound south of the Acropolis, which was adorned with the cross bundles, or *wiintenah* motif, was also flanked with a large limestone *yunque*. A chert example was found nearby next to Str. 10L-32 on the platform above.

THE CONTEXT OF THE CACHES

Strategic excavations at the compound have revealed a series of primary contexts, including tombs, burials, middens, and caches. Within the last category we registered a series of six obsidian blade deposits. Four of these were documented on the west side of the central patio, in alignment with the central axis of Str. 69C (figure 11.3; Op 14 Sub Op 2).

These caches consisted primarily of unused, or little used, third-series pressure-blades ("3s" in the Clark and Bryant 1997 typology), which seem to have been placed according to the cardinal directions. Two other deposits, which contained large quantities of double-notched pressure blades, were located on the northwest corner of the patio under the tribune of Str. 70W (Op 14 Sub Op 1). These deposits do not represent isolated activities but are rather part of larger ceremonial events associated with construction episodes—as part of the economic and ideological systems of the people who lived there. Not all of the blades have been analyzed, since we plan to carry out further residue analyses on some of them.

Before we move to the context of the caches, it is necessary to outline the sequence or evolution of Str. 69C. This structure can be traced back to Stage V, when the first formal structure with cobblestone foundations and adobe walls was built. This building was replaced by a new structure assigned to Sub-stage IVA, which was periodically renovated three times, during Stage IVB, IIIB, and Stage I. During these episodes the structure grew every time it was modified. As Stages IVA, IVB, and IIIB were built, the new stairway on the eastern side was enlarged to reach the higher buildings at the top of the substructure. By this

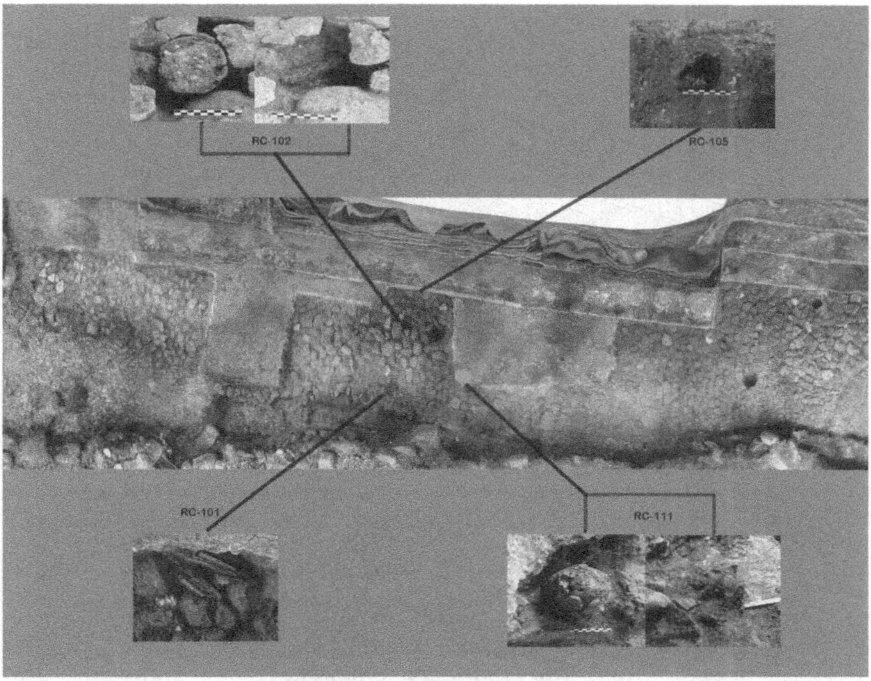

FIGURE 11.3. *Location of blade caches in front of Structure 69C.*

time of Stage IIIB the building consisted of a single large chamber, without any other architectural features. We argue that the scant furniture is evidence of a ceremonial function. During the last episode (Stage I), this room was divided into three individual rooms. The main access to the building was to the central chamber, which was connected to the southern room through a 1 m doorway and to the northern room through a 1.1 m doorway. In contrast to all the other buildings documented at the compound, each room in Sr. 69C was furnished with stucco-surfaced benches. However, it is the only building on the west side of the courtyard with no architectural sculpture. Vertical narrow windows decorate the main facade flanking the doorway, which were also present inside the rooms. At the base of the wall, in alignment with these niches, were small square windows that seem to complete an abstract geometric mask when viewed from the central patio.

The Str. 69C deposits (RC-2-101, RC-2-102, RC-2-105, RC-2-111, and RC-2-117) most probably relate to these renovation events and were perhaps meant to venerate the ancestors buried under the stairway and center of the building (Tombs 14/2-3, 5, 7, and 8). Based on the stratigraphy, we concluded that these tombs were deposited right before or during the Stage IVA, IVB, and IIIB construction

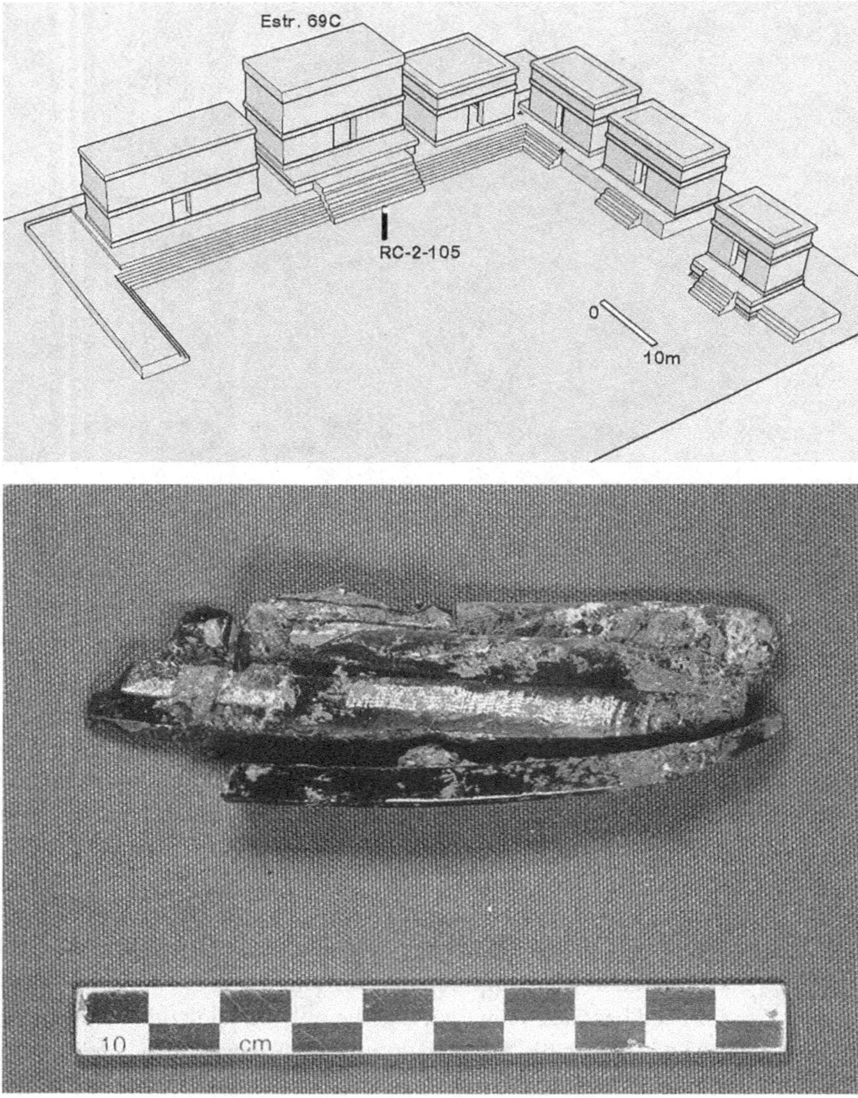

FIGURE 11.4. *(a) Location of blade cache, offering RC-2-105, in relation to Str. 69C; (b) offering RC-2-105.*

events. We discuss the caches temporally, from earliest to latest, and not according to the sequence of their field number. The earliest of these (RC-2-105) was laid out before the construction of Stage IIIB, and therefore Tomb 14/2-3. It consisted of three blades, probably bundled anciently, which were oriented west-east and placed in the fill under the first step of the central staircase (figures 11.4a

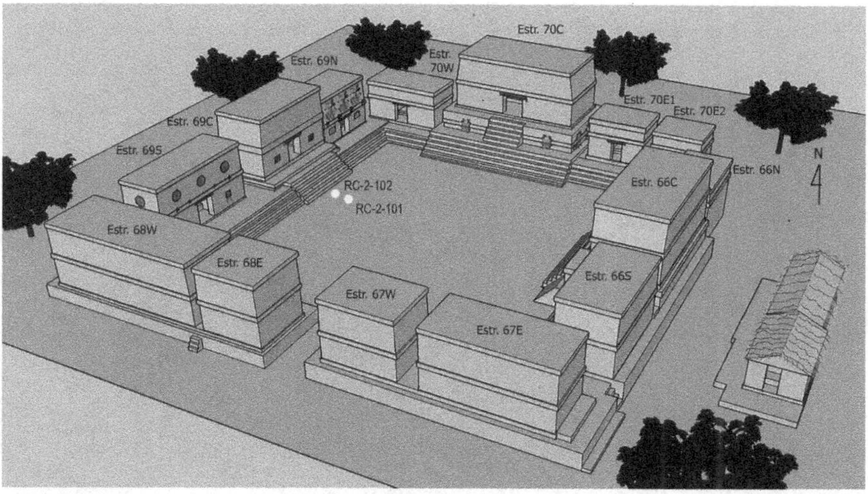

FIGURE 11.5. *Location of blade caches, offering RC-2-101 and RC-2-117 in relation to Str. 69C.*

and 11.4b). It bears mentioning that several of the offerings were deposited in the stucco floor of Sub-stage IVA in front of the stairway. In other words, they were directly related to Sub-stage IVB at the time of the interment of Tomb 14/2-7.

The second and third deposit (RC-2-101 and 117) were interred along with another deposit (RC-2-102) right before laying the floor for Stage I of the patio (figures 11.3 and 11.5). RC-2-101 consisted of two bundles of ten blades each, oriented roughly north-south (close to southwest-northeast) and placed in a circular hole 15 cm in diameter (figure 11.6). What is interesting about RC-101 is how precise the positioning of the blades was and their relationship to the core from which they were removed. In the western bundle, five blades were placed with the proximal ends pointing north, and the other five below had the proximal ends pointing south. In the eastern bundle, which was parallel to the western bundle, all of the blades were pointed with the proximal end pointing north. Refits and band patterns on the blades suggest that they were taken from two separate cores (figures 11.7a and b) and that the series in which they were removed determined the bundle in which they were placed.

This is direct evidence between the craft production of the blades, their symbolism, and the process of depositing the cache goods. They were not random blades collected *ad hoc* for the offering. The other interesting pattern is the focus on the long count units of 5, 10, 15, and 20, which suggests they may be related to katun numerology. RC-2-117 consisted of one complete blade oriented east-west and a fragment of another blade oriented north-south. Both artifacts were placed in front of the northeast corner of the central stairway.

FIGURE 11.6. *Blade offering RC-2-101 in situ.*

The Cache RC-2-102 was not focused on blades per se. A group of five blades were deposited underneath a ceramic vessel before laying out the stucco floor of Stage I (figure 11.8a). In this case, one blade was placed east-west with the proximal end pointing east, two more were laid on top of that blade but both facing the opposite direction, and then two more blades were placed on top with the proximal ends facing east again. On top of that a vessel containing a *Spondylus* shell was placed (figure 11.8b). In this cache, very specific directionality mattered in the placement of the blades. According to the length, color, and banding of the blades, it is likely that they came off of the same core. The last deposit (RC-2-111) relates to Stage IB, at a time when Stage I was still in use. The Maya made a hole 15 cm in diameter on the patio floor and placed a single blade oriented vertically roughly east-west (figure 11.9a). Significantly, they later covered this obsidian blade with a vessel facing down, thus capping it in three-dimensional space (figure 11.9b).

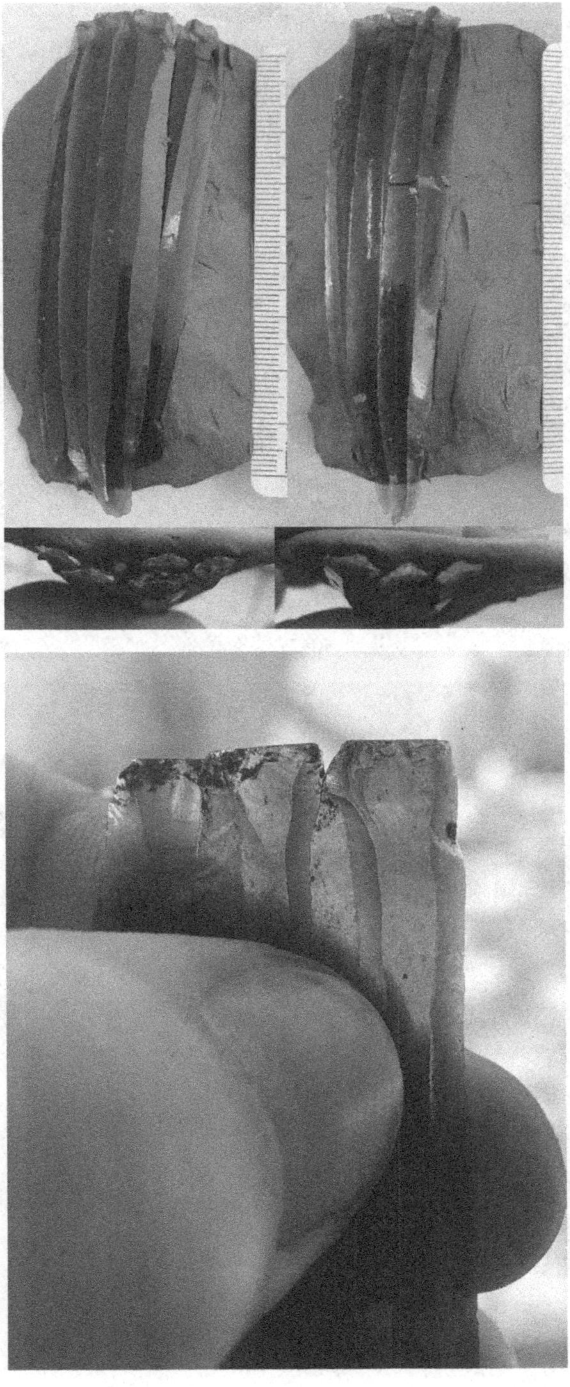

FIGURE 11.7. *Refitting process and example of blade pieces from two cores: (a) dorsal and proximal views of the two sets of refits from RC-2-101; (b) homogeneity of the material in one group of refits.*

FIGURE 11.8. *(a) Ceramic that covered the offering RC-2-102; (b) blades and cache vessel from offering RC-2-102 in situ.*

FIGURE 11.9. *(a) Blade deposit from offering RC-2-111 in situ after vessel was removed from around the blade; (b) ceramic vessel of offering RC-2-111 in situ above blade deposit.*

It may be significant that blades in other Copan caches are placed similarly according to the cardinal directions. The famous Stromsvik Macroblade Cache, which contained over 500 blades and flakes, was stacked according to the cardinal directions (Hruby and Tokovinine n.d.). The proximal and distal ends of the macroblades were directed to east, west, north, and south respectively. Even the basal blade cache (RC-1-224) in the nearby western platform of Str. 70W at 8N-11 was organized into a quadripartite configuration. Clearly directionality was an important factor in using obsidian in ritual deposits at Copan.

METHODS: TECHNOLOGY, USE-WEAR, AND REFIT STUDIES

We examined all the blades first according to technology and manufacture. The primary goal was to understand the stage of reduction in which the blades were removed from the core. The regularity of the blade, lack of percussion scars on the dorsal surface, and extent of platform preparation, as well as the overall size of the blade, allow us to identify when it came off in the process of blade-core reduction. The technological identification of the blades is derived from replication experiments, comparative samples, and technological work previously published by John Clark and Douglas Bryant (1997; Clark 1997). It will be recalled that some of the blades have not been entirely cleaned. Thus, in some ways this chapter does not represent a complete study of the caches. Some of them will stay unwashed indefinitely, perhaps for future researchers to study. We believe, however, that the cleaned sample, and the basic technological traits we were able to glean from the unwashed examples, provide enough information to characterize the cache as a whole.

Use-wear was examined visually, and also microscopically using an illuminated, 60× magnification hand-loupe. Of course, this kind of analysis does not allow us to identify extremely light forms of use-wear that may be visible using a metallurgical microscope of the kind used by Kazuo Aoyama (1999). These kinds of use could ostensibly be the most captivating and include the cutting of flesh, perhaps in bloodletting rituals, but they would exclude heavy uses, such as wood carving and stoneworking. Platforms were also examined with the loupe to better understand the kind of preparations that were used before blades were detached form the core. Platforms were heavily modified elsewhere in the lithic sample, so we were able to compare the cache platforms to others recovered from the group to see whether they fell into a typical range of practice for the blade makers of 8N-11.

Perhaps the most important method for looking at the process of production and the *chaîne opératoire* that went into the creation of these caches is refit study. Refitting is difficult and in the cased of unwashed artifacts impossible. Given the subtractive nature of lithic technology, one is able to match different flakes, blades, cores, and fragments to one another with 100 percent certainty, if they were detached in the same series. Also available to refit specialists in some cases is variability in the color, texture, and banding of colors within materials. These patterns are especially helpful in the case of highly translucent obsidians, since one is able to see through the material and match patterns more easily. One may also guess that two blades may have belonged to the same clast or core even if there is no direct refit. Finally, basic metric similarities between blades, especially in terms of length, betray the length of the core from which the blade was removed. Groups of blades with the same color, pattern, and blade length are more likely to have been detached from the same core.

RESULTS OF ANALYSIS

Cache RC-105

This cache was the first deposited along the centerline of the Str. 69C and probably related to the individual buried right before it (i.e., Tomb 14/2-7). It could also have marked the place for the next phase of construction of the building, Stage IIIB. As mentioned previously, the cache was deposited within the fill under the first stairway block, and therefore at the same time as the stairway or some point before. Only half of each of the interred blades remained in this cache, indicating that they were disturbed before its placement (figure 11.4b). The directionality of the blades was mostly intact, however, which allowed us to determine how the blades were likely placed in the original cache. We have not washed the blade fragments and consequently have not yet collected use-wear information, but we were able to record the technological type and recover basic metric data.

Cache RC-102

The blades of this feature were probably placed in a small hole, much like the later RC-101 cache, which was placed a few meters east of RC-105. The broken ends of the blades were pointing east. It appears that two blades were placed first with the distal ends pointing east, another with the distal end pointing west, and finally another with the proximal end pointing east again. There was another portion of a blade fragment, indicating that the overall number of the blades may have been five, but this has not been conclusively established.

According to the metric analysis of RC-105 and other caches found in 8N-11, some striking differences are revealed. Although the length of the blades is clearly shorter, since they are fragmentary, the average thickness of the blades is markedly greater than other caches. Table 11.1 compares all of the complete blades from the 8N-11 caches to give an idea of average core size and an idea of which blades they were selecting for different deposits. The blades in RC-105 are almost a millimeter thicker than the average blades in the other caches, which, while seemingly insignificant to the layperson, is actually impressive in

TABLE 11.1. Comparison of complete blades from 8N-11 caches that contained obsidian

3s Complete	*Number*	*Length*	*Width*	*Thickness*
RC-101	14	91.89	11.49	3.15
RC-102	5	97.76	10.61	3.13
RC-105	4	75.59	14.19	4.36
RC-111	1	100.20	14.38	3.40
RC-224	12	100.72	11.37	3.41
RC-219	45	104.55	14.49	3.44
Total/average	81	95.12	12.76	3.48

the grander scheme of systematic blade production, wherein blade thicknesses usually vary by fractions of a millimeter. It means that the Maya may have been selecting a more robust blade for these caches earlier in the history of the group than later. It could indicate a slightly different blade removal technique or platform preparation than that of the other cached blades in Str. 69C. Based on their basic morphology, future studies will likely reveal refits between the blades as is the case with RC-101.

Cache RC-101

In terms of number of blades and deposition technique, RC-101 was the most complex of the 69C blade caches, with a total count of twenty (figure 11.10). As mentioned in the "The Context of the Caches" section, the blades were positioned in two different groups (bundles) of ten, in a roughly north-south axis, and directionality varied within each group. Refit studies reveal that the production process was a significant aspect of the cache, since twelve of the twenty blades refit with one of the other blades (see figure 11.7). These refits were discovered without having access to all of the blades in the deposit; two of them, one from each ten-blade group, were not washed so it could be reserved for future residue analyses. Thus, it is possible we may find that all of the blades were in some way related. On the other hand, two of the blades seem to have different metrics than the others: Blades 1F from the first group and 2G from the second feature a different color of obsidian (figure 11.7b). Blade 1F (figure 11.10) was longer than the rest of the blades, and also wider, except for 2G. Both of these blades may have been taken off from a different core but from a similar stage of blade-core reduction.

All of the blades are easily identifiable as 3s, but they were detached somewhat late in the process, or, in other words, they were removed when the core was relatively small for Copan (see Hruby 2024 for coparative blade sizes from the Maya Lowlands). It is possible that these later blades were used as bloodletters, since at other sites late 3s pressure-blades are more commonly found in offering

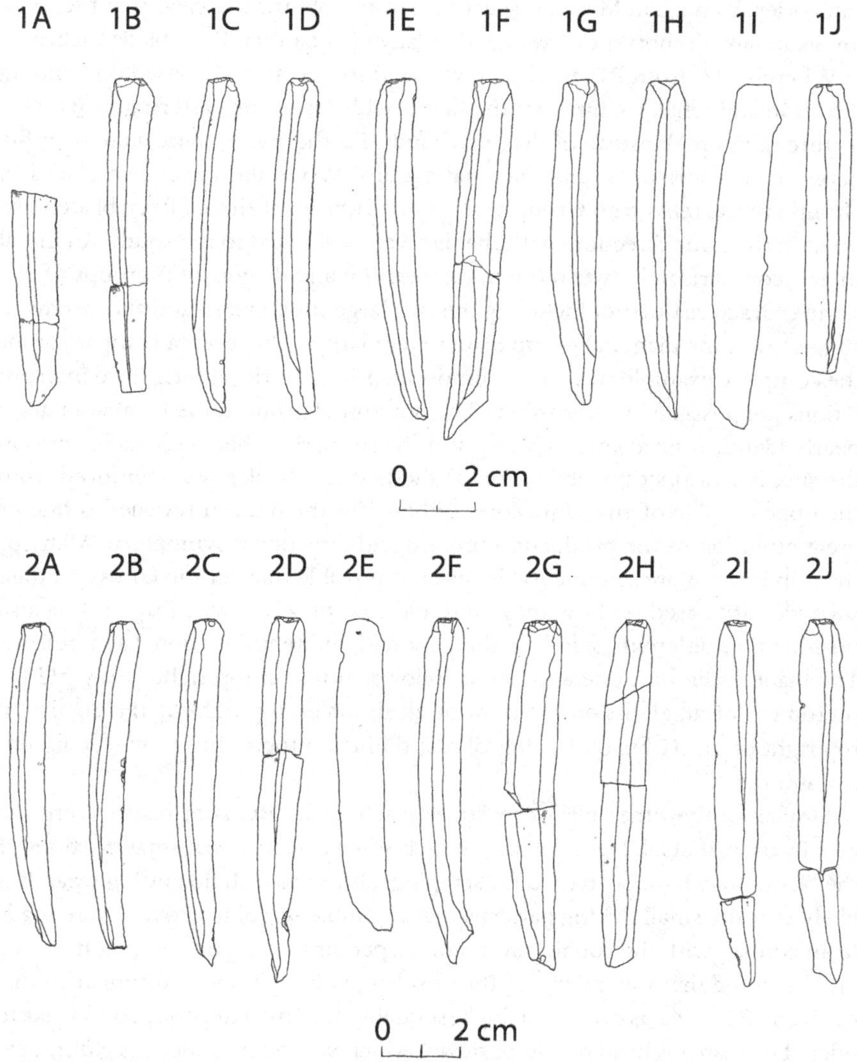

FIGURE 11.10. *Blades from offering RC-2-101: (a) Set 1; and (b) Set 2 (drawings by Lindsay Powell).*

bowls and burials with other bloodletting instruments. It may be significant that the blades in this cache and others of Str. 69C feature either a pointed distal end or a sharp feather termination, which would have been useful for bloodletting activities. The blades in this cache, in typical Copaneco form, are still very large compared to blades from elsewhere in the Maya Lowlands, which may confuse categories from a pan-Maya perspective. From table 11.1 it is clear that the RC-2-101 blades are all shorter on average than any of the other 8N-11 blade caches.

When blades from RC-2-101 are compared to one another based on banding direction and refits, we find that the first five blades of the eastern group (Set 1) feature horizontal instead of diagonal bands. Furthermore, there were no refits between the horizontal and diagonal groups. While the lower five blades in Group 1 reveal refits with Group 2 (see figure 11.7), all of these fifteen blades also point in the same direction, with the distal ends directed to the south. An initial metric comparison between the horizontal (14) and diagonal (4) groups of the same characteristics (not including the two larger outliers: 1F and 2G), seems to suggest that the diagonal group blades were larger. However, a comparison of the complete examples reveals an almost identical metric pattern. Two interpretations are possible in this case: (1) The horizontal group came from a separate nearly identical sized core, which given the similarities between blade cores in the site, is a distinct possibility; or (2) the first five blades were removed from the opposite side of the same core. Either way, the pattern revealed is one of great attention to the production process and directional symbolism. Many of the implications are imponderable, but it is possible that certain cores had their own identity based on how they were reduced or who owned them. It is also possible that different "sides" of the core had different directional symbolisms. The blades refit with one another as follows: 2I fits on top right of 2A; 2H fits on top left of 2J; 2F fits on top right of 2J; 1E fits on top right of 1H; 2D fits on top right of 2I; 1G fits on top left of 2C; 1H fits on top right of 2H (see figures 11.7 and 11.10).

Similar to the other blades in Str. 69 caches, the RC-2-101 blades were not heavily used, if at all. Nevertheless, there is some evidence that small portions of the blades may have been used for scraping, although with no cutting wear. It is likely that the small flaking patterns that resemble scraping activity were made from contact with the cobble pavement, or perhaps from one another. It should also be noted that the quality of the obsidian, although fine and similar to that of Cache RC-2-102, is not of the highest quality that the Ixtepeque source has to offer. The lustrous brown-gold obsidian, which was often traded long distances, was not sought out for the caches. Instead the blade size and type seem to have been of greater importance than the quality of the raw material.

Cache RC-2-102

The primary components of this cache are not obsidian blades but instead a ceramic vessel that contained an articulated *Spondylus* shell. Other caches at Copan feature *Spondylus* shells, such as the famous stone-vessel caches excavated by Seichi Nakamura (2003, 98; see also Taube 2010, 266), which contained large quantities of finely carved jade. *Spondylus* shells can represent aspects of femininity and are a common central element of belt assemblages of royal women, perhaps marking female genitalia. They are somehow related to the Maize God (Taube 2010, 266) and may reference the sea from which the sun rises. In caches *Spondylus* shells are often placed in juxtaposition to jade, much in the same way obsidian and flint objects are placed in opposition to one another in caches and in the iconography of Mesoamerica. Obsidian and flint are often placed in caches to reflect "the whole" universe combined: the eastern chert-bearing land of the Maya and the western obsidian lands of Central Mexico (Hruby 2007; Taube 1991). The opposition of jade and *Spondylus* may also reflect directional symbolism, with jade representing the east, where jade is found geologically, and *Spondylus* representing the west, where, symbolically at least, the largest *Spondylus* shells are collected on the west coast of Mexico. It is also possible, however that jade represents zenith and the heavens, and *Spondylus* nadir and the watery underworld. Thus, some caches represent a cosmic diagram of the Maya universe.

In RC-2-102 there is no flint or jade. Instead, a small ceramic vessel contained the *Spondylus* shell, and underneath were five delicately placed obsidian blades. In materials terms, then, the cache is associated with the concept of the west, the watery underworld, and nighttime symbolism. In the case of a cache excavated at El Peru-Waka', small pieces of obsidian debitage were placed at the base of a ceramic olla, while flint debitage was placed above. This separation of debitage "framed" the olla, which contained jade and *Spondylus* offerings. RC-102 features a similar arrangement, but, as is the case with other obsidian deposits at Copan, the entirety of the directionality was expressed only in obsidian blades, without the addition of flint or chert. Like the RC-2-101 cache, there appears to have been scant use-wear, but it is unclear whether those marks were created during deposition or excavation.

The blade orientations of RC-2-102 are similar to those of RC-2-105 in the sense that they are a smaller group of blades with alternating directionality within the same group and also were oriented roughly east to west. However, in this case the pattern appears to be slightly different: one blade was laid with the proximal end pointing east, two facing west on top of that, and then two more facing east. All of the blades probably came from the same core, but two came from one side and three came from the other. It is likely that A and E refit and come from one side of the core (because they tilt in the opposite direction), and C and D refit with the right portion of C, fitting on top of the left dorsal portion of D. Given the similarity in color between the obsidians, we believe that three blades were removed from one side of the core and two from the other, although it is possible that two came from one core and three from a separate core. In either case, less attention was paid to the sequence in which the blades were detached, as was the case with RC-2-101.

Cache RC-111

At first glance, this cache appears to be the simplest of the deposits, with a single vessel containing a single blade. Upon closer inspection, however, it is slightly more complex, because the undecorated vase was placed inverted over the blade. Furthermore, the blade was not simply laid on its side but found oriented roughly vertically within the chamber of the vessel. In this way the cache resembles other covered caches from the Maya world that contain chipped stone but without the lower vessel. It is possible that a lower vessel was made out of wood, or more likely a gourd. At this point we have not cleaned the blade and consequently we have no direct data on what it may have been used for.

One wonders how and why the obsidian blade was standing near vertically inside the vase. The iconography of heart sacrifice in the Maya area often depicts excised hearts and incense balls centrally placed in a cache vessel, censers, or bowls in ceremonial scenes (Taube 2009, 99, fig. 11). These balls and hearts sometimes display a sacrificial knife protruding vertically from the sacrifice as part of the offering. In Postclassic Central Mexico bloodletters are often shown protruding vertically from woven grass balls, like a Western pincushion. It is possible that RC-2-111 contained one of these objects associated with sacrifice: excised heart, incense ball, or grass ball. The blade itself (table 11.2), a 3s pressure blade, is large compared to the other blades from Str. 69C caches. Only residue analysis will be able to tell us more about its function.

Cache RC-117

Only two blades were included in this deposit. They are perhaps the least complex of the caches. They were deposited near the north corner of the stairway of the structure in a small hole of 12 cm diameter underneath the stucco floor of Stage I. Carrying the simplest of the arrangements, both pieces in this cache were roughly oriented west-east.

SYNTHESIS OF ANALYSES, CONTEXTUAL INFORMATION, AND STRATIGRAPHY

Instead of a collection of obsidian blade deposits, the caches of 8N-11-69C reveal an entire "language" of caching behavior not thoroughly discussed in Mesoamerica. The primary themes in these caches are directionality and centrality, as well as world creation and period completion communicated through the metaphors of craft production and, quite likely, blood sacrifice. What is clear is that in this particular structure, obsidian played a key role in the creation and re-creation, or rebuilding, of the central structure of the western side of the courtyard. By extension, the building and these caches likely reify or apotheosize the individual buried directly in alignment and below the caches. The caches were ceremonially deposited in two distinct episodes, during Stage III and Stage

TABLE 11.2. Comparison of blade metrics from offering RC-2-101 in color groupings, and individually in Set 2

Blade Color	*Count of Site*	*Average of Length*	*Average of Width*	*Average of Thickness*
Fine diagonal banding	14	89.83	11.36	3.15
Fine horizontal banding	4	81.06	11.51	3.20
Grand total	18	87.88	11.39	3.16
Complete	11	91.52	11.41	3.15
Complete	2	91.00	11.49	3.18

Blade Number	*Length*	*Width*	*Thickness*	*Weight*
2B	94.00	10.78	3.10	2.8
2C	98.93	10.69	3.04	3.2
2D	97.47	10.26	3.05	2.8
2E	99.21	10.65	3.22	3.5

I. Years earlier, during Stage IV, the burial chamber was introduced in the central axis of the stairway. The tomb is one of two earliest funerary deposits found at the site, which may possibly belong to the first ancestors of the group. Detailed results, however, will have to be presented in a different forum.

Even though the blades are central to these caches, our analyses reveal another important factor that was *not* materially present: the blade core. It appears that part of the symbolism of each blade and how it was ultimately deposited was derived from the sequence or portion of the core (or cores) from which they were detached. Zachary Hruby (2007) argues that blade cores may have themselves been cosmograms and that certain kinds of debitage may have had different symbolic associations. At some sites, like Piedras Negras and Tikal, there was a correlation between certain deities (i.e., silhouettes or incised obsidians) and debitage types. At Las Sepulturas one could conjecture that different blade cores may have had their own *identification*, which changed over the use-life of the core, similar to a lunar deity waning over time or the sun setting. Thus, the sequence of blade detachment and the portion of the core that was reduced may have themselves been ritually significant. It is a possibility that imported macrocores were considered to be "full" and that then, as they were reduced, they became diminished and hungry. Famished gods are hungry gods, and as the blades become smaller on an exhausted core they become more appropriate for bloodletting, to feed those starving nighttime deities.

It is also interesting that the blades are grouped in fives and tens, perhaps marking completions of the long count calendar, instead of reflecting normal cache groupings of 3, 4, 6, 7, 9, 13, 18, and 21, which are important numbers in Maya numerology and the structure of the universe. On Stela F, for example, the *lahun*, or completion of a ten-year period, is associated with a diminishing of the gods, and one wonders whether there is a similar symbolic load here (Hruby and Tokovinine n.d.). Since the base five and twenty numerology is related to the Long Count Calendar, it is possible that caches are related to or celebrate those kinds of period completion. Although the sacred number of Teotihuacan, eighteen, is not invoked here, there may be some relation to a Mexican system. It may be useful to test for *yauhtli* residue or other substances that are related to Central Mexican New Year ceremonies. In any case, the directionality so common in Petén lithic caches is clearly in evidence here, but using a completely different semiotic system to communicate these directional elements. No matter the final interpretation of these deposits, the caches index the blademaker as an indelible and inherent part of the caching ceremony at 8N-11—crafting and ritual intertwined. Crafting is a common theme in world creation myths, and the Maya may have associated blade production with some aspect of world creation, diminishing, and destruction.

FINAL THOUGHTS

This chapter only begins to introduce these incredible new lithic discoveries. In traditional archaeological terms, they would be described as "various blade deposits." In this study we delved deeper. Detailed technological and contextual analyses reveal more patterns than clearly exist on the surface. We argue that all Mesoamerican caches deserve this level of detailed analysis. These offerings, ritually placed, had a formal agenda: they aimed at reconstructing of the cosmos and a re-creation of creation, and undoubtedly, given the material, they were tied to blood sacrifice. We work from the bottom up to decipher the numerology, directionality, and technology involved in the multivalency of these ritual deposits. At Copan we see a different tradition from that of the eccentrics and eccentric caches found in the Petén, but nevertheless a tradition steeped in as much symbolism, complexity, and meaning as those from the Classic Maya Heartland. Here, only patience and perseverance can lead to greater understanding. Future residue and use-wear analyses can only deepen our knowledge of these sacred remains.

REFERENCES

Agurcia Fasquelle, Ricardo, Payton Sheets, and Karl Andreas Taube. 2016. *Protecting Sacred Space: Rosalila's Eccentric Chert Cache at Copan and Eccentrics Among the Classic Maya*. Monograph 2. San Francisco: Precolumbia Mesoweb Press.

Aoyama, Kazuo. 1999. *Ancient Maya State, Urbanism, Exchange, and Craft Specialization: Chipped Stone Evidence of the Copán Valley and the La Entrada Region, Honduras*. Memoirs in Latin American Archaeology 12. Pittsburgh, PA: University of Pittsburgh Press.

Clark, John E. 1997. "Prismatic Blademaking, Craftsmanship, and Production." *Ancient Mesoamerica* 8(1):137–159.

Clark, John E., and Douglas Donne Bryant. 1997. "A Technological Typology of Prismatic Blades and Debitage from Ojo de Agua, Chiapas, Mexico." *Ancient Mesoamerica* 8(1):111–136.

Fash, L. William, Barbara W. Fash, and Jorge H. Ramos. 2016. "New Approaches to Community Stewardship, Education, and Sustainable Conservation of Cultural Heritage at Rastrojon, Copan, Honduras." In *Finding Solutions for Protecting and Sharing Archaeological Heritage Resources*, edited by Anne P. Underhill and Lucy C. Salazar, 135–152. New York: Springer International Publishing.

Hruby, Zachary X. 2007. "Ritualized Chipped-Stone Production at Piedras Negras, Guatemala." In *Rethinking Craft Specialization in Complex Societies: Archaeological Analysis of the Social Meaning of Production*, edited by Zachary Hruby and Rowan K. Flad, 68–87. Archaeological Papers of the American Anthropological Association, Vol. 17. Arlington, VA: American Anthropological Association.

Hruby, Zachary. 2024. "Toward an Understanding of Obsidian Values Among the Ancient Maya Through a Comparative Approach." In *Realizing Value in Mesoamerica: The Dynamics of Desire and Demand in Ancient Economies, Scott Hutson and Charles Golden*, editors, pp. 139–162. Palgrave McMillan, Cham. Switzerland.

Hruby, Zachary, and Alexandre Tokovinine. N.d. "Exploring the Symbolic Aspects of the Stromsvik Macroblade Cache from the Grand Plaza of Copan, Honduras." In *The Rise, Collapse, and Rebirth of a State*, edited by Cameron McNeil and Alexandre Tokovinine. Gainesville: University of Florida Press.

Inomata, Takeshi. 2001. "The Power and Ideology of Artistic Creation: Elite Craft Specialists in Classic Maya Society." *Current Anthropology* 42(3):321–349.

Kovacevich, Brigitte. 2011. "The Organization of Jade Production at Cancuen, Guatemala." In *The Technology of Maya Civilization: Political Economy and Beyond in Lithic Studies*, edited by Zachary X. Hruby, Geoffrey E. Braswell, and Oswaldo Chinchilla Mazariegos, 149–161. London: Equinox Publishing.

McNeil, Cameron L., Edy Barrios, Zachary Hruby, Nathan Meissner, Alexandre Tokovinine, Mauricio Díaz García, and Walter Burgos. 2022. "Control, Collapse, and Resilience at Rio Amarillo in the Copan Valley, Honduras." *Ancient Mesoamerica* 33(1): 62–82.

Mallory, John K. 1984. "Late Classic Economic Specialization: Evidence from the Copan Obsidian Assemblage." PhD diss., Pennsylvania State University, State College.

McNeil, Cameron L., Edy Barrios, Zachary Hruby, Nathan Meissner, Alexandre Tokovinine, Mauricio Díaz García, and Walter Burgos. 2022. "Control, Collapse, and

Resilience at Rio Amarillo in the Copan Valley, Honduras." *Ancient Mesoamerica* 33(1): 62–82.

Nakamura, Seichi. 2003. "The Rise and Fall of the Copan Dynasty." In *Shipono Ocho Maya Bunmeiton (Dynasties of Mystery: Maya Exhibit Catalog)*, edited by Masahira Ono and Hisao Baba, 89–104. Tokyo: TBS.

Ramos, Jorge H., Barbara W. Fash, and William L. Fash. 2011. "The Portrait of a Royal Ancestor Emerges in the Copan Valley." In *Symbols*, Spring 2011, 4–11. Peabody Museum and Department of Anthropology, Harvard University.

Taube, Karl A. 1991. "Obsidian Polyhedral Cores and Prismatic Blades in the Writing and Art of Ancient Mexico." *Ancient Mesoamerica* 2(1):61–70.

Taube, Karl A. 2009. "The Womb of the World: The Cuauhxicalli and Other Offering Bowls of Ancient and Contemporary Mesoamerica." In *Maya Archaeology*, vol. 1, edited by Charles Golden, Stephen Houston, and Joel Skidmore, 86–106. San Francisco: Precolumbia Mesoweb Press.

Taube, Karl A. 2010. "Catalogue Entry 89." In *Fiery Pool: The Maya and the Mythic Sea*, edited by Daniel Finamore and Stephen D. Houston, 266–267. Salem, MA: Peabody Essex Museum.

Taube, Karl A., Zachary Hruby, and Luis Romero. 2011. "Ancient Maya Jade Workshops: Archaeological Reconnaissance in the Upper Río El Tambor, Guatemala." In *The Technology of Maya Civilization: Political Economy and Beyond in Lithic Studies*, edited by Zachary X. Hruby, Geoffrey E. Braswell, and Oswaldo Chinchilla Mazariegos, 141–148. London: Equinox Publishing.

Webster, David, Barbara Fash, Randolph Widmer, and Scott Zeleznik. 1998. "The Skyband Group: Investigations of a Classic Maya Elite Residential Complex at Copan, Honduras." *Journal of Field Archaeology* 25:319–343.

12

Obsidian for the Dead

The Technology and Patterning of Classic Maya Blade and Debitage in Tomb Contexts at El Peru-Waka'

DAVID RAFAEL McCORMICK ALCORTA, ZACHARY HRUBY, DAMIEN MARKEN, OLIVIA NAVARRO-FARR, MICHELLE RICH, AND KEITH EPPICH

This study investigates the deposition of obsidian excavated from elite tombs at El Peru-Waka' (hereafter Waka'), Guatemala. Although predominantly understood as mundane, obsidian blades and the debitage resulting from their manufacture may be considered "ceremonial lithic artifacts" or ritualized lithics based on their recovery contexts and symbolic usage (chapters 1 and 10). Here, we examine such material recovered above, around, and within burials and suggest that this practice, common at many ancient Maya cities, represents a ritually and ideologically charged action (Hall 1989; Hruby and Rich 2014; Johnson 2016; Moholy-Nagy 1997; Trachman 2002). We argue that specific kinds of blades and debitage were selected and deposited to convey different kinds of meanings associated with burial and termination rites. Although the meanings of these symbolic acts and a highly detailed technological analysis will be presented elsewhere, the data we present here constitute a foundation for those studies. The main goals of the present chapter are to describe the contexts of the tomb obsidians, to demonstrate that tomb obsidians were not homogenous, and to show that certain kinds of debitage may reflect different symbolisms and ritual behaviors associated with

https://doi.org/10.5876/9781646427819.c012

FIGURE 12.1. *Map of the Maya area with detail of Western Central Petén indicating the location of El Peru-Waka' in relation to other Maya sites. Map by Damien Marken.*

inhumation and tomb termination practices. We suspect, given often diminutive quantities of debitage in some contexts and the abundance of usable blades in others, that the deposition of tomb obsidians, generally speaking, does not reflect an elaborate means of waste management or disposal as has been argued elsewhere (Moholy-Nagy 1997; but see Andrieu 2020).

Located in the Western Petén, Waka' was an important Classic Maya center located on top of a natural ridge overlooking the San Juan River, a tributary to the San Pedro Mártir, which was a major artery of commerce and communication

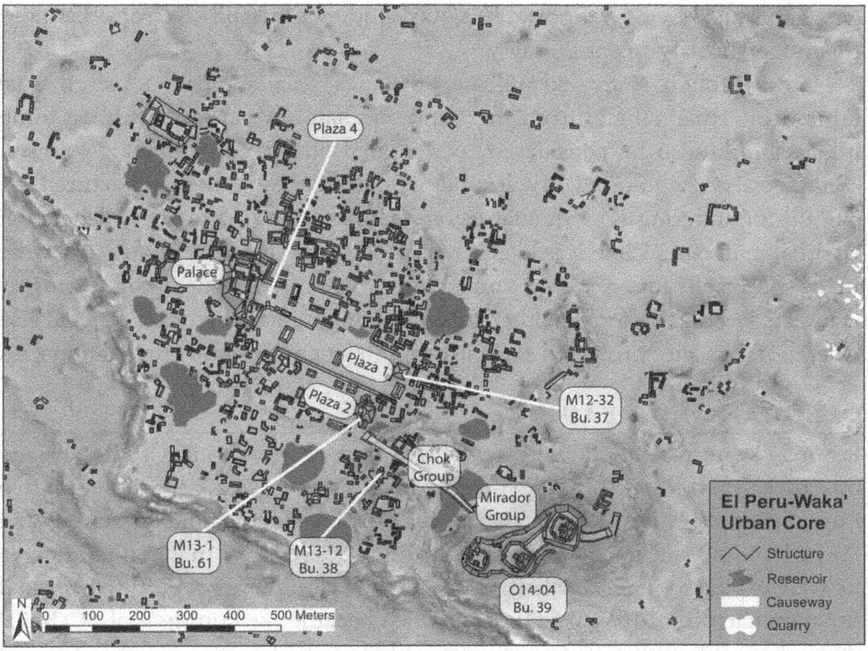

FIGURE 12.2. *Relief map of El Peru-Waka' urban center. Locations of burials indicated. Image by Damien Marken.*

of the Maya Lowlands (figure 12.1). The San Pedro flows east to west, where it joins the Usumacinta River and creates a connection with the western Maya Kingdoms. To the east, this route connected Waka' to Tikal. The city covers some 4 km^2 and is comprised of public plazas, temple pyramids, palaces, and residential neighborhoods. The site's highly nucleated core is surrounded by elite residential groups and temples radiating in all directions (figure 12.2). The burials analyzed here all come from the site core, and based on the frequency of luxury and prestige goods within the tombs, as well as their high-profile contexts, it is likely that the individuals were elites or members of the Waka' royal line.

The study of these tomb obsidians began in 2007, when Hruby analyzed debitage from three different tombs: Burial 38, excavated by Keith Eppich; Burial 39, excavated by Michelle Rich; and Burial 37, excavated by Héctor Escobedo and Juan Carlos Meléndez. This preliminary analysis focused on technology, identifying flake and blade types. This study was followed up by McCormick Alcorta in 2018 and 2019, maintaining a similar technical approach for comparative purposes with more recently discovered tomb obsidians excavated by Olivia Navarro-Farr from Burial 61 (Navarro-Farr et al. 2013; Navarro-Farr et al. 2021). The second phase of analysis included the gathering of metric data and geochemical sourcing via portable X-Ray Fluorescence (pXRF) from the aforementioned burials except Burial 37.

Hattula Moholy-Nagy and William Coe (2008) reserve the term "ceremonial lithics" for "small effigies from chert and obsidian." We argue that most tomb lithics should also be considered "ceremonial lithics," or following Johnson (2016; see also chapter 10), "ritualized lithics," since they were manufactured specifically for or during a ceremonial act or that material was curated for deposition in ritual contexts. While it is likely that some of the blades and debitage were made specifically for one of the tombs (Burial 61), as we argue in the "Recovery Contexts" section below, the others may contain obsidian that was collected elsewhere, perhaps not produced explicitly for interment or tomb termination rites.

OBSIDIAN AND MEANING

From a Western scientific perspective, we understand obsidian to be an igneous rock of volcanic origin whose formation processes contribute to two important properties: (1) the capacity for predictable conchoidal fracture; and (2) particularly sharp edges, which make it a desired raw material and led to its wide exploitation and exchange. In Mesoamerica, these properties propelled the manufacture of prismatic blades, which became the cutting tool of choice throughout the culture area beginning around 600 BCE into the Colonial period (Hirth 2003). Demand for obsidian, its restricted geological distribution, and the considerable skill involved in blade-core reduction provoked the emergence of long-distance exchange networks and craft specialists who became indispensable components of pre-Hispanic economies (i.e., Clark 1997, 2003; Hirth 2012; Hirth and Andrews 2006). Obsidian has been interpreted to have been a prime mover in the development of complex society and the maintenance of political economies of state-level societies in Mesoamerica (i.e., Aoyama 1999, 2005; Clark 1987; Santley 1983, 1989). The Mesoamerican worldview saw obsidian not only as integral to everyday economic and political life but enmeshed in otherworldly concerns as well.

Pre-Hispanic Mesoamerican people understood the natural world as animated by cosmic forces with which they were in constant interaction (i.e., Houston and Taube 2000; Saunders 2001). The landscape itself was a tapestry of physical and symbolic attributes inextricably linking the people, matter, and supernatural forces that constituted it. By extension, the use of a material referenced its origin on the landscape and the cosmic energy of its creation. Obsidian can be understood to have a physical and symbolic value attached to its place of origin, the political-economic relationships generated in securing the materials, and the reduction strategies regarded as appropriate for manufacture (Saunders 2001, 222). It is no surprise, then, that obsidian played a large role in virtually every major Mesoamerican ritual type: caches, burials, calendrical rituals, blood sacrifice, and warfare. The material was coveted not only for what it could do but also for its role in the natural and mythological realms as an inherent aspect of the universe. In general, the material was associated with lightning and celestial fire, the night-time sun, and the west, and rightly so, since obsidian was the most dangerous and transformative element in a world without metallurgy (Saunders 2001). In the burial context, in which tombs often represent cosmograms or microcosms of the universe, obsidian was likely used to frame the burial and the individual within the cosmos (see chapter 1). Obsidian, in a very general way, was used as a marker for one of the five sides of the cosmos, with the individual interred with their riches and jade reflecting the central direction or axis mundi.

Origins of Obsidian in Mesoamerica

The origin of obsidian was of importance to traditional Mesoamerican world-views: it is a geologically restricted commodity requiring substantial labor to bring it to regions where it does not occur naturally, like the Maya Lowlands. Stephen Houston (2014, 25) notes that to the Classic Maya, obsidian was a foreign substance tied with the west, Teotihuacan, night, and darkness (see also Agurcia Fasquelle et al. 2016, 11). Although most Maya obsidian came from Guatemalan sources and not the West per se, it was still an import from distant mountains. More precisely, it came from their interiors—mythical places of abundance in Mesoamerican mythology (i.e., Taube 2004). This underworld place of abundance was the source of obsidian, according to *The Annals of the Cakchiquels*:

> Then the Obsidian Stone (*Ri chay abah. Chay*, The Cakchiquels worshipped this stone as a symbol of divinity) by the wondrous Xibalbay, by the precious Xibalbay. (Xibalbay [is] one of the places endowed with riches and beauty and the cradle of the sacred stone, *Chay Abah*. The implication that the obsidian stone, which, like all minerals, was produced in the interior of the earth, came from the precious Xibalbay shows that the Cakchiquels imagined this place as a subterranean kingdom of great power and magnificence, an idea that coincides in part with the Quiché conception of this place, which is elaborately described in the *Popol Vuh*). (Recinos and Goetz 1953, 45–46)

The Annals of the Cakchiquels also clearly tie the making of humans to the nourishment of the stone through human sacrifice and war (Chinchilla Mazariegos 2022). Although the underworld is a place of plenty, it is also a dangerous place, the realm of jealous spirits (Pastrana and Aithe 2014, 97). Obsidian blades are employed by the miserly Lords of Xibalba in the second trial or punishment of the Hero Twins in the Popol Vuh, in the House of Knives or Razors (Tedlock 1996, 258–259). In Nahua cosmology, the soul must cross nine levels of the underworld, three of which are trials by obsidian, or *iztli*. The fourth and fifth layers are "Obsidian (Knife) Hill," *Iztepetl*, and the "Place of Obsidian Wind,"

Itzehecayan (M. Coe 1988, 232; López Austin 1988, 55–56). *Izmictlan apochcalocan*, "the obsidian place of the dead, with no outlet for smoke," is the final level of the Nahua underworld (López Austin 1988, 55).

Obsidian's role in ancient Mesoamerican life was multivalent, being at once ubiquitous tool and symbolically charged material (Levine and Carballo 2014; Saunders 2001). Both in a geological and mythological sense, it is born of the very innards of the once-volcanic mountains where it was obtained. In this way, the people who utilized it referenced its mountainous underworld and, in the case of the Lowland Maya, its faraway origins. Conversely, the material was also associated with the sky, or at least the aftermath of celestial fire in the form of lightning and meteors (see chapter 1; Taube 2000). Manipulation of this material was an interaction with the supernatural forces that created it, and it had to be done purposefully according to socially prescribed rules. In the case of tomb obsidians, the material was probably used to create a cosmogram, in the same way it was used as such in cache contexts, as an integral aspect of the underworld and heavens alike. The question remains: In what sense was it employed in burial contexts?

Obsidian Manipulation and Ritual

As with all things in the natural world, obsidian was endowed with cosmic essence by the gods during its creation (Friedel et al. 1993). Obsidian came from a dangerous place, and its extraction and use by people were tasks that had to be handled appropriately. Late Postclassic exploitation of the Sierra de las Navajas source in Hidalgo was accompanied by ritual offerings to specific deities at the summit of the mountain, within the mines, and on workshop walls (Pastrana and Athie 2014, 96). The extraction process was a dangerous and delicate procedure, as these were the property of mountain deities whose permission and protection were necessary to penetrate the earth (96, 101–105). Extraction at some sources required entering the mountain itself via tunnels or shafts, and the openings to these mountains were scattered with millions of sharp flakes and blades (Braswell 1996; M. Coe and Flannery 1964; Mejía and Suyuc 1998; Pastrana 2002; Sheets et al. 1990).

Ethnographic and ethnohistoric accounts attest to the ritual nature of stone tool manufacture in Mesoamerica and suggest some level of continuity through time (Clark 1989a, 1989b). Friar Toribio de Benavente Motolinía ([1568] 1941) wrote an oft-cited account of obsidian prismatic-blade production and ritual use at the Temple of Matlaluege, outside of Tlaxcala:

> [Then] they came, the masters/teachers that detached the blades, they too had fasted and prayed, and they detached many knives with which tongues were to be opened, and as they detached the blades they placed them on a clean mantle, and

> if one broke on detachment, they said that they had not fasted properly . . . they will take more than 200 hundred blades from a stone and then turn some into lancets for bloodletting; and having put the blades on a clean mantle, they perfumed them with their incense, and when the sun had set, all the priests there together, four of them sung to the blades with songs of the devil, beating their drums; and when they had sung awhile, both drums and voices quieted, and then the same four priests sang without drums, this time very sad song, to evoke devotion and wept; I think (they wept) for what they would soon suffer. The second song finished, all the same priests were together, and soon a master/teacher dexterous like a surgeon pierced everyone's tongues through the middle, having made a good hole with those blessed blades; and soon the eldest and chiefly of priests pulled through his perforated tongue four hundred and five sticks. (de Benavente Motolinía [1568] 1941, 64–65, translation by McCormick Alcorta)

This passage demonstrates not only the ritual nature of production but also the pageantry of ritual bloodletting in a public setting. The Spanish chroniclers recorded numerous examples describing similar ritual bloodletting (i.e., de Landa in Tozzer 1941; Durán 1971), and it appears in the Madrid Codex (page 19b) as well (Clark 1989b, 314).

John Clark (1989a, 1989b) observed similar rites prior to the knapping of chert among the Lacandon Maya of Chiapas, albeit in a domestic setting. In the Lacandon case, both fasting and chanting were considered integral parts of the technique and were necessary to clear the head of the artisan so they could concentrate (Clark 1989b, 305). Once again, if a blade broke during manufacture, it was considered to be a result of not fasting properly (Clark 1989a, 444). Citing Roberto Bruce (1976, 62), Clark (1989a, 1989b, 1991) notes that prior to the "commercialization" of flintknapping, the craft was carried out in a "god house," where the uninitiated were not allowed. Interestingly, knapping techniques were learned by males twelve to fourteen years old, suggesting this was a rite of passage for adolescent boys. Ethnographically, debitage was carefully collected in a cloth and placed in a receptacle, which, when full, was disposed of some distance from the village, unlike other household trash, a practice said to have been observed when work was done in the "god house" (Clark 1989a). In 1980, Clark observed that one Lacandon knapper discarded his debitage at "the base of a large Pre-Columbian pyramid . . . (where) there are numerous large cavities" (1991, 259). This may be a vestige of pre-Hispanic reverence for the dead, as ruins are seen by the modern Maya as places of the ancestors.

The ritual nature of obsidian procurement and production is supported by the post-Classic archaeological, ethnohistoric, and ethnographic records; it is likely that these practices held similar ritual connotations in the pre-Hispanic era (Hruby 2006, 2007). The collection of debitage at the time of production and the disposal of it in "special purpose dumps" (sensu Moholy-Nagy 1997) ethnographically suggests a continuity of practice in waste curation and discard as well.

THE ASSOCIATION OF "MUNDANE" LITHIC DEPOSITS WITH TOMBS AMONG THE LOWLAND MAYA

Large accumulations of flakes, blades, and other debitage have been associated with signaling the presence of royal and elite burials entombed within structures among the Lowland Maya (Hall 1989). Such deposits are often "paired," consisting of both chert and obsidian, and are found exterior to burial chambers (Hall 1989; Johnson 2016; Moholy-Nagy 1997). These two lithic raw materials may be considered thematic opposites in Mesoamerican cosmology; however, chert (day, east) and obsidian (night, west) should be understood not as diametrically opposed materials but rather as "opposite qualities interact(ing) dynamically and [that] are themselves manifestations of an underlying unity (Jordan 2016, 1430)." Burial 125 at Tikal is the earliest-known such deposit, dating to 173 ± 45 CE (Coe 1965, 1416). Hall's survey of fourteen Maya sites does not detect "debitage layers" associated with tombs again until the fifth century; ten of these twelve examples contained chert only, the exceptions being Tikal, Lamanai, and Dos Hombres (Hall 1989, 259, fig. 79; Trachman 1999, 2002). The comingling of chert and obsidian became more widespread in the Late Classic (Hall 1989) but was always present at Tikal (Moholy-Nagy 1997).

In her discussion of craft production at Tikal, Moholy-Nagy (1997, 304; also Moholy-Nagy 2003, appendix F) contextualizes these features as a specific type of "special deposit," denoting eight such deposits. Seven of these deposits (Burials 10, 22, 23, 24, 125, 116, 196, and 200) were encountered in Group 5D-2, the monumental core of Tikal (Coe 1990). The eighth (Burial 77) was excavated in Group 5D-10, an adjacent elite residential group (W. Coe 1967; Haviland 1963). Unfortunately, many of these eight debitage layers at Tikal were so large that they were neither recovered nor recorded in their entirety. Often, only a count of blades is provided, and in most cases debitage was weighed or simply noted. However, it bears mentioning that these debitage deposits were larger than other "Lithic Production Waste Deposits" uncovered further afield from the city's center (Moholy-Nagy 2020). The relative frequency of chert and obsidian in these assemblages varies case by case, but generally the pattern holds that chert provides the majority of the weight and less overall pieces while obsidian provides the majority of the pieces and less weight (1997). The chert contribution is primarily bifacial thinning flakes, followed by decortication flakes, cores, chunks, and unfinished and occasionally finished artifacts (1997, 304; Moholy-Nagy 2003, 24–25, appendix F). The obsidian components consisted mostly of small percussion flake-blades, some exhausted prismatic cores, unused prismatic blades, macroblade and -flake fragments of polyhedral cores, and transverse flakes characteristic of manufacturing eccentrics from exhausted blade cores (1997, 304; 2003, 26, 31–35, appendix F). Eccentrics were recovered from Burials 10 and 23.

Similar lithic layers have been reported at Altun Ha (Pendergast 1979; Welsh 1988), Altar de Sacrificios (Smith 1972), Uaxactun (Smith 1950), Río Azul (Hall 1986, 1989), Dos Hombres (Trachman 1999, 2002), Lamanai (Pendergast 1981), Caracol (Chase and Chase 1987; Johnson 2016; Johnson et al., chapter 10, this volume; Johnson and Johnson 2021), Yaxchilán (García Moll 2004), and at Waka' (Eppich 2007; Escobedo and Meléndez 2007; Navarro-Farr et al. 2013; Rich 2009). The lithic layers at these sites vary through time and space, with variable frequencies of obsidian and chert, in some cases with one or the other omitted, but share the common association with elite burials, usually along the central axis of buildings. Unfortunately, reporting on these layers is inconsistent. In these reports, obsidian is variably referred to as "chips, flakes, flake blades, and/or prismatic blades." It is unclear in many examples whether the assemblage is dominated by debitage or blades or if one or the other is absent.

Several interpretations of the purpose and function of these deposits have been put forth and fall into two basic camps: practical and symbolic. William Coe (1990, 486) and Grant Hall (1989, 168) suggest that debitage layers could serve to warn future construction workers of the presence of a tomb. Moholy-Nagy (1997; see also Hall 1989) suggests that they may be opportunistic dumps for the discard of waste and exhausted tools generated in architectural renovation and/or the construction of tombs and related to site maintenance. The first of these practical options should not be discarded, since, without metal shovels, deep strata of sharp lithic material may have acted as a physical and symbolic warning to pre-Hispanic looters. It "would also have served as a reliable beacon for ancient grave robbers, just as it does for looters—and archaeologists—today" (Moholy-Nagy 1997, 306).

Symbolic or ritual motivations for the interment of these deposits have also been suggested, generally by their direct or tangential association with richly furnished tombs (i.e., W. Coe 1965). Michael Coe (1988, 232) was the first to utilize ethnohistoric references to suggest that lithic layers were, in fact, a deliberate intention to symbolically mark Classic Maya tombs as existing in the Underworld. Hall (1989, 308; see also Fitzsimmons 2002, 356–357; 2009, 103–104) interprets debitage layers as the teeth of symbolic cave mouths or "Cauac Monsters" in the shafts and formal entries to tombs at Río Azul. Given that caves can be understood as portals into the Underworld and mountain interiors, neither interpretation should be ruled out; in fact, the symbolism may be conflated. In W. James Stemp and Jaime Awe's (2014, 230) discussion of obsidian's role in Maya cave ritual, they posit that "caves, akin to Xibalba, are places of both birth and death. The Maya believed . . . that they would be transported back through caves into the underworld when they died . . . (caves were) places where religious practitioners go to petition the gods and ancestors." Understanding caves as portals to the Underworld or even loci of the *emergence of* or *communication with* the dead allows us to see the use of lithic layers marking these spaces above or within Maya tombs as similar or analogous liminal places and evidence of such a communicatory event.

A possible related symbol set is the iconographic and material equivalency between lithic caches and burials (Mock 1998). It has been noted that there are some symbolic and material similarities between caches and burials, with a central jade figure, usually the sacrificed or reincarnated Maize God, surrounded by instruments of sacrifice (i.e., stingray spines, obsidian, and flint) and placed within an offering vessel. Sacrificial bowls, and by extension most cache vessels, were understood to be places of emergence through the sacrificial act, with the example par excellence of Pakal's tomb at Palenque. Thus, there is an iconic and symbolic connection between the negative spaces of the cave, burial chamber, and cache bowl as a place of *emergence of* and *communication with* the ancestors. These ancestors may be couched in terms of the maize or sun deities, and the sacrifice is a necessary part of the process to keep ancestors revered and the cosmos in order. It is important to note that the cache and burial materials and their colors actually set up the five directions of the cosmos: green (jade), red (*Spondylus*), yellow (pyrite), black (obsidian), and white (flint). In other words, a general symbolic function of tomb debitage is to finish the tomb as an axis mundi, in a similar way that the quadripartite badge as an offering bowl to the sun becomes a point of emergence for the World Tree and the rebirth of the sun and maize.

Many scholars have noted the Maya temples reference mountains and the sacred caves within them (Brady and Ashmore 1999; Chinchilla Mazariegos 2018; Friedel et al. 1993; Stuart 1997; Taube 2004). Temples were also funerary monuments and ancestral shrines (Ashmore 2003; Lucero 2007). Knowing the symbolic linkage of temples to mountains within which the dead entered the underworld, the recovery contexts of "lithic layers" can be understood as evidence of a communicatory event with the dead in a human-made cave, marking a liminal space between realms and possibly a metaphor of the Underworld. The acts that resulted in these deposits may well have been part of prescribed social practices tied to the renovation-renewal of buildings that required the permanent closure of tomb access. As discussed in "Obsidian Manipulation and Ritual," the collection of debitage was likely done at the time of production, and disposal of it may have been saved for such ceremonial acts (Andrieu 2020). Moholy-Nagy (1997, 307) notes that "he incorporation of debitage (above burial chambers) may well have been opportunistic rather than obligatory" but later notes (2020, 15) that "large deposits of chert and obsidian debitage (were) reverentially placed exterior

TABLE 12.1. Comparison of the burials discussed in the text

Location	*Sex*	*Age*	*Interred*	*Reentry*
Burial 37				
Str. M12-32; east side of Plaza 1, city center	Unknown	Adult	570–770 CE	No
Burial 38				
Str. M13-12; south side of Chok Group, city center	Unknown	Unknown	570–700 CE	Yes—final reinternment early 800s
Burial 39				
Str. O14-4; central area of Mirador Group; southeast of city center	Unknown	Adult	600–650 CE	Yes—Reentered and sealed 700–820 CE
Burial 61				
Str. M13-1; east side of Plaza 2, city center	Female	Adult	700–800 CE	No direct evidence

to some royal chamber burials and within standard caches." Fitzsimmons (2009, 104) notes that "sealing a tomb is a transformative act . . . The deceased became part of the funerary structure, even personified by that structure." It is likely the case that the act of interment did not require lithic layers to be incorporated, but perhaps the performance of certain rites associated with the closure of tombs did, at which point either new chipped stones would be produced or curated material would be used.

RECOVERY CONTEXTS

The obsidian recovered from the four tombs discussed here can be understood to have been recovered from three different types of contexts: (1) as part of the primary funerary goods; (2) as part of the tomb fill; and (3) above and around the tomb chamber. The first was restricted to a single example, Burial 37. The second and third types were associated with the final closure of the tombs and recovered in each of the other three. It is also possible that the second type may have resulted from the reuse of type three upon the termination of the tomb for further reentry. A temporal comparison of the burials analyzed in this study, along with tomb entry information can be found in table 12.1.

Burial 37

Excavations into Structure (Str.) M12-32 encountered a carefully infilled room pertaining to an earlier construction. Removal of the fill revealed significant amounts of ash and burned stuccoed floor, suggesting a ritual associated with a burial. This led to the interpretation of this room as a sanctuary for the

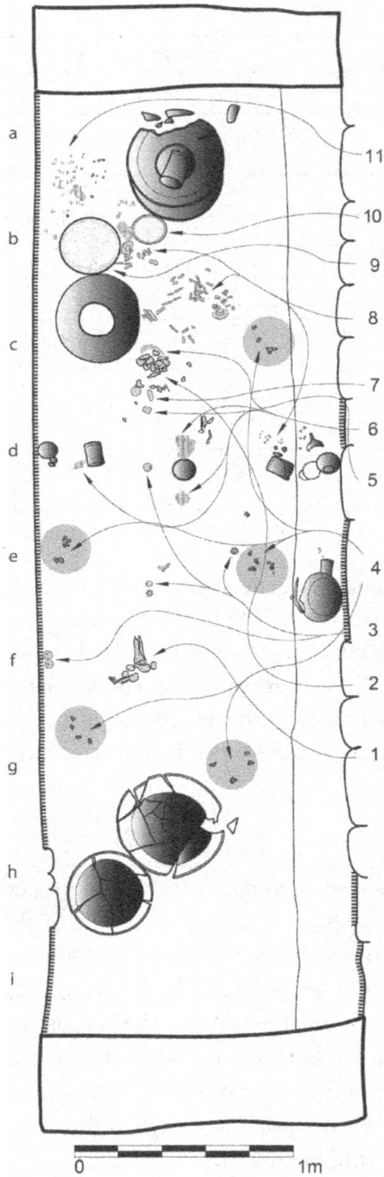

veneration of the individual interred below (Escobedo and Meléndez 2007, 104). The stuccoed floor had an intentional cut in it, granting access to the burial. The burial goods consisted of jade ornaments, sixteen vessels, a shell mosaic, whole shells, and likely two jaguar pelts based on the number of jaguar claws recovered (figure 12.3). A total of 251 obsidian blades, blade fragments, and flakes were recovered from this burial, most of which appear to be associated with primary funerary goods. A total of twenty-four chert flakes were also recovered. The majority of the obsidians (*n* = 248) and all the cherts were collected intermingled or around and beneath the jaguar claws. Based on the location of the tomb in the largest structure of the city's east-central precinct, care to preserve the sanctuary, and the associated grave goods, Héctor Escobedo and Juan Carlos Meléndez (2007, 105–106) interpret this burial as belonging to a king of Waka' entombed during the sixth century CE (Navarro-Farr et al. 2024).

FIGURE 12.3. *Burial 37 plan view. Materials as follows: (1) Human foot bones; (2) cranium fragments; (3) spherical greenstone beads; (4) jaguar Claws and obsidian blades, blade fragments, and flakes; (5) green pendant with crocodile head design; (6)* Spondylus *shells; (7) greenstone; (8) tubular beads and greenstone earspool; (9) pyrite mirror; (10) alabaster vessel; (11) shell beads (*Spondylus*). Area of lithic concentrations indicated by light gray shading. Drawing by Juan Carlos Meléndez, Laura Gamez, and Jennifer Piehl (Escobedo and Meléndez 2007, fig. 9).*

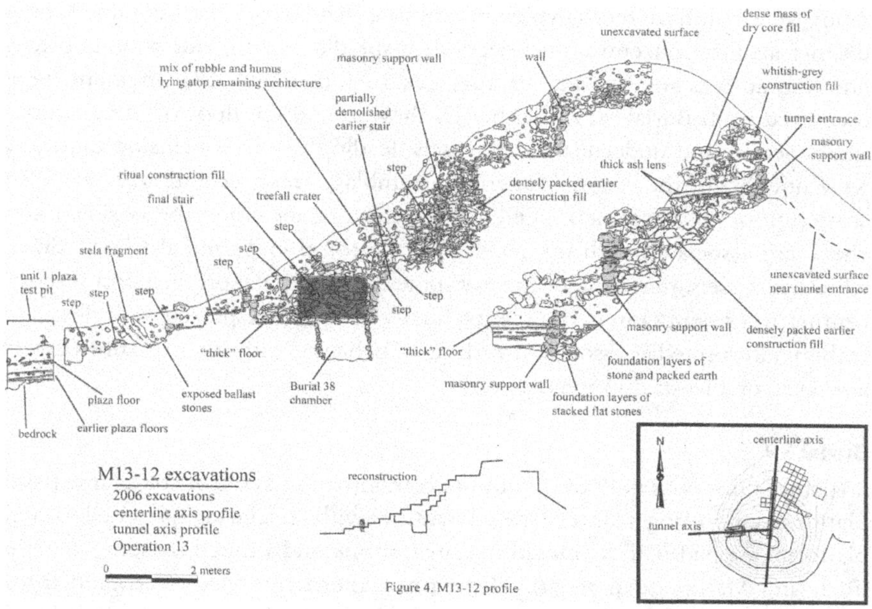

FIGURE 12.4. *M13-12 excavation profile view. Area of "ritual construction fill" above Burial 38 chamber indicated in gray. Drawing by Kieth Eppich (Eppich 2007, fig. 4).*

Burial 38

Excavations along the centerline staircase of M12-12 encountered an earlier partially demolished staircase dating to the mid-seventh century (Rich and Eppich 2020). Between the construction episodes, dense fill with "circular concentrations of flaked obsidian blades and exhausted core fragments," chert flakes, and human remains were placed on top of a thick plaster floor, interpreted as being ritual in intent (Eppich 2007, 12; see also Eppich 2014, 124–127). The presence of eroded fine orange sherds dates this "ritual construction fill" to the Terminal Classic. The plaster floor had been intentionally cut, and the ritual fill continued below it. Obsidian and chert continued to be encountered in the fill, sometimes in lenses but also intermingled with other materials (figure 12.4). Burial 38 was beneath this construction fill, in what is described by Welsh (1988, 18, 351–352) as an "elaborate crypt" or small "stone-lined tomb" (see Eppich 2014, 125 for detailed description).

The funerary assemblage of Burial 38 clearly labeled it as elite; it contained eleven whole vessels, jade beads and mosaic fragments, burned animal bone, a blue stuccoed river stone, a partial slate mirror, shell, and a mass of human remains grouped with six stingray spines. The vessels are similar to those recovered in Burial 39 by Michelle Rich (2011; Rich and Matute 2014; Rich et al. 2007), many of which have hieroglyphic inscriptions. The burial also contained three distinct artifact concentrations: decorative spindle whorls, jade mosaic pieces, and human remains (see Eppich 2014, 125–126). In total, 3,262 obsidians were recovered from Burial 38, and although they littered the floor of the chamber, none were found underneath the vessels, leading to the conclusion that they were not part of the original funerary assemblage (2007, 14; 2014, 126–127). The scattering of the aforementioned artifact clusters, including the obsidian and chert, are associated with the reentry of the tomb in Terminal Classic times, while the vessels, mirror, and stingray spines belong to the original mid-seventh-century interment (2014). The reentry material was interspersed with flecks of carbon, interpreted as associated with ritual behavior with the structure's use as a *muknal*, or ancestral shrine (116, 127).

Burial 39

Structure O14-4 was built on a natural promontory and consists of a large basal platform supporting a large terraced pyramid with an *adosada* (attached frontal platform) topped with a vaulted masonry shrine and vaulted masonry temple (Rich and Matute 2014, 70–71). Three royal interments were recovered from the adosada of O14-4, the most recent of which was Burial 39. The individual in Burial 39 was interred in a vaulted tomb with a rich assortment of prestige goods, including a narrative scene of twenty-three ceramic figurines, a green stone Olmec-style heirloom figurine (Rich et al. 2010), thirty-three ceramic vessels, four mosaics, greenstone ornaments, worked shell ornaments, animal bones, and a juvenile human (figure 12.5; Rich and Matute 2014, 74). During the Late Classic or Terminal Classic, the tomb was reentered by removing the floor of the shrine on the final construction of the *adosada*, with remarkable precision, only disturbing the area directly above the tomb.

During the reentry event, flat stones were placed over the individual, and the western vault and capstones were collapsed into the chamber. The chamber was then infilled with recycled construction fill, midden material, and materials directly related to the tomb reopening (Rich 2009). A total of 2,561 artifacts were recovered in this fill, including ceramic sherds; faunal remains; fragments of sculpture; and decorative elements made of stucco, shell, chert, and obsidian tools and debitage (Rich 2011). Materials directly related to the tomb's revisiting included 1,197 chert artifacts and 287 pieces of obsidian. The chert assemblage consisted primarily of biface fragments, alternate flakes, and rejuvenation flakes, and most were recovered from fill close to the body (see Hruby and Rich 2014, 171–182). The obsidian was primarily deposited in discrete clusters consisting of cores, core fragments, blades, blade fragments, and debitage. There was also evidence of in situ burning associated with the infilling and permanent closure of the tomb.

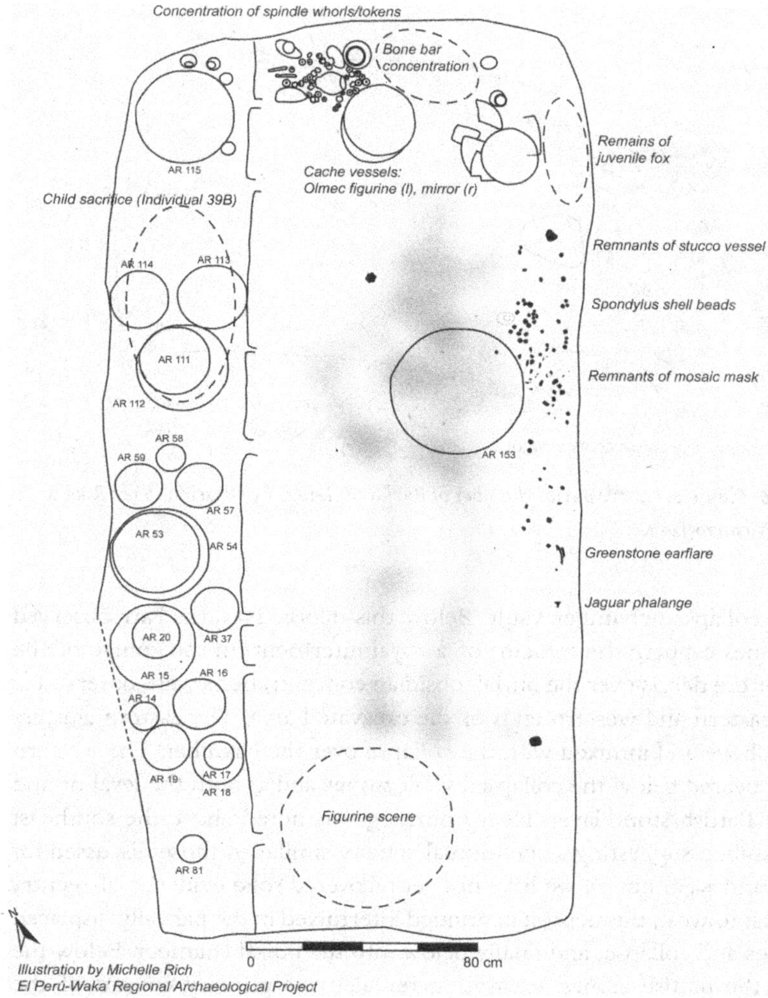

FIGURE 12.5. *Plan view of Burial 39. Shaded image denotes general location of human skeletal remains. Objects in mortuary assemblage obscured by skeletal material are not included. Ceramics are depicted whole but were broken in situ when the tomb was infilled after reentry. Drawing by Michelle Rich, Jennifer Piehl, and Varinia Matute.*

Burial 61

The tomb was discovered during the investigation of the building's *adosada* construction sequence by Olivia Navarro-Farr and colleagues (2013). Excavations first revealed debris from a floor far above the level of the chamber, which had fallen in naturally due to underlying instability (figure 12.6). The collapsed floor debris

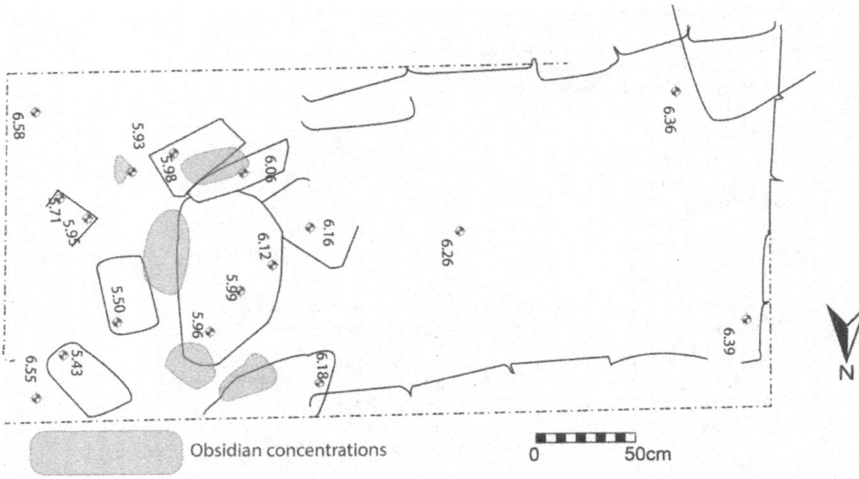

FIGURE 12.6. *Plan view above burial chamber of Burial 61. Image by Griselda Peréz Robles and Olivia Navarro-Farr.*

covered a collapsed chamber vault. Below this debris, Navarro-Farr observed flattish stones capping the remains of a royal interment. In the course of the removal of the debris over the burial, obsidian concentrations were observed at both the eastern and western ends of the excavated area. The eastern clusters appeared above and inmixed with the collapse over the chamber. The western clusters appeared below the collapsed vault stones and at both the level of and below the flattish stone layer. Heavy burning was noted above the southeast of the chamber, suggesting a ceremonial activity similar to those discussed for Burials 38 and 39, although we have not yet recovered solid evidence of reentry. Moving east to west, this deposit continued intermixed in the partially displaced vault stones and collapse, and finally below into the burial chamber. Below the vault and the flattish stones, excavation revealed the primary context of the burial chamber (Navarro-Farr et al. 2013, 77).

The chamber of Burial 61 was well intact. The interred, believed to be Lady K'abel, was found with a variety of funerary goods, including 19 ceramic vessels and 1 alabaster, a pyrite mirror, jade, shell, pearls, wood, stingray spines, stucco, faunal remains, paint, and over 2,000 pieces of obsidian. Lady K'abel was either the daughter or granddaughter of Yuknoom Ch'een II, the great king of Calakmul, and a dynast of the Kaanul regime, or "Snake Dynasty." She ruled over Waka' as *ix kaloomte'* (female overlord, high-king/queen, or emperor/empress), outranking her husband, during the seventh century and possibly into the early eighth century (Eppich and Navarro-Farr 2019, 204; Guenter 2014, 156–157). The title *kaloomte'*, while etymologically problematic, was "restricted to only the strongest dynasties"

(Martin and Grube 2008, 17). There was no clear evidence of reentry to the tomb, such as a floor cut, although the arrangement of the flattish stones suggests they had been placed there following such an event, similar to that described in Burial 39. Therefore, it is not yet certain whether all of the materials recovered were in primary deposition. Nevertheless, because obsidian was found throughout each level of the underlying interment, it can be noted that some of these items may have been part of the original internment event. Ceramic evidence deriving from Eppich's analysis of the ceramic tomb vessels indicates the interment dates to between the early and middle eighth century, during the Late Classic.

ANALYSIS

Aggregate technological analyses were undertaken in two facets: the first by Hruby in 2007 on materials from Burials 37, 38, and 39 and the second on a subset of the materials recovered from Burial 61 by McCormick Alcorta in 2018–2019. Geochemical and attribute analysis—including technological, metric, and macroscopic use-wear analyses—were undertaken by McCormick Alcorta on a subset of the materials from Burials 38, 39, and 61 and were conducted in 2018–2019; material from Burial 37 was not available for this subsequent analysis.

Although the obsidian in these four burials is all derived from blade-core technology, a number of important nonutilitarian patterns stand out. First, the volume of material varies widely across these contexts (figure 12.7). Burial 61 has the most material, with 4,580 pieces, 2,266 of which were analyzed in this study. Burial 38 boasts some 2,527 pieces, 2,278 of which were incorporated in this analysis—all demonstrably from secondary deposition. Obsidian recovered from Burials 39 and 37 total 215 and 245, respectively. A second interesting difference in these contexts is the frequency of prismatic cores and core fragments (figure 12.8). As implied from their absence in earlier descriptions, no cores or core fragments were recovered from Burial 37. Burial 38 contained 113, while Burial 39 had 94 cores and core fragments, which make up 4.9 and 43.7 percent of their respective analyzed totals. A single complete core was recovered from Burial 61 and confirmed to be of El Chayal obsidian by pXRF.

Third, flakes directly relatable to core maintenance (distal and platform rejuvenation and error removal flakes) make up variable frequencies of each deposit (figure 12.9). In this sense, Burials 38 and 39 are once again similar, both with 9.8 percent of these debitage types. Burial 37 has only 1.2 percent of its sample determined as core maintenance. Similarly, Burial 61 contains only 2.1 percent of its sample related to core maintenance.

Fourth, the debitage resulting from the actual blade detachment process differs in frequency from context to context (figure 12.10). In this category, Burials 37 and 38 are most similar; the former is 60 percent, and the latter has 51 percent of its assemblage consisting of initial-series blades (first- and second-series–related

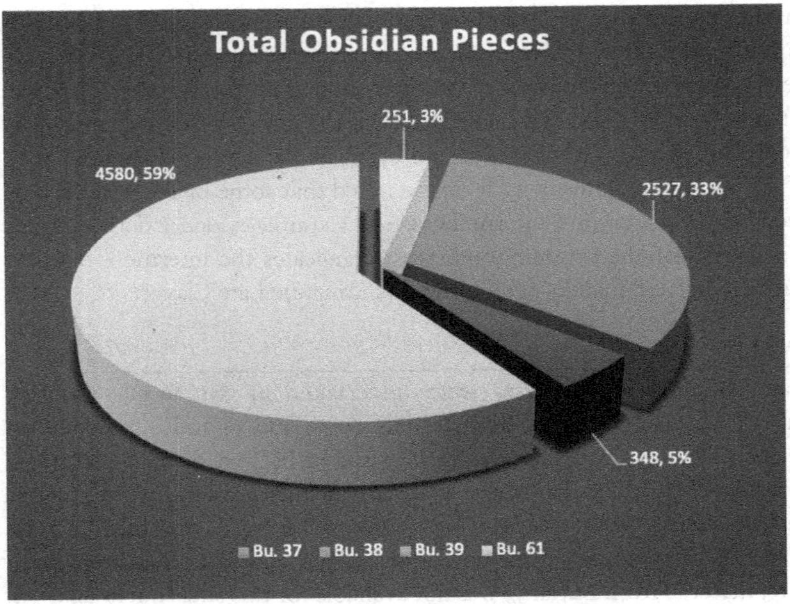

FIGURE 12.7. *Pie chart of obsidians recovered at burials discussed in this study.*

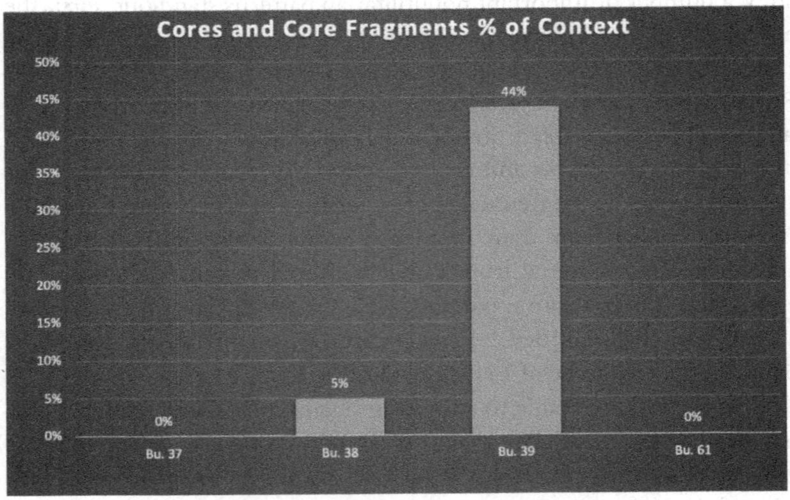

FIGURE 12.8. *Bar graph of prismatic cores and core fragments as a percentage of each context.*

debitage and blades). These two assemblages also consist of approximately one-quarter third-series prismatic blades. In contrast, the blades from Burial 39 have a much lower percentage of third-series blades, with 6 percent, and far more initial-series blades, which make a total of 31 percent of this assemblage. Burial

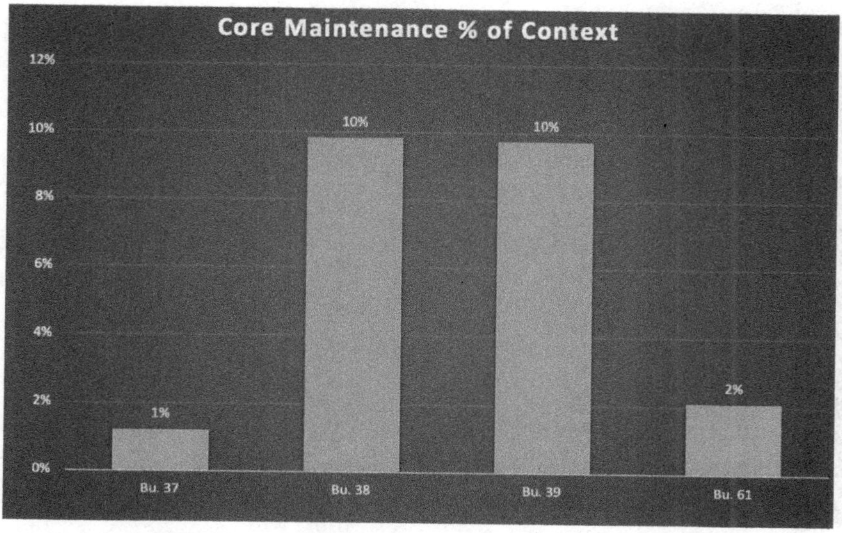

FIGURE 12.9. *Bar graph of flakes directly related to prismatic core maintenance.*

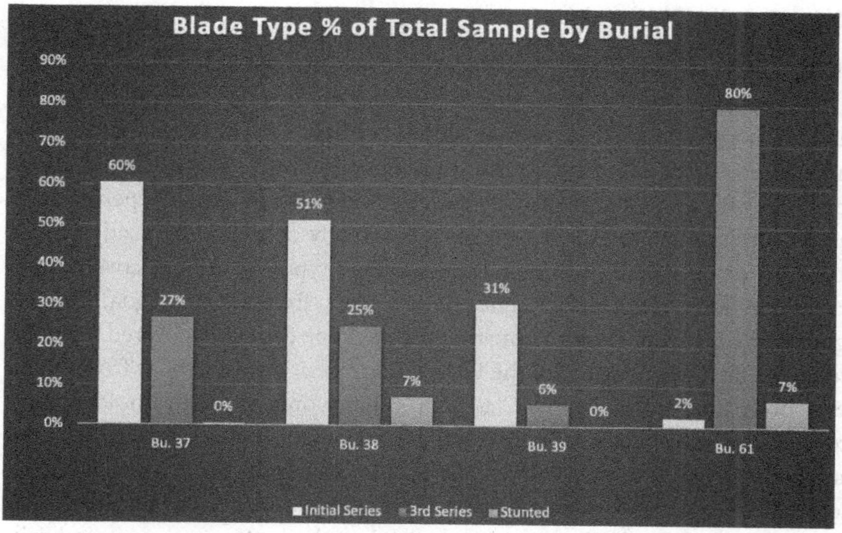

FIGURE 12.10. *Bar graph of each burial's blade assemblage in relation to stage of reduction process.*

61 is a total departure from this and contains 80 percent of its assemblage as third-series blades, with only 2 percent initial series. Additionally, significant percentages of stunted blades were only recovered from Burials 38 and 61; in each case, stunted blades contributed 7 percent of the assemblage.

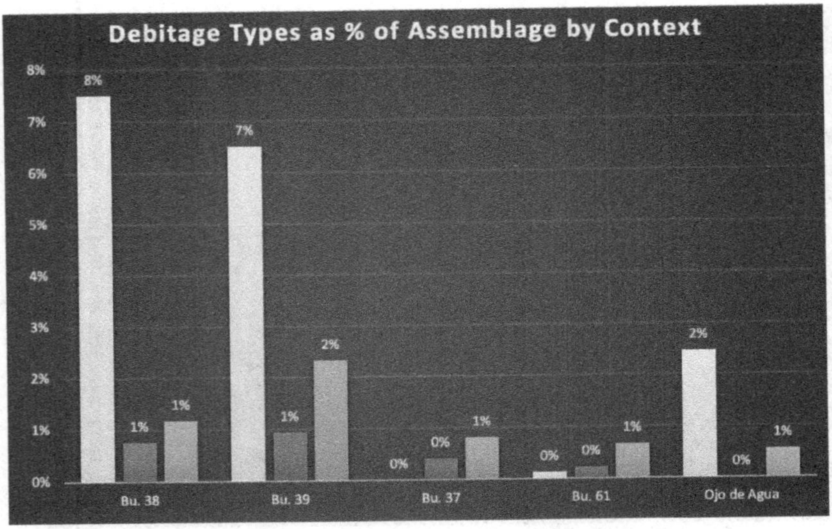

FIGURE 12.11. *Bar graph of core maintenance strategy applied in obsidian burial assemblages at Perú-Waka' compared with that from Ojo de Agua (Drawing by McCormick including data from Clark 1997).*

Fifth, the vast majority of the blades analyzed for use from Burial 61 show little to no sign of use, which suggests that they were created *for* the interment ritual, perhaps with some attendant blood sacrifice. To date, neither the obsidian from above the fill between the tomb chamber and vault nor from above the vault of Burial 61 has been analyzed. A comparison of debitage percentages was also undertaken to elucidate what was actually chosen to be used in tomb rituals. To do so, Hruby compared the debitage types to those recovered and analyzed by John Clark and Douglas Bryant (1997) from Ojo de Agua, Chiapas (figure 12.11). It is important to note that the debitage deposit recovered from the *chultun* at Ojo de Agua dated to the Early Classic period, while those from Waka' are securely Late and Terminal Classic. In other words, some technological differences were expected, since core reduction strategies changed over time in the Maya area (Hruby 200, 2024). For example, at Ojo de Agua, there was a dearth of lateral rejuvenation flakes in contrast to a complete absence at Waka'. Platform rejuvenation and distal rejuvenation seem to have been the more common rejuvenation techniques at Waka'.

Furthermore, although Ojo de Agua contained five times the number of obsidian artifacts, there were only six cores and core fragments in that sample, while 208 were recovered at Waka'. The exhausted or partially reduced cores absent from the chultun at Ojo de Agua likely went to other ritual caches at the site or were traded down the line. The cores from Waka', in some cases, were systematically destroyed in what we argue was a ritualistic and stylized manner.

FIGURE 12.12. *Purposefully broken core from Burial 38. Refit demonstrates systematic destruction of core into five pieces. Photo by Zachary Hruby.*

Refit studies suggest that some of the cores were purposefully destroyed for the ceremony, since they were sectioned in ways that have absolutely nothing to do with core rejuvenation or the normal sequence of blade-core reduction. In fact, the sectioning may have been done in accordance with sacred numerology, the number 5 in this case (figure 12.12). The complete destruction of the cores was somehow an important part of creating the deposit. Interestingly, obsidian and flint eccentrics are exceedingly rare at Waka'. Although it may be that these caches have yet to be discovered, Hruby's unpublished analysis of a debitage cache from the site demonstrates that flint and obsidian were a common facet, along with *Spondylus* and jade, of ritual deposits at the site (Hruby and Rich 2014). They do not seem to have been using exhausted cores to create eccentrics at Waka'. Instead, they seem to have used exhausted cores in certain burial, ancestor veneration, and tomb termination rituals.

Finally, only certain obsidian sources were utilized in the lithic deposits at Waka'. Geochemical data demonstrate that although most of the obsidian at

Waka' comes from El Chayal; however, the city's residents obtained obsidian from no less than five other sources, including Ixtepeque, San Martin Jilotepeque, and San Bartolome Milpas Altas in the Guatemalan Highlands, as well as Pachuca and Zaragoza in Central Mexico. Most obsidian came to Waka' primarily in the form of polyhedral cores and finished prismatic blades (see Sheets 1975, fig. 3). Over 99 percent of the obsidian analyzed by pXRF from the tombs discussed here originated from El Chayal, with a few examples of Ixtepeque, and a single Pachuca flake from Burial 61. This is important insomuch as it indicates that obsidian from San Martin Jilotepeque, San Bartolome Milpas Altas, and Zaragoza likely arrived at Waka' through different networks from the obsidian incorporated in tombs. All the cores from tomb contexts analyzed by pXRF were confirmed to be of El Chayal obsidian, suggesting elite involvement in the procurement of obsidian from this source.

COMPARISON OF TOMB CONTEXTS AND OBSIDIAN TECHNOLOGY

The four elite burials discussed demonstrate variability in the employment of obsidian artifacts to commemorate and entomb the dead at Waka' during roughly the mid-seventh through the tenth centuries CE. Obsidian use in burial practices at Waka' can be differentiated either as part of primary funerary goods or as tomb fill. The latter of these seems to be consistently associated with the final closure of tombs and, in two of the three cases, securely linked to secondary reentry events. Technologically, the material is almost all associated with blade-core reduction, with few examples of flakes related to biface production. Most of the obsidian from these deposits are blades; however, the stage in the reduction process varies considerably by context. Cores are present in all contexts except Burial 37, the only example of obsidian as part of the primary funerary assemblage. Debitage or manufacturing waste occurs in all of the contexts, but Burial 37 is almost devoid of it, with only 3 examples of 245 samples analyzed.

The obsidian associated with Burials 38 and 39 are similar in the types of debitage present and their relative frequencies but are quite different in scale. Both burials contained substantial amounts of obsidian but also included chert debitage related to late-stage biface reduction and rejuvenation. Chert was placed in the fill of the chamber after reentry in antiquity (see Hruby and Rich 2014 for a detailed analysis). The obsidian in Burials 38 and 39 were of a secondary nature and not deposited as part of the original funerary goods. These deposits were not placed in a highly systematic or stylized way, but we interpret them as part of the termination ritual associated with the permanent closure of the tombs. By termination, we mean that while these tombs may have been serially reentered over the years for ancestor veneration, there came a point, perhaps due to remodeling or architectural instability, that tombs were sealed for good. These deposits contain significant amounts of carbon and other evidence of burning, which drew the initial interpretation that the flakes may have been part of an *och k'ahk*, or fire-entering ceremony (Fitzsimmons 2009, 156–160; Stuart 1998). It bears mentioning that carbon was encountered above Burial 37, although it lacked a robust and distinct debitage deposit.

Eppich describes the material above Burial 38 as a ritual construction fill, wherein obsidian occurs in circular lenses above and within the burial chamber. Material from these lenses includes small flakes, complete blades, exhausted core fragments, and microdebitage. Core fragments from this deposit were demonstrated to refit. Thus, the co-occurrence of these artifact classes in lenses is interpreted as in situ, "flaking blades off a core and then snapping the core into pieces during deposition" (Eppich 2014, 124). The absence of macroscopic use-wear on material from these lenses supports this hypothesis. The deposit of obsidian associated with Burial 39 is similar but much smaller in scale, totaling 215 pieces compared with over 2,500 in Burial 38.

The obsidian deposited in Burial 37 departs from this pattern, having been deposited inside the burial as part of the original interment. Technologically, the majority of the blades were initial series (i.e., first and second series) recovered near disarticulated jaguar paws and claws; five such concentrations are found in this burial. Escobedo and Meléndez (2007) hypothesized that these claws are what remained of two jaguar pelts that may have covered the individual interred here, which is supported by the orientation of human remains recovered. It may be relevant that jaguar paws are often depicted with blades and flames emanating from their claws, both in Central Mexico and the Maya area during the Classic period. In other words, the obsidian debitage, in this case, may have either decorated the claws of the pelt or embellished them at the time of interment to emphasize their ideological potency. This departure from the secondary deposition of obsidian encountered in Burials 38 and 39 suggests a very different process or function of obsidian deposition as part of the primary funerary goods.

Burial 61 seems to merge the patterns observed in Burials 38 and 39 with that of Burial 37 but differs from all of them in that it contains no chert. It is possible that although Lady K'abel's remains were covered with flattish stones and sealed by a vault, periodic veneration continued above the vault, and *och k'ahk*–style ceremonies took place there without disturbing the body. Whatever the case, a substantial offering of obsidian was made as part of the original funerary rights and appeared throughout debris associated with the overlying vault as well. Navarro-Farr et al. (2013) suggest that a funerary event may have been accompanied by the partial destruction of M13-1 Sub II, punctuating its ritual termination. Sometime after the interment of Lady K'abel, a stucco floor sealed Sub-Structures I and II, which was part of a remodeling of the structure that included establishing a new centerline axis. This floor later became the site of another dedicatory burial (Burial 60). However, this was a far simpler interment that contained few funerary goods and no obsidian.

Chronologically, these burials all belong to the Classic period and show evidence of burning associated with their permanent closure. Burials 37, 38, and 39 were all originally interred during the mid-seventh century, while Burial 61 dates to the late seventh–early eighth century. The burials all show definitive signs of prolonged use except for Burial 61, although postinterment veneration rites appear likely for a couple of reasons: (1) the possible locus of reentry was never explored; and (2) the discovery of the entombed Stela 45 (Navarro-Farr 2024), also beneath the Sub II stairway, was definitively reentered. The reentry of Stela 45 strongly increases the likelihood that Burial 61 was also reentered at that same time. Prolonged use at Burial 37 resulted in the accumulation of carbon but not lithic layers as was the case in the other burials. Reentry and the final closure of Burials 38 and 39 resulted in two similar, albeit disproportionate, lithic layer deposits of obsidian and chert. Burial 61's lithic layer incorporated only obsidian and was primarily of third-series prismatic blades, unlike all of the other deposits.

CONCLUSIONS

The deposition of obsidian blades and core-blade debitage was a symbolically charged act utilized in a variety of termination rituals across the Maya world. In the Maya Lowlands, lithic layers incorporating chert and obsidian were often used in the rites associated with the sealing of royal tombs during the Classic period. Although this practice is largely confined to the Central Petén (Fitzsimmons 2009, 103), it has been observed in the Western Petén at Waka' and at Yaxchilán in Chiapas (García Moll 2004) and at several sites in Belize (Johnson 2016; Trachman 1999, 2002). The motivations for the deposition of this material are ritual in nature and likely reference multiple mythological beliefs held by the Classic Maya. Differences among the deposits discussed here may have to do with shifting or changing traditions over time but also increased wealth and the conspicuous consumption of obsidian goods, as is the case with Burial 61. Furthermore, the incorporation of broken, exhausted cores in Burials 38 and 39 may have been explicitly involved in tomb termination rights that signified the end of the production sequence, the complete exhaustion of the resource. Burial 61 reflects a different stage in the production process: the reduction of the most fruitful portion of the core and the inclusion of complete, usable, and still economically valuable third-series blades.

Obsidian was used in a variety of ways in burial contexts at Waka', both as primary and secondary deposits. In Burial 37, it was incorporated into the primary funerary assemblage as small concentrations of blades mixed with chert flakes and jaguar claws; the obsidian recovered here contained almost no debitage. Burials 38 and 39 are utilized differently as part of a secondary offering associated with the permanent closure of these tombs. Evidence of in situ, perhaps *ad hoc* reduction of exhausted obsidian debitage is present at Burials 38 and 39 in the form of systematically destroyed cores in conjunction with primarily initial series blades and debitage related to the rejuvenation of cores. Burial 61 has very little debitage and is the only case where third-series blades dominate the assemblage, denoting a different type of deposition. This is similar to a deposit encountered above what is believed to be the tomb of another *ix kaloomte'* in the Western Petén at Yaxchilán (García Moll 2004).

The deposition of cores in Burials 38 and 39, as well as that of third-series blades in Burial 61, represents the removal of a substantial amount of wealth from circulation. In these three cases, it is notable that little to no macroscopic wear was detected, indicating that the material was manufactured for the express purpose of depositing it in the act of sealing these tombs. These, therefore, cannot be opportunistic acts related to site maintenance as has been suggested for similar deposits at Tikal (Moholy-Nagy 1997, 2003) but rather purposeful ritual acts.

Burial 61 is different from Burials 38 and 39 in a few aspects that warrant attention. First, this was the largest deposit of obsidian by number but did not contain evidence of in situ production. Second, substantial lenses of chert were not recovered here. Third, the predominance of prismatic blades is in opposition to the initial series of blades recovered from Burials 38 and 39. The obsidian assemblage from this burial resembles that of Tomb 3 at Yaxchilán, believed to house the remains of Ix K'abal Xook (aka Lady Xook), wife of Shield Jaguar, King of Yaxchilán (García Moll 2004; Martin and Grube 2008, 125). Between a plaster floor and Tomb 3, element 50: "containing a large quantity of obsidian chips, dominated by prismatic blades and totaling 16.5 lb(s) on top of ample stone fill. Smaller amounts of polychrome ceramics and flint chips covered two layers of stone slabs, which in turn covered the tomb" (García Moll 2004, 267). Unlike Burial 61, Lady Xook's lithic layer contained flint, a much larger deposit of obsidian, and an associated deposit where 146 of the blades were transformed into double-notched bloodletters (element 21). These burials are similar in that they both contained important nonlocal female personages, likely the wives of rulers, and the assemblage of the lithic layer is predominantly made up of prismatic blades. At their respective sites, stelae and carved stone lintels depict these *ix kaloomte'* as queens in full warrior regalia belonging to powerful foreign dynasties (Navarro-Farr et al., 2021). Unfortunately, Lady Xook's tomb goes largely unpublished, and a comparison of grave goods may elucidate more similarities.

Importantly, the association of obsidian deposits with carbon and other burned material associated with tomb-sealing events at Burials 38, 39, and 61 suggests that these deposits are part of termination rituals. Such events were transformative acts changing the way people could interact with the "bones and grave goods of their ancestors or perform rites within funerary chambers" (Fitzsimmons 2009, 104). Burning was often part of both the termination rituals of tombs among the Maya elite and the dedication of structures. The sealing of tombs involving burning and lithic concentrations may be seen as both the termination of interaction with an ancestor and an act of renewal associated with the dedication of a structure (101). The placement of the lithic layers at Waka' can be understood to symbolize multiple things: marking an entrance, possibly a cave maw to or within the Underworld; trials that had to be undertaken by the interred on their Underworld travels to ensure renewal and rebirth. This transformative act made the structure the permanent house of the interred and embedded in the human-made sacred landscape of the Maya city.

REFERENCES

Agurcia Fasquelle, Ricardo, Payton Sheets, and Karl Andreas Taube. 2016. *Protecting Sacred Space: Rosalila's Eccentric Chert Cache at Copan and Eccentrics Among the Classic Maya*. Monograph 2. San Francisco: Precolumbia Mesoweb Press.

Andrieu, Chloé. 2020. "A Stone Duty? Flake Ritual Deposits and the Missing Workshops of the Maya Lowlands." In *The Real Business of Ancient Maya Economies: From Farmers' Fields to Rulers' Realms*, edited by Marilyn Masson, David A. Freidel and Arthur A. Demarest, 418–433. Gainesville: University Press of Florida.

Aoyama, Kazuo. 1999. *Ancient Maya State, Urbanism, Exchange, and Craft Specialization: Chipped Stone Artifacts of the Copan Valley and the La Entrada Region, Memoirs in Latin American Archaeology* 12. Pittsburgh, PA.: University of Pittsburgh.

Aoyama, Kazuo. 2005. "Classic Maya lithic production at Copan, Honduras." *Mexicon* 27 (2/3):30–37.

Ashmore, Wendy. 2003. "Ancient Maya Landscapes." In *Continuities and Changes in Maya Archaeology: Perspectives at the Millennium*, edited by Charles W. Golden and Greg Borgstede, 88–100. New York: Routledge.

Brady, James E., and Wendy Ashmore. 1999. "Mountains, Caves, and Water: Ideational Landscapes of the Ancient Maya." In *Archaeologies of Landscape Contemporary Perspectives*, edited by Wendy Ashmore and A. Bernard. Knapp, 124–145. Oxford: Blackwell.

Braswell, Geoffrey E. 1996. "A Maya Obsidian Source: The Geoarchaeology, Settlement History, and Ancient Economy of San Martin Jilotepeque, Guatemala." PhD diss., Department of Anthropology, Tulane University, New Orleans.

Bruce, Roberto D. 1976. *Textos y dibujos lacondones de Naja: Colección científica linguística*. Mexico City: INAH-SEP.

Chase, Arlen F., and Diane Z. Chase. 1987. *Investigations at the Classic Maya City of Caracol, Belize: 1985–1987*. Institute Monograph 3. San Francisco: Pre-Columbian Art Research.

Chinchilla Mazariegos, Oswaldo. 2018. "Moradas oscuras: Un ensayo de interpretación de los templos mesoamericanos." In *Tiempo detenido, tiempo suficiente: Ensayos y narraciones mesoamericanistas en homenaje a Alfonso Lacadena García-Gallo*, edited by Harri Kettunen, Verónica Amellali Vázquez López, Felix Kupprat, Cristina Vidal Lorenzo, Gaspar Muñoz Cosme, and María Josefa Iglesias Ponce de León, 831–843. Couvin, Belgium: Wayeb Publications.

Chinchilla Mazariegos, Oswaldo. 2022. "The Sustenance Providers: War, Sacrifice, and the Origin of People in Ancient Mesoamerica." In *3,000 Years of War and Peace in the Maya Lowlands*, edited by Geoffrey Braswell, 304–329. New York: Routledge.

Clark, John E. 1987. "Politics, Prismatic Blades, and Mesoamerican Civilization." In *The Organization of Core Technology*, edited by J. K. Johnson and C. A. Morrow, 259–284. Boulder, CO: Westview Press.

Clark, John E. 1989a. "La Técnica de Talla de los Lacandones de Chiapas." In *La obsidiana en Mesoamérica*, edited by Margarita Gaxiola, John E. Clark, 443–448. Mexico City: Instituto de Antropología e Historia.

Clark, John E. 1989b. "Obsidian: The Primary Mesoamerican Sources." In *La obsidiana en Mesoamérica*, edited by Margarita Gaxiola G., John E. Clark, 299–319. Mexico City: Instituto Nacional de Antropología e Historia.

Clark, John E. 1991. "Modern Lacandon Lithic Technology and Blade Workshops." In *Maya Stone Tools: Selected Papers from the Second Maya Lithic Conference*, edited by Thomas Hester and Harry J. Shafer, 251–265. Madison, WI: Prehistory Press.

Clark, John E. 1997. "Prismatic Blademaking, Craftsmanship, and Production." *Ancient Mesoamerica* 8(1):137–159.

Clark, John E. 2003. "Craftsmanship and Craft Specialization." In *Mesoamerican Lithic Technology: Experimentation and Interpretation*, edited by Kenneth G. Hirth, 220–233. Provo: University of Utah Press.

Clark, John E., and Douglas Donne Bryant. 1997. "A Technological Typology of Prismatic Blades and Debitage from Ojo de Agua, Chiapas, Mexico." *Ancient Mesoamerica* 8(1):111–136.

Coe, Michael D. 1988. "Ideology of the Maya Tomb." In *Maya Iconography*, edited by Elizabeth P. Benson and Gillett G. Griffin, 222–235. Princeton, NJ: Princeton University Press.

Coe, Michael D., and Kent V. Flannery. 1964. "The Pre-Columbian Obsidian Industry of El Chayal, Guatemala." *American Antiquity* 30(1):43–49.

Coe, William R. 1965. "Tikal, Guatemala, and Emergent Maya Civilization: Excavations Reveal Evidence of Early Complex-Living at a Prime Maya Indian Site." *Science* 147(3664):1401–1419.

Coe, William R. 1967. *Tikal: A Handbook of Ancient Maya Ruins*. Philadelphia: University Museum, University of Pennsylvania.

Coe, William R. 1990. *Excavations in the Great Plaza, North Terrace, and North Acropolis of Tikal: Tikal Report 14*. University Museum Monograph 61. Philadelphia: University of Pennsylvania Museum of Archaeology and Anthropology.

Coe, William R. 1967. *Tikal: A Handbook of Ancient Maya Ruins*. Philadelphia: University Museum, University of Pennsylvania.

de Benavente Motolinía, Fr. Toribio. (1568) 1941. Historia de los Indios de la Nueva España. Reprint, Mexico City: Salvador Chávez Hayoe.

de Landa, Fr. Diego. 1941. *Landa's Relación de las Cosas de Yucatán: A Translation by Alfred Tozzer, Papers of the Peabody Museum of Archaeology and Ethnology*. Cambridge, MA: Peabody Museum of Archaeology and Ethnology, Harvard University.

Durán, Diego. 1971. *Book of the Gods and Rites of the Ancient Calendar: F. Horacasitas and D. Heyden (trans)*. Norman: University of Oklahoma Press.

Eppich, Keith E. 2007. "Investigations in the Grupo Chok." (English translation of "WK-13: Excavaciones en el Grupo Chok.") In *Proyecto Arqueológico El Perú-Waka': Informe no. 4, temporada 2006*, edited by Héctor L. Escobedo and David A. Freidel, 259–316. Guatemala City: Instituto de Antropología e Historia.

Eppich, Keith E. 2014. "Ritual Narratives from El Perú-Waka': Ceremonial Deposits in Non-Royal, Elite Contexts." In *Archaeology at El Perú-Waka'*, edited by Olivia C. Navarro-Farr and Michelle Rich, 112–133. Tucson: University of Arizona Press.

Eppich, Keith E., and Olivia C. Navarro-Farr. 2019. "The Curious Case of Lady K'abel's Snuff Bottle: The Functionality, Form, Context, and Distribution of Classic Maya Tobacco Flasks." In *Breath and Smoke: Tobacco Use Among the Maya*, edited by Jennifer Loughmiller-Cardinal and Keith E. Eppich, 203–326. Albuquerque: University of New Mexico Press.

Escobedo, Héctor L., and Juan Carlos Meléndez. 2007. "Excavaciones en la Estructura M12-32." In *Proyecto Arqueológico El Perú-Waka': Informe no. 4, temporada 2006*, edited by Héctor L. Escobedo and David A. Friedel, 89–107. Guatemala City: Instituto de Antropología e Historia.

Freidel, David A., Linda Schele, and Joy Parker. 1993. *Maya Cosmos: Three Thousand Years on the Shaman's Path*. New York: William Morrow and Company.

Fitzsimmons, James L. 2002. "Death and the Maya: Language and Archaeology in Classic Maya Mortuary Ceremonialism." PhD diss., Department of Anthropology, Harvard University, Cambridge, MA.

Fitzsimmons, James L. 2009. *Death and the Classic Maya Kings*. Austin: University of Texas Press.

García Moll, Roberto. 2004. "Shield Jaguar and Structure 23 at Yaxchilan." In *Courtly Art of the Ancient Maya*, edited by Mary E. Miller, Simon Martin and Kathleen Berrin, 268–270. New York: Fine Arts Museums of San Francisco, Thames and Hudson.

Guenter, Stanley. 2014. "The Epigraphy of El Perú-Waka'." In *Archaeology at El Perú-Waka'*, edited by Olivia C. Navarro-Farr and Michelle Rich, 147–166. Tucson: University of Arizona Press.

Hall, Grant D. 1986. "Results of Tomb Excavations at Rio Azul, Season of 1984." In *Rio Azul Reports Number 2: The 1984 Season*, edited by E. W. Adams Richard, 69–110. San Antonio: Center for Archaeological Research, University of Texas at San Antonio.

Hall, Grant D. 1989. "Realm of Death: Mortuary Customs and Polity Interaction in the Classic Maya Lowlands." PhD diss., Department of Anthropology, Harvard University, Cambridge, MA.

Haviland, William A. 1963. "Excavation of Small Structures in the Northeast Quadrant of Tikal, Guatemala." PhD diss., Department of Anthropology, University of Pennsylvania, Philadelphia.

Hirth, Kenneth G. 2003. "Experimentation and Interpretation in Mesoamerican Lithic Technology." In *Mesoamerican Lithic Technology: Experimentation and Interpretation*, edited by Kenneth Hirth, 3–9. Salt Lake City: University of Utah Press.

Hirth, Kenneth G. 2012. "The Organizational Structures of Mesoamerican Obsidian Prismatic Blade Technology." In *The Emergence of Pressure Blade Making*, edited by Pierre M. Desrosiers, 401–415. New York: Springer.

Hirth, Kenneth G., and Bradford W. Andrews. 2006. "Craft Specialization and Craftsman Skill." In *Obsidian Craft Production in Ancient Central Mexico*, edited by Kenneth L. Hirth, 258–274. Salt Lake City: University of Utah Press.

Houston, Stephen D. 2014. *The Life Within: Classic Maya and the Matter of Permanence*. New Haven, CT: Yale University Press.

Houston, Stephen D., and Karl A. Taube. 2000. "An Archaeology of the Senses: Perception and Cultural Expression in Ancient Mesoamerica." *Cambridge Archaeological Journal* 10(2):261–294.

Hruby, Zachary. 2006. "The Organization of Chipped-Stone Economies at Piedras Negras, Guatemala." PhD diss., University of California, Riverside.

Hruby, Zachary. 2007. "Ritualized Chipped-Stone Production at Piedras Negras, Guatemala." In *Rethinking Craft Specialization in Complex Societies: Archaeological Analysis of the Social Meaning of Production*, edited by Zachary Hruby and Rowan K. Flad, 68–87. Washington, DC: Archeological Papers of the American Anthropological Association.

Hruby, Zachary. 2024. "Toward an Understanding of Obsidian Values Among the Ancient Maya Through a Comparative Approach." In *Realizing Value in Mesoamerica: The Dynamics of Desire and Demand in Ancient Economies, Scott Hutson and Charles Golden*, editors, pp. 139–162. Palgrave McMillan, Cham. Switzerland.

Hruby, Zachary, and Michelle Rich. 2014. "Flint for the Dead: Ritual Deposition of Production Debitage from El Perú-Waka', Burial 39." In *Archaeology at El Perú-Waka':*

Ancient Maya Performances of Ritual, Memory, and Power, edited by Olivia C. Navarro-Farr and Rich Michelle, 167–183. Tucson: University of Arizona Press.

Johnson, Lucas R. 2016. "Toward an Itinerary of Stone: Investigating the Movement, Crafting, and Use of Obsidian from Caracol, Belize." PhD diss., Department of Anthropology, University of Florida, Gainesville.

Johnson, Lisa M, and Lucas R. Martindale Johnson. 2021. "Sealing with Stone: Assessing an Assemblage of Lithic Debitage from a Funerary Context at the Lowland Maya City of Caracol, Belize." *Latin American Antiquity* 32(1):39–56.

Jordan, Keith. 2016. "Cosmology in Mesoamerica." In *Encyclopaedia of the History of Science, Technology, and Medicine in Non-Western Cultures*, edited by Helaine Selin, 1429–1449. Dordrecht: Springer Netherlands.

Levine, Mark, and David M. Carballo, eds. 2014. *Obsidian Reflections: Symbolic Dimensions of Obsidian in Mesoamerica*. Boulder: University Press of Colorado.

López Austin, Alfredo. 1988. *The Human Body and Ideology: Concepts of the Ancient Nahuas*. Translated by Thelma Ortiz de Montellano and Bernard Ortiz de Montellano. Vol. 1. Salt Lake City: University of Utah Press.

Lucero, Lisa J. 2007. "Classic Maya Temples, Politics, and the Voice of the People." *Latin American Antiquity* 18(4):407–427.

Martin, Simon, and Nikolai Grube. 2008. *Chronicle of the Maya Kings and Queens: Deciphering the Dynasties of the Ancient Maya*. London: Thames and Hudson.

Mejía, Héctor E., and Edgar Suyuc. 1998. "La industria de obsidiana de El Chayal." In *XI Simposio de Investigaciones Arqueológicas en Guatemala, 1997*, edited by Juan P. Laporte and Héctor Escobedo, 660–673. Guatemala City: Museo Nacional de Arqueología y Etnología, Guatemala.

Mock, Shirley B. 1998. *The Sowing and the Dawning: Termination, Dedication, and Transformation in the Archaeological and Ethnographic Record of Mesoamerica*. Albuquerque: University of New Mexico Press.

Moholy-Nagy, Hattula. 1997. "Middens, Construction Fill, and Offerings: Evidence for the Organization of Classic Period Craft Production at Tikal, Guatemala." *Journal of Field Archaeology* 24:293–313.

Moholy-Nagy, Hattula. 2003. *The Artifacts of Tikal: Utilitarian Artifacts and Unworked Material, Tikal Report 27B*. University Museum Monograph 118. Philadelphia: University of Pennsylvania Museum of Archaeology and Anthropology.

Moholy-Nagy, Hattula. 2020. "Problematical Deposits at Tikal, Guatemala: Content, Context, and Intent." *Ancient Mesoamerica* 31(1):47–63.

Moholy-Nagy, Hattula, and William R Coe. 2008. *The Artifacts of Tikal: Ornamental and Ceremonial Artifacts and Unworked Material, Tikal Report 27A*. University Museum Monograph 127. Philadelphia: University of Pennsylvania Museum of Archaeology and Anthropology.

Navarro-Farr, Olivia C., Griselda Pérez, and Dámaris Menéndez. 2013. "Operación WK-1: Excavaciones en la Estructura M13-1." In *Proyecto Regional Arqueológico El Perú-Waká, Informe No. 10 Temporada 2012*, edited by Juan Carlos Pérez, 3–91. Guatemala City: Instituto de Antropología e Historia.

Navarro-Farr, Olivia C, Griselda Pérez Robles, Juan Carlos Pérez Calderón, Elsa Damaris Menéndez Bolaños, Erin E. Patterson, Keith Eppich, and Mary Kate Kelly. 2021. "Burial 61 at El Perú-Waka's Structure M13-1." *Latin American Antiquity* 32(1):188–200.

Navarro-Farr, Olivia, Juan Carlos Pérez, and Griselda Pérez Robles. 2024. "Lady K'abel and the City's Temple: Reinforcing Cosmic Order Through Sacred Architecture." In *El Peru-Waka': New Archaeological Perspectives on the Kingdom of the Centipede*, edited by Keith Eppich, Damien Marken and David Freidel, 157–177. Gainesville: University Press of Florida.

Pastrana, Alejandro. 2002. "Variation at the Source: Obsidian Exploitation at Sierra de las Navajas, Mexico." In *Pathways to Prismatic Blades: A Study in Mesoamerican Obsidian Core-Blade Technology*, edited by Kenneth G. Hirth and Bradford Andrews, 15–28. Los Angeles: Cotsen Institute of Archaeology at UCLA.

Pastrana, Alejandro, and Ivonne Aithe. 2014. "The Symbolism of Obsidian in Postclassic Central Mexico." In *Obsidian Reflections: Symbolic Dimensions of Obsidian in Mesoamerica*, edited by David M. Carballo and Marc N. Levine, 75–110. Louisville: University Press of Colorado.

Pendergast, David M. 1979. *Excavations at Altun Ha*. Vol. 1. Toronto: Royal Ontario Museum.

Pendergast, David M. 1981. "Lamanai, Belize: Summary of Excavation Results, 1974–1980." *Journal of Field Archaeology* 8:29–53.

Recinos, Adrian, and Delia Goetz. 1953. *The Annals of the Cakchiquels*. Norman: University of Oklahoma Press.

Rich, Michelle. 2009. "Refuse or Ritual? An Examination of the Problematic Deposit of Artifacts in the Tomb Fill of Burial 39, El Perú-Waka', Guatemala." 74th Annual Meeting of the Society for American Archaeology, Atlanta, GA.

Rich, Michelle. 2011. "Ritual, Royalty and Classic Period Politics: The Archaeology of the Mirador Group at El Perú-Waka.' " PhD diss., Southern Methodist University, Dallas.

Rich, Michelle, and Keith Eppich. 2020. "Statecraft in the City of the Centipede: Burials 38, 39 and Internal Alliance Building at El Perú-Waka', Guatemala." In *A Forest of History: The Maya After the Emergence of Divine Kingship*, edited by Travis W. Stanton and M. Kathryn Brown, 88–106. Louisville: University Press of Colorado.

Rich, Michelle, David A. Friedel, Kent Reilly, and Keith E. Eppich. 2010. "An Olmec-Style Figurine from El Perú-Waka', Petén, Guatemala: A Preliminary Report." *Mexicon* 17(5):115–122.

Rich, Michelle, and Varinia Matute. 2014. "The Power of the Past: Crafting Meaning at a Royal Funerary Pyramid." In *Archaeology at El Perú-Waka'*, edited by Olivia C. Navarro-Farr and Michelle Rich, 66–84. Tucson: University of Arizona Press.

Rich, Michelle, Varinia Matute, and Jennifer C. Piehl. 2007. "WK-11: Excavaciones en la Estructura O14-04." In *Proyecto Arqueológico El Perú-Waka': Informe no. 4, temporada 2006*, edited by Héctor L Escobedo and David A. Friedel, 217–257. Guatemala City: Instituto de Antropología e Historia.

Santley, Robert S. 1983. "Obsidian Trade and Teotihuacan Influence in Mesoamerica." In *Highland-Lowland Interaction in Mesoamerica: Interdisciplinary Approaches*, edited by Arthur G. Miller, 69–124. Washington, DC: Dumbarton Oaks.

Santley, Robert S. 1989. "Obsidian Working, Long-Distance Exchange, and the Teotihuacan Presence on the South Gulf Coast." In *Mesoamerica After the Decline of Teotihuacan: AD 700–900*, edited by Richard A. Diehl and Janet Catherine Berlo, 131–151. Washington, DC: Dumbarton Oaks.

Saunders, Nicholas J. 2001. "A Dark Light: Reflections on Obsidian in Mesoamerica." *World Archaeology* 33(2, "Archaeology and Aesthetics"):220–236.

Sheets, Payson. 1975. "Behavioral Analysis and the Structure of a Prehistoric Industry." *Current Anthropology* 16(3):369–391.

Sheets, Payson, Kenneth G. Hirth, Fred Lange, Fred H. Stross, Frank Asaro, and Helen Michel. 1990. "Obsidian Sources and Elemental Analyses of Artifacts in Southern Mesoamerica and the Northern Intermediate Area." *American Antiquity* 55(1):144–158.

Smith, A. Ledyard. 1950. *Uaxactun, Guatemala: Excavations of 1931–1937, Publication 588.* Washington, DC: Carnegie Institution of Washington.

Smith, A. Ledyard. 1972. *Excavations at Altar de Sacrificios: Architecture, Settlement, Burials, and Caches, Papers of the Peabody Museum* 62(2). Cambridge, MA: Harvard University.

Stemp, W. James, and Jaime Awe. 2014. "Ritual Use of Obsidian from Maya Caves in Belize: A Functional and Symbolic Analysis." In *Obsidian Reflections: Symbolic Dimensions of Obsidian in Mesoamerica*, edited by David M. Carballo and Mark Levine, 223–254. Boulder: University Press of Colorado.

Stuart, David. 1997. "The Hills Are Alive." *Symbols* (Spring):13–17.

Stuart, David. 1998. "'The Fire Enters His House': Architecture and Ritual in Classic Maya Texts." In *Function and Meaning in Classic Maya Architecture*, edited by Stephen D. Houston, 373–425. Washington, DC: Dumbarton Oaks Research Library and Collection.

Taube, Karl A. 2000. "Lightning Celts and Corn Fetishes: The Formative Olmec and the Development of Maize Symbolism in Mesoamerica and the American Southwest." In *Olmec Art and Archaeology: Social Complexity in the Formative Period*, edited by John E. Clark and Mary E. Pye, 297–337. Washington, DC: National Gallery of Art.

Taube, Karl A. 2004. "Flower Mountain: Concepts of Life, Beauty, and Paradise Among the Classic Maya." *RES: Anthropology and Aesthetics* 45(1):69–98.

Tedlock, Dennis. 1996. *Popol Vuh: the Mayan Book of the Dawn of Life. With commentary Based on the Ancient Knowledge of the Modern Quiché Maya*. New York: Simon and Schuster.

Trachman, Rissa M. 1999. "An Additional Technological Perspective on Obsidian Polyhedral Core Platform Rejuvenation." *Lithic Technology* 24(2):119–125.

Trachman, Rissa M. 2002. "Early Classic Obsidian Core-Blade Production: An Example from the Site of Dos Hombres, Belize." In *Pathways to Prismatic Blades: A Study in Mesoamerican Obsidian Core-Blade Technology*, edited by Kenneth Hirth and Branford Andrews 105–120. Los Angeles: Costen Institute of Archaeology at UCLA.

Welsh, W. Bruce M. 1988. *An Analysis of Classic Lowland Maya Burials*. BAR International Series 409. Oxford: British Archaeological Reports.

13

Volume Summary

Keeping Stone Alive, Active, and Attractive

LUCAS R. M. JOHNSON

Mesoamerica offers one the best places to study flaked or chipped stone for several reasons. The most germane, as stated in the introduction, is that pre-Contact Mesoamerican peoples used stone to make their world. Given that most Mesoamerican ontologies seem to frame the natural and manmade world as animate, and to varying degrees sacred (e.g., mountains and temples), it is to be expected that stone occupied a central role in ceremony and ritual. In my view, rituals that involved stone were a direct way to symbolically situate, recognize, and affirm stone as essential for the production life and the re-creation of origin mythologies. It should always occupy our intellectual pursuits as well. It was essential for daily tasks, for representing complex ideologies, and for indexing known places and ideas. This volume directs our attention to those flaked technologies that likely accrued special meaning through production procedures, eventful displays, and necessary depositional actions. Since these three aspects are interrelated, the goals of this volume are ambitious and situated within frames of actions. I think it is safe to say that the contributors see stone as active and processual in reduction and use and that most of us are curious about its

https://doi.org/10.5876/9781646427819.c013

study and what it *does*. In the opening chapter, Zachary Hruby provides extensive background to ceremonial flaked or chipped stone, including descriptions of chapters and their thematic aims. Therefore, this closing chapter focuses brief attention to several themes that are particularly useful for future studies. These include the operationalization of communities and constellations of practice versus chipped-stone traditions, the ongoing debate over terminology to classify ceremonial and symbolically charged chipped-stone objects, the importance of the identification of indirect percussion and the continued insights replication experiments provide, the importance of refitting studies, and finally the need for detailed descriptions of archaeological contexts that support symbolic or other kinds of meaning (i.e., iconic, indexical, bundled).

To begin, there are several volumes and individual publications available for those ready to engage complex reduction sequences through experimentation, as well as to learn how to do chipped-stone analyses based on technology, distribution, and use (Aoyama 2001; Clark 1987; Hirth 1998). These publications are punctuated moments in Mesoamerican archaeological history. In other words, the study of chipped stone is not as cohesive as ceramic studies and those of epigraphic texts, but rather collaborators produce unique volumes describing the state of the field at moments of inflection and reflection. For example, John Clark, in association with other notable scholars, assembled a cohesive summary of work several years ago to bring everyone into a current state of obsidian studies in Mesoamerica while asking for a massive database be constructed on the topic (Clark 2003). This volume represents a start in assembling some aspects of that database beyond the description of prismatic obsidian blades. Hruby begins with an outline for such a collective work by explaining problems with names and moving to a broader summary of ritualized chipped-stone, highlighting when some specific terms should be used over others (i.e., *silhouettes* vs. *ritualized*). This executive summary and the chapters that follow are a start to a massive effort by lithic analysts to establish a cohesive, yet adaptive, language for use by both specialists and nonspecialists. Project directors from across Mesoamerican archaeology should take notice of the terminology used herein when thinking about information management and broader interpretations beyond "we have eccentrics from a cache." Tabular databases include descriptive fields that are largely made up. As such one entry that likely occupies all chipped-stone databases is *eccentric*, potentially followed by wordy descriptions of morphology and context. From my work summarizing and reevaluating some thirty years of obsidian collections at Caracol, Belize, descriptions of ritualized stone are so cumbersome to code, *eccentric* is often shorthand employed during cataloging and for summaries and interpretations. To be sure, and as several authors point out, this term obscures virtually everything contemporary researchers are interested in, including assumptions regarding material type.

Furthermore, this volume works to overturn long-established norms in ceremonial chipped-stone description to reinvent and reinvigorate the field. It combines the experience of the older guard with the often-younger apprentices in much the same way Hruby advocates we use communities of practice to understand the regional participation in the ceremonial and ritual production and use of chipped stone. This book reflects that community, once imagined, and now real as expressed in shared ideas and practices. But like all communities, the members are diverse and spatially distant at times. It is up to us to understand what we and those in the past share through our research and actions. It is important to reflect on the important underpinnings of why such a volume exists. With this brief introduction and slight reflection on the importance of this volume, numerous aspects are of note for discussion and conclusion.

COMMUNITIES OR CONSTELLATIONS OF PRACTICE

The use of communities and constellations of practice is of relatively recent use in archaeology (Roddick and Stahl 2016). Its use is largely to relate individuals, as represented by their actions in creating crafted object as observed in artifact analysis, to others doing the same task but not directly related to the agents under analysis. In other words, how does the analysis of material from my excavated household, community, site, or other area relate to another where the same type of practices occurred? Various authors (Hruby, chapter 1; Hruby et al., chapter 8) use the communities of practice scheme to do just that with great effectiveness. Because communities of practice extend from models of situated learning—an education theory—there is a leap in its archaeological application when people are absent, and the master-apprentice dynamic is inferred. The leap is assisted through skilled delivery of physical analytical data. John Clark (chapter 4), James Woods (chapter 5), and Payson Sheets (chapter 6) present detailed analysis of ceremonial lithics. These studies form a model of analysis Hruby argues is a prerequisite in determining whether communities of practice existed in ancient Mesoamerican cities. While they are not explicitly trying to identify communities, the sequential movements they describe through a type of forensic reconstruction (Clark, chapter 4) allow for the inference of shared practices including the procurement of stone and the movement of near-finished to finished objects ready to use in singular eventful ceremonies.

Hruby (chapter 1), through the review of ceremonial lithics, shows how the act of using lithics in ceremonies represents a community of practice. Several chapters (Hruby and colleagues, [chapter 8], Joshua Kwoka [chapter 9], Lucas R. M. Johnson et al. [chapter 10]) provide a review of technological forms and contexts for ceremonial lithics. At least three examples are of note and include the use of modified laurel-leaf bifaces, retouched exhausted blade cores, and obsidian blade caches. The comparison between the Late and Terminal Classic eastern lowlands is described in great detail to demonstrate how techniques of lithic production and depositional practice were shared against the backdrop of political disunity (Hruby et al., chapter 8). Furthermore, Hruby (chapter 8) reminds archaeologists that communities of practice are not necessarily discoverable through detailed analytical techniques but are also a way to envision relationships (citing Lave 2008; Lave and Wenger 1991). This finding is important as archaeologists cannot necessarily apply a theory or model of doing directly through analytical comparison, but they adapt the model ontologically, endeavoring to reconstruct relationships between places and regions. For example, Johnson et al. (chapter 10) presents a broad survey of ritualized obsidian from Caracol, Belize. While data from Caracol clearly demonstrate its connection to the Central Petén area and Northern Belize, few ceremonial chert artifacts have been found at the site, with the exception of four large chert silhouettes recovered from household excavations (Chase and Chase 2008, 2010). Three of these are from the same cache and are accompanied by a suite of materials including obsidian and several stingray skeletons that are likely from the Belizean coast (Conningham-Smith et al. 2014). Two of these are notched laurel-leaf bifaces (Chase and Chase 2008, fig. 34a-c). The other chert silhouette is a scepter-like form with a hole in the center recovered from a Terminal Classic household burial (Chase and Chase 2010, fig. 55). The lack of ceremonial chert objects perhaps indicates that Caracol did not directly participate in certain relationships that brought in cherts from the Northern Belize and instead used obsidian almost exclusively.

With a volume of this scope available to new and current scholars interested in ceremonially used chipped stone, regional relationships can be reconstructed in addition to those defined by ceramics spheres and the like. If some lithic techniques were guarded or otherwise restricted but the distribution of ceremonial lithics is regional, then the teaching of ritualized chipped-stone reduction was an intentional act that transcended political boundaries. Since knappers were creating powerful effigies or supernatural forms, they themselves were ritual practitioners and thus operated in exclusive circles and shared knowledge with select people who (re)created physical deities. Hruby (chapter 8) likens this to learning to writing Maya hieroglyphs, which was only done by a select few.

CLASSIFICATION TERMINOLOGY

For many studying Mesoamerican chipped-stone objects recovered from obvious ritual contexts (e.g., burials and caches), the term *eccentric* is too broad and obscures nuance in both material, reduction techniques, shape, and perhaps even context. By context alone any chipped-stone object could be labeled as an eccentric or a ritualized stone object (i.e., an object used in rituals or ceremonies and given special significance). In chapter 10, I use the phrase *ritualized obsidian* to be more specific than *eccentric*, but even this phrase does not allow for the description of form, shape, or technological classification (e.g., blade core). Technological classification comes next. The bottom line is that a single word or even phrase cannot necessarily accomplish the breadth of items presented in this volume. Therefore, a particular strength of this volume introduces further operational definitions that allow for the study of chipped stone in new ways, advancing the field. There are several authors that show the effectiveness of the *silhouette* classification. Understanding objects as silhouettes repositions objects as two- and three-dimensional representative shapes that did not necessarily require huge time investments but did require knowledgeable actions. Examples of organizing eccentrics into shapes used their outlines to show continuity between deposits (Moholy-Nagy and Coe 2008). This classification scheme is easily applicable across Mesoamerica, as several authors have done (Alejandro Pastrana and Hruby, chapter 2). In terms of classification terminology, the volume justifies that future work across Mesoamerica on ceremonial lithics include a hierarchy of classifications and descriptions starting with understanding the full expression of ceremonial used chipped-stone objects extending to their shapes and the flaking techniques required.

REPLICATION STUDIES, INDIRECT PERCUSSION, AND REFITTING

The shapes of ceremonial lithics are expressions of skilled knowledge, some of which includes detailed iconographic conventions usually carried out by elites and high-status individuals. The understanding that not everyone in the past possessed such knowledge is inferred from experimental studies or those studies that aim to reproduce what forms the archaeological record. While restricted knowledge may not apply only to ceremonial lithics, possibly extending to certain projectile points across the Americas, ceremonial lithic production in Mesoamerica seems to be an extreme case. Several chapters, by Clark (chapter 4), Woods (chapter 5), and Sheets (chapter 6), show how detailed analysis and replication experiments can reveal differences in skill level. The steps inferred involved intimate knowledge of various tools and the physical properties of stone. This knowledge is gained over a lifetime, and those involved are clearly masters of their craft.

One interesting technological flourish in the Maya tradition is the remnant cortex that was intentionally left on the distal or proximal portions of finished bifaces. Hruby (chapter 1), Woods (chapter 5), and Sheets (chapter 6) each describe the presence of cortex to be intentional and perhaps a signature of a workshop—a calling card. Kwoka (chapter 9) reports this too in his description of Northern Belize caches. This observation transcends sites to be present across a broad region of the Maya Lowlands. It even shows up on one of the chert silhouettes from Caracol (Chase and Chase 2010, fig. 55). To intentionally leave a cortical "maker's mark" underscores just how knowledgeable knappers were. The same has been argued for craft specialization in the producing lanceolate projectile points from Napa Valley, which could be from a singular workshop (see Carpenter and Mikkelson 2005). Leaving markers of cortex also signifies you, as the producer, have access to the primary source rock and thus occupy a privileged status among knappers.

Although preceding the experimental studies, Benjamin Eble and Hruby (chapter 3) describe a critical aspect of knapping that is argued to be largely missing from analysis. The review of indirect percussion from the Americas is extraordinary, providing documented accounts backed up with archaeological data. A significant takeaway from this contribution is that indirect percussion can involve multiple people, according to ethnohistoric and ethnographic accounts from the Americas. Assertions of multiperson crafting is not new per se and is implicated in Kenneth Hirth's multicrafting approach (2009; see also Shimada 2007). However, in chipped-stone reduction, which is dangerous, trust between holder and striker is important. Both must communicate to stay safe and execute the task at hand. This is one likely context of learning, which would be essential to a master-apprentice relationship in a community of practice model. Spoken word cadence and rhythm are an interesting aspect that supports the ritualized nature of ceremonial lithic production. A review of indirect percussion early in the volume sets up a deeper understanding of how chipped-stone reduction most likely occurred, even if the subsequent chapters do not necessarily reference it.

Refitting in lithic studies is extremely time consuming but can be a fruitful endeavor. While refitting is not explicitly a topic covered in the volume, several authors show how determining refits adds to the interpretation of actions during blade removal in terms of overall core size and the movement of the core during blade removal. This removal can be during pressure reduction (Jorge Ramos et al., chapter 11) and also percussion reduction (Kelsey Sullivan and Jaime Awe, chapter 7). Refit studies do not necessarily relate only to blade removal, a highly skilled process, but also to the systematic destruction of stone objects, such as exhausted blade cores and other percussion techniques (David McCormick et al., chapter 12; and Johnson et al., chapter 10). Refit studies such as those presented allow for the inference of time between blade production and deposition, especially when blades remain adjacent to their refit neighbor. Jorge Ramos et al.'s (chapter 11) refit studies moreover include analysis of color and texture as well as physically refitting two or more blades.

CONTEXTUAL ANALYSIS

In Mesoamerica, ceremonial lithics are associated with specific contexts; they are usually directly deposited with inhumations or other kinds of interments, be they dedication or termination rituals. Their presence signals the deposit. For example, a cache of obsidian at the base of a stela signals the ritual use of stone in the dedication of a monument. Other ritual contexts include a suite of materials, of which obsidian or chert is only a small portion. Furthermore, as Hattula Moholy-Nagy (1997) reported several decades ago, chipped stone—both obsidian and chert alike—are sometimes associated with burials of high-status individuals at Tikal. These layered deposits of sometimes tens of thousands are found elsewhere above the burial chamber and were interpreted as convenient places to dump dangerous production waste. While Moholy-Nagy (2011) has suggested otherwise, that these layered deposits are part of the interment process, others have fully invested in this interpretation (McCormick et al, chapter 12; see also Johnson and Johnson 2021). The last main chapter of the volume, by McCormick and colleagues, understands the deposition of blades, cores, and debitage to be a symbolic act present at various types of Maya termination and burial rituals. It and a few other studies (Johnson and Johnson 2021) should be a model for (re)interpretation.

SYMBOLIC CONCERNS

Lars Fogelin (2007, 63) stated, "The question still remains . . . how best to study symbolism within the confines of archaeological research?" Cited several times throughout the volume, and underscoring that conversations within are part of an ongoing research program regarding Mesoamerican ritualized and ceremonial stone technologies (see Levine 2014), is a need to explicitly focus on symbolism (Hruby in prep). Symbolic analysis is not the primary concern of this volume. However, the contributors actively engage the material concerns structuring symbolic forms (chapters 2, 9, 11, and 12). Furthermore, many of the authors operationalize or apply a symbolic approach to how objects mean, or to the semiology of ritual through material and technological analysis. It is important to state that a symbol is a high-level abstract construct that can be used too frequently and thus becomes difficult to operationalize on the past. What this volume does is attempt to remove the abstractness of a symbol (i.e., eccentric) and give it meaning through material and contextual analysis as well as link forms to other referents. The enthusiasm for examining silhouettes is therefore movement toward greater meaning (chapter 2). Based on this positioning of symbolic analysis within archaeology and chipped stone in particular, several points are important to address. After, I propose returning to statements by Marc Levine (2014) that advocated archaeologists concerned with (obsidian) chipped stone think through the work of Charles Peirce (1991; see also Preucel 2006), who applied *symbol* with *icon* and *index* relationally in understanding the ways objects can mean in differing ways at the same time (see Joyce 2007 for ceramic figures).

Religion, Ritual, and Elite-ness

Many contributors herein implicitly position their analysis within expectations pursuant to either or both religion and/or ritual. In other words, religion is expressed through analyses of final forms with cosmological linkages (chapters 1, 2, 8, and 11), while ritual is concerned with sequences of events—production and deposition—that may not have an overt religious connection (chapters 10 and 12). While a dialect is apparent between the two, wherein Fogelin (2007, 56) states, "Ritual is the action or the cutting edge of religion," not everyone is concerned with the reconstruction of ancient religion. Material expressions of religion broadly may be studied outside the confines of production and context per se by examining the overall shape embracing the silhouette or outline of the thing across a broadly studied area (chapter 2). Ritual is analyzed often without the need to address the shape of things, as several chapters point out (chapters 10, 11, and 12). This is especially true when the form is unmodified debitage, blades, bifaces, or the like (chapter 10). Clearly the presence of refitting in several well-examined contexts points toward the recognition of ritualized production (chapters 7, 10, and 11).

Power and prestige are intractable constituents of silhouettes or eccentrics, or similar objects (Robb 1998, 329). Most herein are examinations of rituals that signal elite political power (chapters 1, 4, 8, and 11; cf. chapter 10). How much is elite-ness a priori in our analysis? How is the elite-ness we associate with eccentrics biasing our understanding of ancient Mesoamerican religions? In other words, what or who is being implicitly privileged through a symbolic analysis, contextual analysis, or production analysis (see Fogelin 2007, 66)? Many researchers focus on ritual production and analysis of context, while others situate ceremonial chipped-stone within a religious framework.

Practice, Symbolism, Material, and Context

It is clear throughout the volume that a practice approach underlies much of the discussion (chapters 1, 2, 8, 10, and 12). The need to be practice oriented is because rituals are rule based and are carried out through performative acts of production (Fogelin 2007, 58)—both in the making of symbols and the ritual space—but can be creative and diverge from convention seen at other sites (chapter 10). John Robb (1998, 336), citing several seminal works, reminds us that "technological knowledge is an integral part of the symbolic world." Furthermore, the ways we do something incorporates cosmological beliefs about tools, materials, qualities, and other processes (336). There are hints of this melding throughout the volume (chapters 2, 3, 4, 5, and 7) and in the argument about why blades are cached objects (chapter 11). Melding technological practice and symbolic expressions is evident in several authors' explicit linkages to ritualized production and the support for esoteric production of ceremonial lithics (chapters 5 and

9). Hruby (chapter 1) states this point at the outset of the volume, and it is supported throughout.

If symbolic meaning is tied to context, then a simple unshaped flake in the right place can be meaningful (Robb 1998, 334). This point is especially examined through analyses of layered stone deposits from significant burial contexts (chapter 12; see also Johnson and Johnson 2021). McCormick and colleagues (chapter 12; see also Hruby and Rich 2014) argue that specific kinds of debitage and intact, unused blades were intentionally selected and deposited above or around specific tombs to complete the eventful interment and be understood to likely mark an entrance to a cave opening. This placement effectively transformed the space by further emphasizing the physicality of the sacred landscape. With this recent discussion and the history of interpretation on the matter (Moholy-Nagy 2011), future examinations should consider these deposits as symbolic or otherwise referring to a layering event intended to seal the interred as well as indexing the larger chipped-stone community and its fundamental role in ceremonial and daily life (Johnson and Johnson 2021). Layered deposits emphasize context and the formation of ritual space (Fogelin 2007, 59). To be sure, the space functions to remove dangerous material as well, but more to the point, stone without modification can still index or represent something else (i.e., lightning, directionality, etc.); thus several authors point out its affordances without direct reference (chapters 10, and 12). Similarly, Robb (1998, 331) reminds us that symbolic analysis (symbolism) helps to reinforce a meaning beyond the material, although the material—its properties—is inexorably embedded in its shape and form.

FINAL THOUGHTS: SEMIOLOGY IN CEREMONIAL CHIPPED-STONE

Interpretation of ceremonial stone may be conditioned on which perspectives we employ. This volume engages several integral research topics that are addressed above and introduced by Hruby (chapter 1), and the use of *symbol*, *symbolic*, or the like is inserted throughout to bolster the importance of the research beyond the descriptive. The blending of technological and contextual descriptions with symbolic analysis is compelling as always and shows meaning making is essential to studying chipped stone. Because semiology is central, I advise future scholars to continue to return to foundational works that relate symbol to other interpretive devices. In that regard, the substitution of symbol with index can add greater meaning within a relational system of signs or semiotics (Peirce cited in Preucel 2006; see also Levine 2014, 20; Johnson 2016). Meaning can be bundled in the event (sensu Keen 2005), and this volume clearly examines stone used to create and execute such events with or without iconic silhouette form (chapters 10, 12). This volume demonstrates that meaning is inferred through material properties, production action, object history, shape (e.g., silhouette), and context situated with an overarching religious and ritualistic understanding. Finally, the volume allows particularistic and highly technical analyses from unique examples to be interpreted through a generalizable understanding of how chipped stone was ritualized across Mesoamerica.

REFERENCES

Aoyama, Kazuo. 2001. "Classic Maya State, Urbanism, and Exchange: Chipped Stone Evidence from the Copan Valley and Its Hinterlands." *American Anthropologist* 103:346–360.

Carpenter, Kimberley, and Pat Mikkelsen. 2005. Lithic Production and Craft Specialization in the Middle Period: Data Recovery Excavations at CA-NAP-172. Far Western Anthropological Research Group, Inc., Davis, California. Submitted to City of Calistoga, California, and State Water Resource Control Board, Sacramento.

Chase, Arlen F., and Diane Z. Chase. 2008. *Household Composition and Ritual Patterning: Continued Investigation of Residential Groups near Caracol's Epicenter*. 2008 Field Report of the Caracol Archaeological Project. Report prepared and submitted to the Belize Institute of Archaeology.

Chase, Arlen F., and Diane Z. Chase. 2010. *Household Patterning, the "Collapse," and LiDAR Ground-Checks: Continued Investigations of Structures in and near Caracol's Epicenter*. Report prepared and submitted to the Belize Institute of Archaeology.

Clark, John E. 1987. "Politics, Prismatic Blades, and Mesoamerican Civilization." In *The Organization of Core Technology*, edited by Jay K. Johnson and Carol A. Morrow, 259–284. Boulder, CO: Westview Press.

Clark, John E. 2003. "A Review of Twentieth-Century Mesoamerican Obsidian Studies." In *Mesoamerican Lithic Technology: Experimentation and Interpretation*, edited by K. G. Hirth, 15–54. Salt Lake City: University of Utah Press.

Conningham-Smith, Petra, Arlen F. Chase, and Diane Z. Chase. 2014. "Fish from Afar: Marine Use at Caracol, Belize." *Research Reports in Belizean Archaeology* 11:43–53.

Fogelin, Lars. 2007. "The Archaeology of Religious Ritual." *Annual Review of Anthropology* 36:55–71.

Hirth, Kenneth G. 1998. "The Distributional Approach: A New Way to Identify Market Exchange in the Archaeological Record." *Current Anthropology* 39(4):451–476.

Hirth, Kenneth G. 2009. "Housework and Domestic Craft Production: An Introduction." In *Housework: Craft Production and Domestic Economy in Ancient Mesoamerica*, edited by K. G. Hirth, 13–32. Archaeological Papers for the American Anthropological Association 19(1):1–252.

Hruby, Zachary X., and Michelle Rich. 2014. "Flint for the Dead: Ritual Deposition of Production Debitage from El Perú-Waka', Burial 39." In *Archaeology at El Perú-Waka': Ancient Maya Performances of Ritual, Memory, and Power*, edited by O. C. Navarro-Far and M. Rich, 167–183. Tucson: University of Arizona Press.

Johnson, Lucas Martindale. 2016. "Toward an Itinerary of Stone: An Investigation of Obsidian from Caracol, Belize." PhD diss., University of Florida, Gainesville.

Johnson, Lucas R. M., and Lisa M. Johnson. 2021. "Sealing with Stone: Assessing an Assemblage of Lithic Debitage from a Funerary Context at the Lowland Maya City of Caracol, Belize." *Latin American Antiquity* 32(1):39–56.

Joyce, Rosemary A. 2007. "Figurines, Meaning, and Meaning-Making in Early Mesoamerica." In *Material Beginnings: A Global Prehistory of Figurative Representation*, edited by Colin Renfrew and Iain Morley, 107–116. Cambridge: McDonald Institute for Archaeological Research.

Keen, Webb. 2005. "Signs Are Not the Garb of Meaning: On the Social Analysis of Material Things." In *Materiality: Politics, History, and Culture*, edited by D. Miller, 182–205. Durham, N.C.: Duke University Press.

Lave, Jean. 2008. "Epilogue: Situated Learning and Changing Practice." In *Community, Economic Creativity, and Organization*, edited by Ash Amin and Joanne Roberts, 283–296. Oxford: Oxford University Press.

Lave, Jean, and Etienne Wenger. 1991. *Situated Learning: Legitimate Peripheral Participation*. Cambridge: Cambridge University Press.

Levine, Marc N. 2014. "Reflections on Obsidian Studies in Mesoamerica: Past, Present, and Future." In *Obsidian Reflections: Symbolic Dimensions of Obsidian in Mesoamerica*, edited by David M. Carballo and Marc N. Levine, 3–44. Boulder: University Press of Colorado.

Moholy-Nagy, Hattula. 1997. "Middens, Construction Fill, and Offerings: Evidence for the Organization of Classic Period Craft Production at Tikal, Guatemala." *Journal of Field Archaeology* 24:293–313.

Moholy-Nagy, Hattula. 2011. "Observations on the Use-Life Trajectory of Lithic Artifacts at Tikal, Guatemala." In *The Technology of Maya Civilization*, edited by Zachary X. Hruby, Geoffrey E. Braswell, and Oswaldo Chinchilla Mazariegos, 30–36. Sheffield: Equinox Publishing.

Moholy-Nagy, Hattula, and William R. Coe. 2008. *Tikal Report No. 27 Part A. The Artifacts of Tikal: Ornamental and Ceremonial Artifacts and Unworked Material*. University of Philadelphia: Pennsylvania Museum of Archaeology and Anthropology.

Peirce, Charles S. 1991. *Peirce on Signs: Writings on Semiotics by Charles Sanders Peirce*, edited by James Hoppes. Chapel Hill: University of North Carolina Press.

Preucel, Robert W. 2006. *Archaeological Semiotics*. Malden, MA: Blackwell Publishers.

Robb, John E. 1998. The Archaeology of Symbols. *Annual Review of Anthropology* 27:329–346.

Shimada, Izumi, ed. 2007. *Craft Production in Complex Societies: Multicrafting and Producer Perspectives*. Salt Lake City: University of Utah Press.

Index

Actuncan, Belize, 188
Adney, Edwin Tappan, 100
aggregate analysis, 335
Aguateca, Guatemala, 299
Agurcia Fasquelle, Ricardo, 169, 174, 181, 183, 223
Akan complex (deity), 257
alliance, 211, 214
Altar de Sacrificios, Guatemala, 13, 17, 27, 29, 30, 32, 36–39, 225, 227, 233, 327
altar, 14, 17, 28–30, 33, 36–39, 182, 190, 219, 225, 227, 233, 257, 276, 327
alternate flaking, 112, 141, 332
Altun Ha, Belize, 28, 33, 36, 203, 224, 253, 258, 327
anthropology, 64, 295
anthropomorph (silhouette), 20, 63, 65, 66
anthropomorphic (shape), 7, 58, 60, 65, 159, 177, 180, 186, 192, 197, 198, 233, 241, 254, 282
antler, 10, 79–85, 88, 93, 94, 98, 100–102, 106–108, 110, 111, 113–119, 132, 136, 145, 150, 217
Aoyama, Kazuo, 308
apprentice (novice), 11, 175, 179, 213, 214, 354, 357
Archaic period, 4, 16, 78, 81, 84, 115, 117
architecture, 5, 35, 181, 243, 273, 276, 300
arrow (arrowhead), 77, 89–91, 95, 96, 98–102, 105, 106, 108, 109, 111–113, 247
artifact genealogy, 270
artisan, 5, 11, 12, 70, 109, 111, 132, 140, 142, 145, 146, 149, 151, 186, 212–214, 249–252, 254–256, 260, 325
artisanship, 198–200, 255
artistry, 153, 233
atlatl, 16–18, 20, 57, 62
Australian Aborigines, 87
Awe, Jaime, 26, 189, 289, 327
axe, 14, 15, 20, 28, 29, 34, 37, 153, 240, 299

Bajo Alacranes, 243, 245
bajo, 216, 245
Baking Pot, Belize, 32, 188, 230

Barnes, Alfred, 84, 85, 87, 88
Barrett, Jason, 34
Barton Ramie, Belize, 188
Beauchamp, William Martin, 77, 109
Becan, Mexico, 36
Beckwith, Thomas, 98, 99
Belcher, Edward, 90, 92, 94, 95, 96
belief, 178, 215, 273, 342, 359
Belize (formerly British Honduras), 12, 20, 27, 29, 30, 33–38, 40, 53, 78, 81, 83, 118, 132, 181, 185–189, 193, 196, 198, 202, 204, 205, 209, 210, 213, 216, 224, 227, 240–242, 251–253, 255, 256, 260, 269, 272, 289, 342, 353, 355, 356
Belize River (Valley), 33, 186–189, 193, 196, 198, 202, 204, 205, 213, 217, 232
Benque Viejo, Belize, 187, 189
biface technology: biface, 8, 10, 14, 15, 17, 20, 23, 25, 27, 28, 34, 35, 36, 38, 40, 60, 67, 68, 79, 82, 83, 94, 97, 106, 115, 119, 131–144, 146, 148–153, 163, 191, 196, 198–200, 217, 218, 245, 246, 248, 251–257, 275, 332, 340; chert biface, 131, 146, 156, 178; obsidian biface, 23, 24, 106, 137, 140, 148, 157, 285
bifurcate (silhouette shape), 198
billet, 81, 83, 84, 86, 88, 105, 111, 148
blade technology: chert blade, 4, 14, 34, 190; chert macroblade, 13, 14, 18, 34, 39, 145, 251, 252; core, 17, 18, 23, 24, 25, 39, 40, 57, 109, 110, 185, 190, 191, 200–203, 227, 258, 270, 272, 274, 276–288, 296, 308, 310, 312, 315, 322, 327, 335, 339, 340, 354, 356, 357; microblade, 136; obsidian macroblade, 13, 14, 17–19, 39, 60, 200–204, 284, 285, 308, 326; macrocore, 24, 276, 277, 280, 281, 315; polyhedral core, 201, 204, 327, 340; primary (1s) pressure blade, 61; prismatic blade core (3s core), 296; secondary (2s) pressure blade, 61; third series (3s) pressure blade (prismatic blade), 16, 61, 300, 310, 314
Blank, Patrick (Jack Crafty), 83
bloodletter, 4, 7, 9, 282, 312, 314, 343
bloodletting, 14, 219, 308, 312, 316, 325
Blue Creek, Belize, 33, 34, 242–246, 251, 253–256, 260
Boggs, Stanley, 177
Bonney, A. F., 77, 105
boulder, 35, 91, 92, 132, 298
boundary object, 11, 212, 213, 215, 217, 233
broker, 11, 212–215, 233
Buenavista del Cayo, 188, 193
bulb, 63, 85, 86, 117, 140, 201, 253
bundle (ritual bundling, fabric wrapping), 227, 228, 270, 303, 304
bundled (qualities), 22, 145, 270, 288, 303, 360
burial, 4, 13, 14, 33, 34, 54, 82, 83, 88, 93, 113, 114, 133, 145, 178, 183, 191, 192, 220, 258, 273, 274, 281, 285, 299, 315, 319, 321–324, 326, 328–343, 355, 358, 360
burning, 4, 257, 260, 332, 334, 341, 342, 344

Cacaxtla, Mexico, 8, 15, 18, 19
cache goods, 12, 304
cache vessel, 211, 278, 314, 328
cache, 4, 6, 7, 10, 13, 15, 16, 20, 28–30, 34, 35, 38, 40, 54, 55, 63, 82, 157, 165, 174–176, 178–181, 183, 189–194, 196–205, 211, 212, 217, 219–225, 227–232, 234, 242, 244–251, 254–256, 258–261, 273, 276, 277, 279–286, 295, 296, 299, 302, 304, 306–314, 316, 324, 328, 339, 353, 355, 358
Cahal Pech, Belize, 32, 188, 189
Cakchiquel, 323
Calakmul, Mexico, 26, 31, 36, 37, 233, 245, 334
Campeche, Mexico, 12, 27, 30, 31, 34–38, 227
Cancuen, Guatemala, 31, 37, 299
capital (wealth), 11, 25, 213, 214, 229
Caracol, Belize, 12, 24, 34, 40, 211, 233, 269–280, 282–289, 327, 353, 355, 357
Carballo, David, 15
Card, Jeb, 178, 181
carving, 4, 14, 219, 299, 308
Catlin, George, 91–96, 102, 103
Ceibal, Guatemala, 31, 37
celestial bodies, 5, 7, 55, 225, 233
celt (axe head), 28, 35, 100, 162, 168, 172
celtiform (shape), 15, 27, 218, 299
centipede, 14, 15, 224, 225
Central Mexico, 3, 5, 8, 9, 13–15, 18, 20, 21, 28, 52, 55–58, 61, 64, 70, 200, 313, 314, 340, 341
ceremony, 3, 4, 8–11, 13, 15, 34, 35, 40, 51, 55, 90, 131–133, 135, 140, 186, 187, 189, 204, 209, 211, 212, 220, 223, 241, 274, 275, 300, 301, 314, 316, 319, 322, 328, 334, 339, 341, 352–360
ch'ajoom (priest), 257, 260
Chahk (deity), 178, 188, 228
chaîne opératoire, 10, 12, 70, 83, 84, 88, 117, 212, 215, 216, 232, 235, 309
chalchihuitl, 64
Chalchuapa, El Salvador, 176, 178, 181, 183
chanting, 79, 219, 325
chemical analysis (NAA, XRF), 203, 216, 286, 322, 335, 340
chert (agate, chalcedony, flint): blank, 27, 36, 83, 98, 109, 155, 156, 195, 217, 218, 230–232, 252; nodule, 26–30, 32–38, 100, 118, 132, 146, 217, 218, 231, 232, 234, 235, 251–253, 298

Chert Bearing Zone (CBZ), Belize, 29, 33
Chiapas, Mexico, 25, 29, 35, 109, 131–133, 325, 338, 342
Chichen Itza, Mexico, 19
Childe, V. Gordon, 55
Chixoy River, Guatemala, 30
Clark, John, 10, 18, 35, 39, 70, 77, 120, 157, 159, 172, 198, 308, 325, 338, 353, 354, 356
classification, 8, 9, 55, 56, 59, 61, 63, 64, 68, 212, 241, 280, 355, 356
cobble, 35, 57, 185, 198, 298–300, 312
Coe, Michael, 327
Coe, William, 9, 13, 32, 54, 212, 225–229, 240, 241, 249, 258, 322, 327
coefficient of variation (CV), 254, 264
Colha, Belize, 10, 20, 33, 39, 81–83, 118, 119, 216, 251–253
Collins, W. H., 77
Colonial, 16, 215, 240, 322
color, 6, 26, 30, 32, 51, 56, 64, 132, 142, 143, 145, 178, 183, 194, 217, 231–233, 251, 256, 270, 298, 304, 309, 310, 313, 315, 328, 357
community, 11, 12, 209, 213–215, 235, 354, 357, 360
community of practice, 11, 12, 209, 211–215, 217, 232–235, 242, 354, 355, 357, 360
concave, 203
conchoidal fracture pattern, 10, 59, 61, 136, 148, 253, 322
conflation (of forms), 16, 37, 38, 63, 223–225, 227, 228
constellation of practice, 11, 12, 13, 209, 212, 214, 215, 233
construction fill, 248, 274, 276, 331, 332, 341
convex, 199
Copan, Honduras, 10, 12–14, 25, 27, 29, 34, 35, 36, 39, 40, 81, 137–139, 142–145, 147–149, 151–153, 155, 156, 170–172, 174–183, 186, 203, 215, 233, 254, 255, 295, 297–299, 308, 310, 311–313, 316; Acropolis, 298, 300; Hieroglyphic Stairway (Temple 26), 176, 181–183; Rosalila Temple, 30, 40, 157, 165, 169, 174–183, 254, 255; Sepulturas, 35, 40, 295, 315
Copan River, 35, 298
Copan ruler, 12, 176, 178, 181, 183
core stabilization (holding method), 93, 117, 151, 195
cortex, 26, 27, 36, 60, 132, 142, 143, 177, 180, 197, 201, 202, 251, 280, 298, 356, 357
cosmogram, 315, 323, 324
cosmology, 270, 297, 323, 326
Coutier, Leon, 85–88, 111
craftsperson, 4, 15, 58, 171, 183, 195
crescent, 7, 14, 16, 21, 22, 37, 39, 63, 68, 190, 198, 199, 202, 224, 225, 228, 229, 231–233
crypt, 274, 281, 297, 331
cubit, 132, 133, 134
Cushing, Frank Hamilton, 77, 105, 109
cut tine flaker, 82, 119
cylinder, 108, 111, 112, 114–117

debitage (blades, chunks, flakes, and shatter as a group), 8, 9, 10, 13, 18, 25, 33, 40, 51, 57, 60, 70, 80, 81, 84, 116, 176, 182, 185, 202, 216, 231, 234, 248, 275–281, 283–285, 298, 313, 315, 316, 319–322, 325–328, 332, 335, 336, 338–343, 358–360; dump, 9, 33, 216, 299, 326, 327, 358; macrodebitage, 25, 275, 279, 280, 281, 283; microdebitage, 39, 136, 137, 298, 299, 308, 313, 341, percussion debitage, 18, 276–278, 280, 281, 283–285
decapitation, 156, 257
decortication, 25, 36, 202, 217, 235, 252, 253, 326
dedication, 182, 194, 256, 344, 357, 358
deer, 80, 81, 90, 98, 100, 102, 104, 106, 108, 109, 112, 115, 119, 132, 150, 177, 180
deity, deities (god, gods), 3, 4, 5, 7, 9, 13, 14, 16, 21, 22, 37, 58, 61, 64, 156, 197, 212, 225, 226, 228, 229, 233, 257, 258, 271, 297, 313, 316, 324, 325, 328, 355
delta flake pattern, 136, 141, 150
Deming, Edwin and Therese, 96, 97, 102
denticulate, 187, 190, 191, 193, 196–200, 202, 204
de Pourtalès, Frank, 90
direct percussion, 10, 23, 59, 60, 61, 64–68, 70, 76, 77, 79, 81, 84, 85, 90, 95, 98–100, 109, 110, 115, 117–119, 145–148, 151, 196, 198, 203, 217, 233, 253
direct regular notches, 157, 158
disc, 21, 22, 63, 68, 226, 228, 229, 231, 232
discoidal (shape), 217, 218, 224, 231
dismemberment, 257
distributional analysis, 269, 273, 274
divination, 5, 7, 212
domestic ritual, 273, 274, 287
Dos Pilas, Guatemala, 37
drift (drift pin), 111, 113, 114, 117
drilling, 186, 198, 253
Duff, Wilson, 109
Dulce River, Guatemala, 35
Dzibanche, Mexico, 28

Early Classic period, 8, 13, 15, 18, 20, 21, 33, 53, 61, 62, 225, 230, 243, 248, 275, 282, 284, 286, 287, 338
economic, 18, 21, 27, 34, 39, 57, 186, 260, 271, 274, 277, 288, 296, 297, 300, 322

EF (eccentric flint in the Coe typology), 32, 225–227, 249
El Chayal, Guatemala, 195, 245, 335, 340
element design, 156, 163, 164
element isolation notch, 156, 163, 164, 165, 167, 168, 169, 253
elite ritual (royal ritual), 204
Ellis, H. Holmes, 88
El Maestro (hypothetical Maya knapper), 174, 175, 178, 179, 181
El Mirador, Guatemala, 36
El Palmar, Mexico, 33, 35
El Peru-Waka' (Waka'), Guatemala, 29, 31, 35, 36, 37, 40, 313, 319–321, 327, 330, 334, 338–340, 342, 344
El Salvador, 29, 35, 174–176, 181, 183
El Tigre (Itzamkanac), Mexico, 29, 30, 33, 36–39, 227
El Zotz, Guatemala, 32, 38
embodied knowledge (embodiment), 78, 209, 233
emic, 13, 27, 29
EO (eccentric obsidian in the Coe typology), 32, 54, 225
epigraphy, 61, 214, 223, 353
eraillure flake, 200, 281
error (knapping), 60, 61, 141, 156, 157, 159, 160, 162, 163, 168, 169, 171, 172, 177, 180, 183, 218, 254, 255, 335
Escobedo, Hector, 35, 321, 330, 341
esoteric knowledge, 5, 11, 61, 185, 196, 197, 204, 209, 211–213, 234, 359
ethnohistory, 7, 77, 78, 90, 117, 215, 324, 325, 327, 357
etic, 174
exchange, 29, 30, 53, 57, 58, 211, 234, 235, 260, 272, 322
exhausted (tools and cores), 23–25, 182, 185, 191, 200, 202, 227, 275–278, 280, 283, 284, 286–288, 315, 326, 327, 331, 338, 339, 341–343, 354, 357

face cache, 271, 277
facial feature, 159, 161, 277
fathom (*brazada*), 132, 133
feather termination, 136, 311
fire, 4, 5, 6, 15, 63, 64, 102, 104, 142, 323, 324, 341
flake: biface thinning flake, 25, 28, 82, 119, 196, 217, 256; flake scar (negative bulb), 10, 26, 27, 88, 115, 134–140, 142, 145–151, 155, 159, 163, 164, 170–172, 194, 196, 200, 252–255, 280, 282; macroflake, 24, 27, 35, 57, 59, 60, 65, 67, 79, 217, 252, 253, 280, 284; Type 1 flake, 135–139, 143, 149, 151; Type 2 flake, 136, 137, 139, 143, 150; Type 3 flake, 136, 143, 149, 150; Type 4 flake, 136, 149, 150; Type 5 flake, 138, 143, 149, 150
flaker, 77, 79–83, 90, 94, 99, 103, 104, 106–109, 112–114, 117, 119, 162, 163, 179
float (nodules in *bajos*), 216
flower (petaloid), 225, 258
Fogelin, Lars, 358, 359
Fowke, Gerard, 77, 109
Freidel, David, 259
Frison, George, 115
fulgurite, 223, 258
Funkhouser, William Delbert, 77

Gann, Thomas, 53, 189, 241
Geib, Phil, 115, 116
general utility biface (GUB), 250
geometric silhouette, 6, 13, 55, 60, 63, 68, 69, 185
Gibson, Eric, 19, 20
Gila River, 91
god effigy, 4, 7–10, 12, 14, 16, 21, 27, 29, 30, 33, 35, 37, 39, 40, 54, 58, 62, 80, 120, 133, 155, 177, 212, 295, 322, 355
grinding (ground), 101, 109, 116, 200, 279, 282, 284, 353
Grinnell, George Byrd, 77, 99, 100, 102
Grube, Nikolai, 32
Guatemala, 14, 28–30, 33, 34, 40, 53, 144, 172, 179, 181, 187, 195, 198, 199, 209, 212, 295, 299, 300, 319, 323, 340
Gunflint, 84, 85, 95, 96, 111

ha'al, 258, 259
Hally, David, 116
hammer (Western), 84, 85, 111
hammerstone (hard percussor), 53, 79, 84, 85, 92, 96–100, 102, 105, 106, 108, 109, 112, 114, 117, 119, 150, 191, 253, 298, 299
Harrington, Mark, 77, 108, 109
Haury, Emil, 12
headdress, 58, 70, 133, 156, 157, 162–165, 168, 172, 177, 180, 259
heat treatment (heat altered, thermal alteration), 32, 101, 231
Hellmuth, Nicholas, 259
hematite, 191, 200, 223
Hertzian cone, 10, 85, 88
hexis, 209, 216
hieroglyph, 6, 176, 181–183, 189, 211, 214, 332, 355
hinge termination, 137, 138, 141–143, 145, 151, 177, 179, 255, 283
Hirth, Kenneth, 357
Hodge, Frederick, 77

Holmes, William H., 53, 76, 77, 83, 106, 109, 111, 118
Honduras, 10, 14, 40, 155, 174, 175, 179, 182, 286, 295
Hooten, Ernest A., 77
horizon, 13, 212, 213
horn, 84, 90, 95, 100, 102, 105, 108, 109, 111, 112, 115, 116, 168
Houston, Stephen D., 257, 323
Hruby, Zachary, 6, 8, 9, 11, 16, 21, 23, 26, 28, 53, 61, 119, 143, 172, 182, 183, 186, 204, 213, 215, 217, 223, 227, 232, 234, 315, 321, 335, 338, 339, 353–360
human sacrifice (ritual killing), 3, 4, 5, 7, 13, 14, 17, 20, 27, 29, 55, 56, 62, 68, 132, 134, 201, 257, 259, 260, 314, 316, 322, 323, 327, 328, 338
hypertrophic object, 62, 134, 142, 149, 153

Iannone, Giles, 19, 20, 39
Ice Age (Pleistocene), 85
icon, 7, 15, 17, 270–272, 279, 328, 353, 358, 360
iconography, 7, 8, 11, 12, 14, 15, 17, 21, 29, 36, 61, 212, 213, 223, 225, 226, 229, 258, 313, 314, 328, 356
ideology, 5, 7, 21, 64, 71, 79, 185, 186, 203, 211, 213, 242, 297, 300, 319, 341, 352
idiosyncratic, 227, 252
impersonation (of deity or ancestor), 4, 13, 213
incense burner, 133, 247
incising, 24, 32, 54, 61, 196, 228, 315
index, 17, 271, 288, 316, 352, 353, 358, 360
indirect percussion, 9, 10, 22, 23, 24, 26, 27, 33, 34, 35, 39, 58, 59, 65–67, 70, 76–84, 86, 88–90, 95–103, 105, 106, 109–111, 113–120, 135–138, 141–143, 145–152, 155, 156, 163–168, 182, 195, 198, 200, 202, 217, 218, 233, 253, 254, 277, 282, 353, 356, 357
interaction sphere, 211, 213
intermarriage, 37, 211
IO (incised obsidian in the Coe typology), 24, 32, 54, 228, 315
Ishi, 94, 106
Ixtepeque, Guatemala, 192, 312, 340
Izabal, Guatemala (department), 34
Izapa, Mexico, 133

jade (jadeite), 4, 5, 8, 14, 20, 59, 64, 190, 191, 200, 222, 223, 243, 244, 271, 277, 278, 282, 297–299, 312, 313, 323, 328, 330–332, 334, 339
jaguar, 5, 7, 20, 133, 178, 257, 330, 341, 343
Jasaw Chan K'awiil (Tikal ruler), 32, 257
jasper, 25, 30, 34, 35, 100, 112, 298
Johnson, Frederick, 114
Johnson, Lucas, 9, 40, 204, 280, 322, 354
Jones, Paul A., 96, 102
Jordan, Jilian, 217
Joyce, Thomas, 241
jute snail, 192, 223

K'an cross, 197
K'awiil (deity, God K), 22, 79, 178–180, 225, 232
K'awiil Maize God, 13, 22, 233
Kaanul regime (Snake Dynasty), 334
kaloomte', 334, 343
Kaminaljuyu, Guatemala, 14, 20, 23, 39
Kidder, A. V., 144
Kidder, Tristram, 20
K'in cross, 197
knotted bundle (three dot jewel), 227, 228
Knowles, Francis, 85, 87
Kovacevich, Brigitte, 7

Lacandon, 109, 110, 120, 325
La Corona, Guatemala, 31, 37
Lady K'abel (Waka' ruler), 334, 341
La Laguna, Mexico, 15
Lamanai, Belize, 29, 33, 203, 253, 326, 327
lance, 133, 240
lancehead, 131–134, 140, 200, 240
Late Classic period, 4, 18, 23, 29, 39, 119, 133, 175, 176, 177, 181, 182, 186–188, 192, 204, 212, 217, 220, 221, 227, 230, 240, 242, 244, 246, 248, 257, 273–275, 280, 281, 282, 284, 286, 287, 297, 298, 326, 332, 335
Late Preclassic (Formative) period, 7, 13, 19, 220, 243, 260, 269, 275, 285
laurel leaf (lanceolate, lenticular shape), 15–17, 19, 21, 22, 27, 29, 30, 33–35, 37–39, 60, 62, 63, 83, 132, 134, 140, 144, 197, 198, 200, 218, 224, 225, 227, 228, 231, 232, 245, 246, 248, 251–257, 354, 355, 357
Lave, Jean, 11, 209, 213
layered deposit, 358, 360
Leakey, L. S. B., 85, 87, 111
Levine, Marc, 358
life history, 270
lightning, 4, 5, 7, 14, 22, 55, 58, 60–63, 66, 67, 156, 178–180, 217, 223, 225, 240, 258, 260, 270, 271, 323, 324, 360
limestone, 5, 30, 34, 35, 103, 190, 216, 230, 277, 299, 300
lineage (genealogy), 181, 183, 211, 213
looting (illegal excavation), 174, 177, 327
Lubaantun, Belize, 34
luxury, 213, 321

Macalister, Robert Alexander Stewart, 77, 109
maize, 3, 328
Maize God, 7, 22, 297, 313, 328
margin (biface, blade, flake), 16, 26, 27, 61, 62, 64, 94, 100, 132, 136, 138, 150, 151, 156–159, 162, 164, 167, 168, 172, 189, 196, 198, 200, 202, 243, 245, 252–256, 280, 283, 284, 299
Martin, Simon, 32
Mason, Otis Tufton, 77, 109
master (craftsperson), 4, 11, 12, 18, 27, 38, 59, 65, 69, 79, 91, 132, 145, 152, 171, 175, 213, 214, 324, 325, 354, 356, 257
mat (silhouette shape), 225
Maya Lowlands (Central, Northern, Southern), 12, 21, 25, 32–34, 37, 53, 189, 193, 195, 198, 199, 202, 209, 213, 227, 250, 255, 310, 311, 321, 323, 342, 356
McAnany, Patricia, 215
Mejia, Jorge, 177
Melendez, Juan Carlos, 36, 321, 330, 341
Mercer, Henry C., 77, 109
Mesoamerica, 3–5, 7–16, 18–20, 27, 36, 51–53, 58, 60, 61, 68, 71, 77, 78, 81, 120, 132, 144, 145, 153, 199, 200, 214–217, 260, 261, 269, 270, 313, 314, 316, 322–324, 326, 352–359
meteor (meteorite), 324
Mexico, 4, 5, 8, 9, 13–21, 28, 40, 52, 53, 55–58, 61, 64, 70, 131, 199, 313, 314, 316, 340
Michaels, George Herbert, 119
Middle Preclassic (Formative) period, 4, 14, 18, 188, 246, 275
Milky Way, 225
Mills, William, 112
mirror (hematite, obsidian, or pyrite), 59, 156, 213, 330–332, 334
model, 11, 84, 194, 213, 216, 250, 254, 287, 354, 355, 357, 358
Moholy-Nagy, Hattula, 29, 54, 172, 212, 225, 322, 326–328, 358
monothetic, 241
Monte Alban, Mexico, 14
monument, 4, 7, 13, 32, 36, 53, 55, 78, 181, 190–194, 211, 214, 218–222, 229, 243, 247–249, 256, 257, 273, 276, 296, 316, 326, 328, 342, 343, 358
moon, 5, 6, 7, 62, 63, 225
Mopan River, Guatemala and Belize, 211
Motagua River, Guatemala, 34, 299
Motolinía, Friar Toribio de Benavente, 324, 325
Mount Maloney black slipped (ceramic type), 211
mythology, 3, 7, 20, 104, 211, 213, 214, 297, 316, 323, 324, 342, 352
naab', 258, 259
Naranjo, Guatemala, 32, 211, 212, 225–227, 229, 230, 232, 233, 235
naturalistic style, 5, 6, 13, 21, 22, 24, 26, 223, 228
Navarro-Farr, Olivia, 321, 333, 334, 341
Nicholson, Grace, 106
Nimli Punit, Belize, 34
Northern Yucatan, 29, 30, 36
notch (biface, uniface, blade), 10, 14, 16, 17, 23, 24, 26, 34, 39, 40, 61, 63, 65–68, 90, 105, 109, 120, 156–169, 177–181, 192, 195, 198, 200, 202, 218, 225, 227–231, 233, 247, 253, 254, 277, 278, 280, 282, 284, 287, 297, 300, 343, 255; deep decorative notch, 162; double notched blade, 16, 297, 300, 343; oblique notch, 157, 158, 180

O'Callaghan, P., 95
object biography, 145
obsidian: blank, 23, 24, 59, 60, 65, 69, 200; clast (nodule), 18, 57, 60, 65, 309; slab (tabular clast), 18, 57, 60, 65
oficio, 23, 213, 215
Ojo de Agua, Mexico, 338
Olsen, Larry Dean, 97, 102
ontology (ontologies), 78, 186, 215, 216, 352, 355
Orcutt, Theodore, 90, 106, 107
Otumba, Mexico, 57, 64
oval (ovoid shaped), 63, 68, 197, 198, 250, 251–253
overshot (outre passe), 164, 277, 278, 281, 282, 284

package (raw material format), 251
Palenque, Mexico, 17, 26, 30, 215, 233, 295, 328
Paleoindian period, 4, 8, 78, 111, 117
Paredon, Mexico, 57
Parker, Arthur Caswell, 77, 98, 109
Pasion River, Guatemala, 30, 37
Pastrana, Alejandro, 5, 9, 18, 39, 51, 59, 156, 172
pecking, 58–60, 64, 68, 149, 200, 216, 299
Pedernales River, Guatemala, 34
Peirce, Charles, 358
perforation (stone), 13, 29, 34, 59, 60, 63, 64, 68, 186, 197, 198, 252–255, 282
Peten, Guatemala (department), 17, 25, 26, 29, 30, 32, 34, 35, 37, 181, 210, 272, 316, 320, 342, 343, 355
Piedras Negras, Guatemala, 12, 13, 23, 24, 28–30, 32, 37, 39, 196, 212, 228, 229, 240, 315
Pine Ridge carbonate (ceramic type), 211
platform (architectural), 190, 192, 219–221, 296, 298–300, 308, 332

platform preparation, 255, 277, 282, 283, 308, 310
plaza, 7, 177, 187, 189, 190, 193, 200, 220–222, 243, 246–248, 256, 258, 259, 276, 282, 295, 296, 321, 329; *plazuela*, 244, 276
political center, 211, 213, 214; city state, 214; polity, 11, 25, 32, 38, 209, 211, 214, 243
Popol Vuh, 7, 323
portrait (facial profile), 21, 37, 156, 159, 177, 179, 196, 226, 228, 229, 231, 232, 258
Postclassic period, 18, 58, 64, 81, 83, 119, 247, 314, 324
Prado, Antonio, 135, 198
Pre-Contact, 215, 352
pressure flaking, 10, 15–19, 22, 24, 26, 35, 58, 64–68, 70, 76, 79, 90, 92, 95, 99, 100, 109, 113, 114, 117, 118, 136, 145, 155, 156, 163, 165, 166, 168, 180–182, 195, 196, 198, 200, 202, 233, 253, 254, 256, 270
prestige (prestige goods), 5, 19, 28, 39, 188, 204, 260, 273, 297, 321, 332, 359
priest (ritual specialist), 58, 61, 95, 132, 204, 257, 325
Principal Bird Deity (PBD), 228
profile (view), 136, 139, 140, 142, 252
projectile point, 16, 18, 29, 35, 59, 60, 62, 63, 67, 105, 109, 114, 117, 214, 225, 356, 357
Proto-historic period, 78, 117, 118
punch (pitching tool), 10, 26, 34, 76, 77, 79–83, 87, 88, 90–92, 94–102, 104–111, 113–119, 132, 142, 145, 148, 150, 152, 155, 163, 164, 217, 218, 254
pyrite, 59, 191, 200, 213, 223, 328, 330, 334. *See also* drift

quadripartite, 197, 308, 328
quadruped (silhouette shape), 53, 62, 67
qualitative measure, 255
quantitative measure, 254, 255, 274
quetzalitzli, 64
Quintana Roo, Mexico, 30, 33, 36
Quirigua, Guatemala, 35, 186

Ramos, Jorge, 40, 300, 357
Rau, Charles, 94
raw material, 5, 6, 26–29, 34, 55, 57, 60, 64, 65, 69, 70, 112, 146, 156, 186, 195, 203, 215, 216, 230, 233, 241, 242, 244, 249–251, 255, 260, 298, 312, 322, 326
Ray, P. H., 77, 101–102
reduction sequence, 187, 195, 274, 280, 282, 353
refitting studies, 200–204, 256, 274, 277, 278, 283, 284, 299, 303, 305, 308–310, 312, 313, 339, 341, 353, 356, 357, 359
rejuvenation, 202, 275–285, 287, 288, 332, 335, 338–340, 343
replication (lithic), 10, 17, 20, 22, 33, 40, 51, 59, 77, 78, 80, 81, 85, 115, 120, 144–146, 150, 151, 155–162, 164, 165, 167–169, 171, 186, 193, 194, 198, 216, 254, 260, 299, 308, 353, 356
residential settlement, 219, 243, 244, 269, 270–278, 280, 284, 285, 287, 297, 298, 300, 321, 326
retouch, 59, 61, 64, 132, 136, 138, 139, 143, 145–147, 150, 151, 153, 172, 195, 198, 200, 201, 253, 254, 274, 278, 282, 354
ribbon flake (or blade), 136, 146
ring (silhouette shape), 21, 33, 63, 225, 229, 231, 232
Ritchie, William, 77, 114
ritual proscription, 10, 219, 220
ritual, 4, 5, 7–9, 13, 19–21, 27, 28, 30, 34, 37, 39, 53, 55, 56, 58, 78, 83, 153, 187, 188, 193, 199, 200, 203, 204, 212, 219, 228, 235, 241, 242, 245, 257, 260, 269–278, 280, 282, 284, 287, 288, 298, 308, 315, 319, 322, 324, 325, 327, 329, 331, 332, 338–344, 352, 355, 358–360
ritualization, 5, 14, 36, 94, 120, 187, 196, 197, 271, 286–288, 316, 325, 354, 357, 359, 361
ritualized object, 7, 9, 10, 14, 21, 22, 26–28, 33, 34, 39, 185–188, 196, 198, 201, 203, 204, 211, 214, 218, 241, 260, 269, 271, 272, 274, 277, 280, 282–284, 286–288, 319, 322, 324, 353, 355, 358
Robb, John, 359, 360
Roddick, Andrew, 217
royal, 5, 11, 14, 28, 32, 37, 133, 153, 181, 211, 212, 272, 297–300, 313, 321, 326, 329, 332, 334, 342
ruler (king, queen), 4, 5, 32, 176, 178, 181–183, 188, 257, 259, 343, 344
Russel, Virgil, 54

sacbe, 297
sacrifice (blood), 3, 4, 5, 7, 13, 14, 62, 201, 257, 259, 260, 314, 316, 322, 323, 328, 338
San Andres, El Salvador, 174–183
San Jose Mogote, Mexico, 18
San Martin Jilotepeque, Guatemala, 340
San Pedro Martir River, Guatemala and Mexico, 30, 37, 38, 320, 321
Sapir, Edward, 77, 104, 109
scepter, 13, 20, 21, 29, 61, 131–133, 355
scorpion, 21, 197, 198, 225, 228, 229, 231, 232, 271, 284, 288
Scorpius (constellation), 225
scraper, 55, 58
scraping, 58, 136, 312
scratching, 194, 200, 202

Seeman, Mark F., 77, 109
Sellers, George Escol, 103
sequence (production and deposition), 8, 21, 24, 30, 40, 64, 139, 140, 143–145, 149, 150, 152, 153, 155, 172, 185–187, 194, 195, 198, 201, 216, 232, 250, 252, 274, 280, 282, 297, 299, 300, 302, 313, 315, 339, 342, 353, 359
serpent (serpentiform silhouette), 14–16, 18, 58, 59, 62–67, 69, 156, 217, 225, 231, 258, 259, 270, 282
serration, 14, 16, 62, 158–160, 197, 198, 224–226, 228, 229, 231, 232, 278, 282, 284
settlement, 57, 187, 216, 243, 244, 260, 272–274, 276, 287
Shafer, Harry, 10, 78, 81–84, 109, 119
Sheets, Payson, 35, 39, 142, 143, 145, 147, 152, 172, 176, 179, 183, 254, 354, 356
shell, 20, 26, 58, 59, 61, 68, 78, 113, 153, 177, 178, 190–192, 200, 223, 225, 271, 277, 278, 304, 312, 313, 330, 331, 332, 334; spondylus, 61, 177, 190, 225, 304, 312, 313, 328, 330, 339
Sierra de las Navajas, Mexico, 5, 51, 56, 57, 59, 63, 64, 70, 324
silhouette (eccentric): definition, 6; 53–56, 68, 70, 356; fully bifaced (chert), 16–18, 25–28, 33, 37, 218, 219, 227, 231, 232; fully bifaced blade core (obsidian), 23, 24; infixation, 223–225, 227, 228; modified core, 25, 284; modified (blade core, flake, blade, debitage), 8, 14, 18, 19, 23–26, 29, 37, 60, 64, 200, 212, 218, 219, 227, 231, 232, 234, 248, 251, 252, 272, 276–279, 282, 284, 295, 354, 359; mostly bifaced (chert), 18, 25, 27, 218, 219, 227, 231; mostly bifaced blade core (obsidian), 23; notched blade core (obsidian), 23, 227, 284; partially bifaced (chert), 19, 25, 26, 218, 219, 227, 231, 232; partially bifaced blade cores (obsidian), 23; superimposition, 223, 224, 227
Simpson, Edward (Flint Jack), 84–86, 95
situated learning (legitimate peripheral participation), 11, 12, 213, 216, 354
skill, 5, 11, 24, 27, 36, 60, 61, 65, 69, 70, 89, 92, 101, 103, 111, 114, 141, 142, 146, 152, 153, 156, 171, 172, 175, 176, 179, 181, 195, 204, 213, 218, 231, 249–251, 254, 255, 260, 271, 322, 354, 356, 357
Skinner, Alanson, 112
sling mallet, 118
Smith, Harlan Ingersoll, 77, 113
smoking celt, 159, 162, 168, 172, 180
soft hammer (soft percussor), 85, 87, 88, 119, 195, 198
Solola, Guatemala, 35
source (geological), 18, 27, 29, 30, 31, 33, 34, 36, 37, 38, 51, 56, 57, 78, 79, 84, 182, 192, 195, 199, 211, 216, 217, 221, 230, 286, 312, 323, 324, 339, 340, 357
spatial layout, 13, 35, 270, 272, 274, 286, 287, 354
spear (spearhead), 21, 29, 33, 35, 62, 89, 91, 95, 98, 106, 112, 113, 133, 228
special deposit, 276, 280, 281, 286, 326
specialized, 53, 58, 59, 60, 70, 234
stage (of reduction), 10, 12, 26, 27, 57, 59–61, 64, 70, 79, 84, 92, 109, 111, 118, 120, 143, 146, 149, 152, 153, 155, 156, 180, 212, 216–219, 231, 234, 235, 276, 279, 287, 288, 298, 299, 308, 310, 337, 340, 342
standardization, 16, 186, 229, 254–256, 271, 272, 277, 282, 287
stemmed (biface or uniface), 13, 16, 19, 33, 39, 228, 245, 246, 251–254, 256
stingray spine, 14, 15, 177, 178, 192, 271, 277, 278, 282, 288, 328, 331, 332, 334, 355
strap-handle, 118
strategy, 33, 55, 57, 65, 149, 195, 216, 217, 230, 243, 245, 287, 322, 338
Stromsvik, Gustav, 182, 296, 308
Stuart, David, 182
stucco, 285, 301, 303, 304, 314, 329–332, 334, 341
stylistic style (stylized), 4–7, 15, 21, 22, 24, 37, 63, 176, 178, 209, 217, 223, 225, 228, 269, 271, 287, 288, 338, 340
subsidiary site, 25, 37, 214, 227
Sugiyama, Saburo, 63
sun (the center of our solar system), 4–7, 15, 58, 63, 64, 179, 217, 225, 228, 229, 257, 313, 315, 323, 325, 328
symbol (symbolism), 4–6, 8, 9, 11, 13, 15–17, 19, 21, 22, 24, 26–33, 37, 40, 52, 54, 55, 57, 58, 60, 63, 71, 133, 153, 162, 171, 179, 180, 186, 187, 203, 213, 214, 218, 223, 225, 228–230, 232–235, 240, 241, 250, 256, 258–261, 269–272, 279, 287, 288, 295–297, 299, 303, 312, 313, 315, 316, 319, 322–324, 327, 328, 342, 344, 352, 353, 358–360
symbolic capital, 11, 213, 229
symmetry, 6, 24, 26, 27, 33, 54, 61, 65, 66, 103, 132, 142, 144, 186, 226, 228, 230, 240, 241, 278, 283, 284

Taube, Karl A., 223
Tazumal, El Salvador, 181
technological analysis, 217, 242, 250, 252, 280, 319, 335, 358
temple (pyramid), 14–16, 20, 28, 52, 58, 61, 62, 177, 189, 190, 220, 221, 243, 257, 287, 321, 324, 325, 328, 332, 352

Teotihuacan, Mexico, 4, 8, 12, 14–22, 28, 40, 51–53, 56–71, 149, 234, 297, 316, 323; Pyramid of the Moon, 52, 58, 62; Pyramid of the Sun, 15, 16, 58; Temple of the Feathered Serpent, 16, 58, 62
Terminal Classic period, 8, 18, 23, 34, 186, 187, 192, 209, 211, 212, 225, 229, 235, 246, 248, 260, 269, 275, 284, 286, 287, 332, 338, 354, 355
termination (flake and blade), 10, 136, 142, 177, 255, 278, 311
termination (ritual or deposit), 13, 256, 319, 320, 322, 329, 339–344, 358
terrace, 276, 297, 332
texture (stone), 51, 132, 142, 143, 232, 251, 256, 270, 309, 357
Thompson, J. Eric S., 54, 258
Three Rivers Region, 242, 243, 245
Tikal, Guatemala, 7, 12, 13, 20, 21, 24, 28–30, 32, 37, 52, 53, 189, 196, 212, 225–230, 232, 233–235, 245, 257, 260, 286, 315, 321, 326, 343, 358
Titmus, Gene, 10, 70, 151, 155, 156, 172, 254
Tlaloc (deity), 5, 64
Tlaxcala, Mexico, 15, 324
tomb, 8, 10, 14, 40, 178, 189, 248, 274, 275, 296, 297, 300–303, 309, 315, 319–324, 326–333, 335, 338–340, 342–344, 360
tomb reentry, 329, 332–335, 340, 342
toolkit, 10, 81, 82, 101, 107, 118
tradition, 4, 6, 8, 11–21, 23–25, 29, 31, 32–40, 53, 54, 58, 60, 81, 88, 96, 117, 119, 120, 183, 186, 195, 209, 213, 214, 216, 227, 229, 233, 234, 240, 295, 316, 323, 342, 353, 356
translucent (translucency), 7, 26, 51, 132, 220, 230–232, 309
trefoil (shape), 16, 29, 34, 37, 68
tridentate (shape), 21, 22, 244–46, 249, 251–253, 255–258, 260
trilobed (shape), 21, 228–230, 232, 282
tripartite, 231, 234
tuff, 58, 298
Tula, Mexico, 18, 19
Tulancingo, Mexico, 57
tusk, 63, 103
typology, 8, 13, 16, 32, 51–53, 56, 114, 185, 212, 249, 286, 296, 300
Tz'unun, Belize, 242, 243, 245–251, 255, 256, 258–260; K'in T'an *plazuela* group, 243, 244

Uaxactun, Guatemala, 24, 32, 144, 327
Ucanal, Guatemala, 12, 27, 32, 209, 211, 212, 219–221, 225–235
underworld (Xibalba), 6, 7, 179, 257, 259, 313, 323, 324, 327, 328, 344
undulation (shape), 14, 197, 217, 249, 254, 258
uniface (unifacial), 14, 55, 59–61, 64–68, 138, 187, 191, 196, 249, 284, 286, 287
unifacial flaking, 61, 138, 196, 277
Usumacinta River, Guatemala and Mexico, 30, 37, 214, 321
utilitarian bifaces, 209

visual analysis, 216
volcanic, 5, 19, 35, 58, 91, 298, 322, 324

Water Lily Serpent (deity), 259
water lily, 7, 240, 258–260
weapon, 4, 9, 13, 28, 29, 51, 55, 57, 58, 62, 63, 91, 95, 156
Webb, William Snyder, 77, 113
Wenger, Etienne, 11, 213, 217
width to thickness ratio, 19, 85, 135, 180
Wilhelm, Paul, 106
Willey, Gordon, 12
Williams, Fredrick, 77, 109, 111
Wissler, Clark, 77
Wood, James, 10, 18, 35, 39, 70, 186, 198, 254, 354, 356
workshop (knapping), 5, 25, 29, 30, 35, 36, 51, 57–59, 64, 70, 81, 92, 95, 143, 168, 169, 177, 216, 217, 234, 251, 252, 255, 256, 270, 274, 296, 324, 356, 357

Xochicalco, Mexico, 18
Xultun, Guatemala, 32, 33, 262
Xunantunich, Belize, 12, 32, 185–195, 197–204, 209, 211, 212, 219, 221–223, 225–235

yax, 225
Yaxchilan, Mexico, 15, 23, 37, 38, 327, 342, 343
Young, Fred S., 89, 90, 98
Yucatan Peninsula, 5, 36, 181
yunque, 35, 298–300

Zapotec, 4, 14, 17, 57
Zaragoza, Mexico, 340
Zender, Marc, 178, 181
zoomorph, 7, 16, 20, 53, 62, 66, 67, 185, 197, 200, 241

About the Authors

This volume is largely the product of a symposium entitled *Ceremonial Lithics of Mesoamerica: New Understandings of Technologies, Distributions, and Symbolisms of Eccentrics and Ritual Caches in the Maya World and Beyond*, which Zachary Hruby organized for the Eighty-fourth Annual Meetings of the Society for American Archaeology, Albuquerque, New Mexico, 2019.

Jaime J. Awe, Northern Arizona University

Arlen F. Chase, University of Houston

Diane Z. Chase, University of Houston

John E. Clark, Brigham Young University

Benjamin Eble, Independent Researcher

Keith Eppich, Tyler Junior College

Christina Halperin, Université de Montréal

Zachary Hruby, Northern Kentucky University

Lucas R. M. Johnson, University of Nevada, Las Vegas and Nellis Air Force Base

Joshua J. Kwoka, Georgia State University

Xinwei Li, Beijing, Chinese Academy of Social Sciences

Damien Marken, Commonwealth University of Pennsylvania

David Rafael McCormick Alcorta, Yale University

Olivia Navarro-Farr, The College of Wooster

Alejandro Pastrana, INAH-DEA

Jorge Ramos, Co-Director of Proyecto Instituto de Arqueología de la Academia de las Ciencias Sociales de China

Michelle Rich, Dallas Museum of Art

Catharina Santasilia, University of California at Riverside

Payson Sheets, University of Colorado Boulder

Kelsey J. Sullivan, University of California at Riverside

James C. Woods, College of Southern Idaho